T0332594

RELATIVISTIC QUANTUM PHYSICS
From Advanced Quantum Mechanics to Introductory Quantum Field Theory

Quantum physics and special relativity theory were two of the greatest break-throughs in physics during the twentieth century and contributed to paradigm shifts in physics. This book combines these two discoveries to provide a complete description of the fundamentals of relativistic quantum physics, guiding the reader effortlessly from relativistic quantum mechanics to basic quantum field theory.

The book gives a thorough and detailed treatment of the subject, beginning with the classification of particles, the Klein–Gordon equation and the Dirac equation. It then moves on to the canonical quantization procedure of the Klein–Gordon, Dirac, and electromagnetic fields. Classical Yang–Mills theory, the LSZ formalism, perturbation theory and elementary processes in QED are introduced, and regularization, renormalization, and radiative corrections are explored. With exercises scattered through the text and problems at the end of most chapters, the book is ideal for advanced undergraduate and graduate students in theoretical physics.

TOMMY OHLSSON is Professor of Theoretical Physics at the Royal Institute of Technology (KTH), Sweden. His main research field is theoretical particle physics, especially neutrino physics and physics beyond the Standard Model.

RELATIVISTIC QUANTUM PHYSICS

From Advanced Quantum Mechanics to Introductory
Quantum Field Theory

TOMMY OHLSSON

Royal Institute of Technology (KTH), Sweden

CAMBRIDGE
UNIVERSITY PRESS

CAMBRIDGE
UNIVERSITY PRESS

Shaftesbury Road, Cambridge CB2 8EA, United Kingdom

One Liberty Plaza, 20th Floor, New York, NY 10006, USA

477 Williamstown Road, Port Melbourne, VIC 3207, Australia

314–321, 3rd Floor, Plot 3, Splendor Forum, Jasola District Centre, New Delhi – 110025, India

103 Penang Road, #05–06/07, Visioncrest Commercial, Singapore 238467

Cambridge University Press is part of Cambridge University Press & Assessment,
a department of the University of Cambridge.

We share the University's mission to contribute to society through the pursuit of
education, learning and research at the highest international levels of excellence.

www.cambridge.org
Information on this title: www.cambridge.org/9780521767262

First published 2011

A catalogue record for this publication is available from the British Library

Library of Congress Cataloging-in-Publication data
Ohlsson, Tommy, 1973–
Relativistic quantum physics : from advanced quantum mechanics to introductory
quantum field theory / Tommy Ohlsson.
p.   cm.
ISBN 978-0-521-76726-2 (Hardback)
1. Quantum theory.   I. Title.
QC174.12.O35 2011
530.12–dc23

2011018860

ISBN   978-0-521-76726-2   Hardback

In memory of my father Dick

Contents

Preface

This book is based on my lectures in the course 'Relativistic Quantum Physics' at the Royal Institute of Technology (KTH) in Stockholm, Sweden. These lectures have been given four times during the academic years 2006–2007, 2007–2008, 2008–2009, and 2009–2010. The main sources of inspiration for the lectures were the books A. Z. Capri, *Relativistic Quantum Mechanics and Introduction to Quantum Field Theory*, World Scientific (2002) and M. E. Peskin and D. V. Schroeder, *An Introduction to Quantum Field Theory*, Addison-Wesley (1995), and indeed, this book serves as a textbook for relativistic quantum mechanics with continuation to basic quantum field theory. The book is mainly intended for final-year undergraduate students in physics or first-year graduate students in physics and/or theoretical physics, who want to learn relativistic quantum mechanics, the basics of quantum field theory, and the techniques of calculating cross-sections for elementary reactions in quantum electrodynamics. Thus, the book should be suitable for any course on relativistic quantum mechanics as well as it might be suitable for a beginners' course on quantum field theory. In summary, the book is a self-contained technical treatment on relativistic quantum mechanics, introductory quantum field theory, and the step in between, i.e. it should fill the gap between advanced quantum mechanics and quantum field theory, which I have called *relativistic quantum physics*. It contains a thorough and detailed mathematical treatment of the subject with smaller exercises throughout the whole text and larger problems at the end of most chapters.

I am deeply grateful to Johannes Bergström, Jonas de Woul, and Dr Jens Wirstam for careful proof-reading of earlier versions of the manuscript of this book and for useful comments, discussions, and suggestions how to improve the book. I am indebted to my former Ph.D. supervisor Professor emeritus Håkan Snellman for teaching me that physics is a descriptive science, which indeed does not explain anything. I would also like to thank my two friends Björn Sjödin and Jens Wirstam, who left science for 'industry', but never lost interest in it, and with whom I obtained many inspiring ideas how to develop this book further. Discussions with Dr Mattias Blennow, Dr Tomas Hällgren, Henrik Melbéus, and Dr He Zhang have been helpful in the process of development. In addition, I would like to thank Professor Mats Wallin, who suggested to me to include the topic 'graphene' in this book.

The author gratefully acknowledges financial support from the degree program 'Engineering Physics' (especially, Professor Leif Kari) at KTH for the development of this book.

Finally, last but not least, I would like to thank my family and friends for always being there for me. This applies particularly to my wife Linda, but also to my mother Inga-Lill and my sister Therése.

1

Introduction to relativistic quantum mechanics

1.1 Tensor notation

In this book, we will most often use so-called *natural units*, which means that we have set $c = 1$ and $\hbar = 1$. Furthermore, a general 4-vector will be written in terms of its contravariant index, i.e.

$$A = (A^\mu) = (A^0, \mathbf{A}),\tag{1.1}$$

where A^0 is the time component and the 3-vector $\mathbf{A}$ contains the three spatial components such that $\mathbf{A} = (A^i) = (A^1, A^2, A^3)$.[1] Thus, the contravariant components A^1, A^2, and A^3 are the *physical* components, i.e. A_x, A_y, and A_z, respectively, whereas the covariant components A_1, A_2, and A_3 will be related to the contravariant components.[2] Specifically, the 4-position vector (or spacetime point) is given by

$$x = (x^\mu) = (x^0, \mathbf{x}) = (x^0, x^1, x^2, x^3) = (ct, x, y, z) = \{c = 1\} = (t, x, y, z).\tag{1.2}$$

Note the 'abuse of notation', which means that we will use the symbol x for both representing the 4-position vector as well as its first spatial component. In addition, we introduce the metric tensor as

$$g = (g_{\mu\nu}) = \mathrm{diag}(1, -1, -1, -1),\tag{1.3}$$

which is called the *Minkowski metric*. In this case, the inverse of the metric tensor is trivially given by

$$g^{-1} = (g^{\mu\nu}) = \mathrm{diag}(1, -1, -1, -1).\tag{1.4}$$

---

[1] Note that we will use the convention that Greek indices take the values 0, 1, 2, or 3, whereas Latin (or Roman) indices take the values 1, 2, or 3.

[2] In order for a vector to be invariant under transformations of coordinate systems, the components of the vector have to contra-vary (i.e. vary in the 'opposite' direction) with a change of basis to compensate for that change. Therefore, vectors (e.g. position, velocity, and acceleration) are contravariant, whereas so-called *dual vectors* (e.g. gradients) are covariant.

Thus, for the Minkowski metric, we have that $g_{\mu\nu} = g^{\mu\nu}$, i.e. the covariant and contravariant components are equal to each other, which does not hold for a general metric. Owing to the choice of the Minkowski metric, it also holds for the 4-vector in Eq. (1.1) that $A^0 = A_0$, $A^1 = -A_1 = A_x$, $A^2 = -A_2 = A_y$, and $A^3 = -A_3 = A_z$. In fact, in order to raise and lower indices of vectors and tensors, we can use the Minkowski metric tensor and its inverse in the following way:

$$A_\mu = g_{\mu\nu} A^\nu, \quad A^\mu = g^{\mu\nu} A_\nu, \tag{1.5}$$

$$T_\mu{}^\nu = g_{\mu\lambda} T^{\lambda\nu}, \quad T_{\mu\nu} = g_{\mu\lambda} g_{\nu\omega} T^{\lambda\omega}, \quad T^\mu{}_\nu = g^{\mu\lambda} T_{\lambda\nu}, \quad T^{\mu\nu} = g^{\mu\lambda} g^{\nu\omega} T_{\lambda\omega}. \tag{1.6}$$

Normally, in tensor notation à la Einstein,[3] upper indices (or superscripts) of vectors and tensors are called contravariant indices, whereas lower indices (or subscripts) are called covariant indices. In addition, note that in Eqs. (1.5) and (1.6), we have used the so-called *Einstein summation convention*, which means that when an index appears twice in a single term, once as an upper index and once as a lower index, it implies that all its possible values are to be summed over.

Using the Minkowski metric, we can also introduce the inner product between two 4-vectors A and B such that

$$g(A, B) = A^T g B = A \cdot B = A^\mu g_{\mu\nu} B^\nu = A^\mu B_\mu = A^0 B_0 + A^i B_i = A^0 B^0 - \mathbf{A} \cdot \mathbf{B}, \tag{1.7}$$

which is **not** positive definite.[4] Therefore, the 'length' of a 4-vector A is given by

$$A^2 = A \cdot A = (A^0)^2 - \mathbf{A}^2, \tag{1.8}$$

where we have again used an abuse of notation, since the symbol A^2 denotes both the second spatial contravariant component of the vector A and the 'length' of the vector A. Nevertheless, note that the 'length' is indefinite, i.e. it can be either positive or negative. One says that A is *time-like* if $A^2 > 0$, *light-like* if $A^2 = 0$, and *space-like* if $A^2 < 0$. In particular, it holds for a 4-position vector x that

$$x^2 = (x^0)^2 - \mathbf{x}^2 = t^2 - x^2 - y^2 - z^2. \tag{1.9}$$

Next, we introduce the Minkowski spacetime M such that $M = (\mathbb{R}^4, g)$, which is the set of all 4-position vectors [cf. Eq. (1.2)]. Note that the metric tensor g is a bilinear form $g : M \times M \to \mathbb{R}$ such that $g(x, y) = g_{\mu\nu} x^\mu y^\nu$, where $g_{\mu\nu}$ are strictly the components of the metric g, which are usually identified with the tensor itself.

---

[3] In 1921, A. Einstein was awarded the Nobel Prize in physics 'for his services to Theoretical Physics, and especially for his discovery of the law of the photoelectric effect'.

[4] In general, note that the superscript T denotes the transpose of a matrix.

Finally, we introduce the totally antisymmetric *Levi-Civita (pseudo)tensor* in three spatial dimensions

$$
\epsilon^{ijk} = \epsilon_{ijk} = \begin{cases} 1 & \text{if } ijk \text{ are even permutations of } 1,2,3 \\ -1 & \text{if } ijk \text{ are odd permutations of } 1,2,3 \\ 0 & \text{if any two indices are equal} \end{cases} \tag{1.10}
$$

as well as in four spacetime dimensions

$$
\epsilon^{\mu\nu\lambda\omega} = \begin{cases} 1 & \text{if } \mu\nu\lambda\omega \text{ are even permutations of } 0,1,2,3 \\ -1 & \text{if } \mu\nu\lambda\omega \text{ are odd permutations of } 0,1,2,3 \ , \\ 0 & \text{if any two indices are equal} \end{cases} \tag{1.11}
$$

which we define such that $\epsilon^{0123} = 1$, which implies that $\epsilon_{0123} = -1$. In addition, in three dimensions, the following contractions hold for the Levi-Civita tensor:

$$
\epsilon^{ijk}\epsilon^{ijk} = 6, \tag{1.12}
$$

$$
\epsilon^{ijk}\epsilon^{k\ell m} = \delta^{i\ell}\delta^{jm} - \delta^{im}\delta^{j\ell}, \tag{1.13}
$$

$$
\epsilon^{ijk}\epsilon^{\ell mn} = \delta^{i\ell}\left(\delta^{jm}\delta^{kn} - \delta^{jn}\delta^{km}\right) - \delta^{im}\left(\delta^{j\ell}\delta^{kn} - \delta^{jn}\delta^{k\ell}\right)
$$
$$
+ \delta^{in}\left(\delta^{j\ell}\delta^{km} - \delta^{jm}\delta^{k\ell}\right), \tag{1.14}
$$

where $\delta^{ij} = \delta_{ij}$ is the *Kronecker delta* such that $\delta^{ij} = 1$ if $i = j$ and $\delta^{ij} = 0$ if $i \neq j$, whereas, in four dimensions, the corresponding relations are:

$$
\epsilon^{\alpha\beta\gamma\delta}\epsilon_{\alpha\beta\gamma\delta} = -24, \tag{1.15}
$$

$$
\epsilon^{\alpha\beta\gamma\mu}\epsilon_{\alpha\beta\gamma\nu} = -6\delta^{\mu}_{\nu}, \tag{1.16}
$$

$$
\epsilon^{\alpha\beta\mu\nu}\epsilon_{\alpha\beta\lambda\omega} = -2\left(\delta^{\mu}_{\lambda}\delta^{\nu}_{\omega} - \delta^{\mu}_{\omega}\delta^{\nu}_{\lambda}\right), \tag{1.17}
$$

$$
\epsilon^{\alpha\mu\nu\lambda}\epsilon_{\alpha\omega\rho\sigma} = -\delta^{\mu}_{\omega}\delta^{\nu}_{\rho}\delta^{\lambda}_{\sigma} + \delta^{\mu}_{\rho}\delta^{\nu}_{\omega}\delta^{\lambda}_{\sigma} + \delta^{\mu}_{\omega}\delta^{\nu}_{\sigma}\delta^{\lambda}_{\rho} - \delta^{\mu}_{\rho}\delta^{\nu}_{\sigma}\delta^{\lambda}_{\omega} - \delta^{\mu}_{\sigma}\delta^{\nu}_{\omega}\delta^{\lambda}_{\rho} + \delta^{\mu}_{\sigma}\delta^{\nu}_{\rho}\delta^{\lambda}_{\omega}, \tag{1.18}
$$

where $\delta^{\mu}_{\nu} \equiv g^{\mu}{}_{\nu} = g_{\nu}{}^{\mu}$ such that $\delta^{\mu}_{\nu} = 1$ if $\mu = \nu$ and $\delta^{\mu}_{\nu} = 0$ if $\mu \neq \nu$.

1.2 The Lorentz group

A *Lorentz transformation* Λ is a linear mapping of M onto itself, $\Lambda : M \to M$, $x \mapsto x' = \Lambda x$,[5] which preserves the inner product, i.e. the inner product is *invariant* under Lorentz transformations:

$$
x' \cdot y' = x \cdot y, \quad \text{where } x' = \Lambda x \text{ and } y' = \Lambda y. \tag{1.19}
$$

---

[5] We will denote the set of Lorentz transformations by $\mathcal{L}$, which is called the *Lorentz group*. See Appendix A for the definition of a group as well as a short general discussion on group theory.

In component form, we have $x'^\mu = (\Lambda x)^\mu = \Lambda^\mu{}_\nu x^\nu$ and $y'_\mu = (\Lambda y)_\mu = \Lambda_\mu{}^\lambda y_\lambda = \Lambda_{\mu\lambda} y^\lambda$, which means that

$$\Lambda^\mu{}_\nu \Lambda_{\mu\lambda} = g_{\nu\lambda} \quad \Leftrightarrow \quad (\Lambda^T)_\nu{}^\mu g_{\mu\omega} \Lambda^\omega{}_\lambda = g_{\nu\lambda}. \tag{1.20}$$

Thus, the Lorentz group is given by

$$\mathcal{L} = \{\Lambda : M \to M : x' \cdot y' = x \cdot y, \quad \text{where } x' = \Lambda x, \ y' = \Lambda y, \text{ and } x, y \in M\}, \tag{1.21}$$

i.e. it consists of real, linear transformations that leave the inner product invariant, and hence, one says that the Lorentz group is an *invariance group*.

An explicit example of a Lorentz transformation relating two *inertial frames* (or *inertial coordinate systems*)[6] S and S' (with 4-position vectors x and x', respectively) that move along the positive x^1-axis is given by

$$x' = \begin{pmatrix} x'^0 \\ x'^1 \\ x'^2 \\ x'^3 \end{pmatrix} = \begin{pmatrix} \cosh\xi & -\sinh\xi & 0 & 0 \\ -\sinh\xi & \cosh\xi & 0 & 0 \\ 0 & 0 & 1 & 0 \\ 0 & 0 & 0 & 1 \end{pmatrix} \begin{pmatrix} x^0 \\ x^1 \\ x^2 \\ x^3 \end{pmatrix} = \Lambda^{(01)} x, \tag{1.22}$$

where ξ is any real number and $\Lambda^{(01)} = \Lambda^{(01)}(\xi)$ denotes the Lorentz transformation. Note that $\Lambda^{(01)}$ is often called the *standard configuration Lorentz transformation* and constitutes an example of a *boost (or standard transformation)*. The parameter ξ is called the *rapidity (or boost parameter)*. Using the hyperbolic identity $\cosh^2\xi - \sinh^2\xi = 1$, one easily observes that indeed $x'^2 = x^2$. By direct computation, one finds that $\Lambda^{(01)}(\xi+\xi') = \Lambda^{(01)}(\xi)\Lambda^{(01)}(\xi')$. Similar to Eq. (1.22), one could define the Lorentz transformations $\Lambda^{(02)}$ and $\Lambda^{(03)}$. Another way of writing the Lorentz transformation $\Lambda^{(01)}$ is

$$\Lambda^{(01)} = \begin{pmatrix} \gamma & -\beta\gamma & 0 & 0 \\ -\beta\gamma & \gamma & 0 & 0 \\ 0 & 0 & 1 & 0 \\ 0 & 0 & 0 & 1 \end{pmatrix}, \tag{1.23}$$

where the two parameters β and γ are introduced as

$$\beta \equiv \frac{v}{c} = \{c = 1\} = v, \tag{1.24}$$

$$\gamma \equiv \frac{1}{\sqrt{1-\beta^2}} = \frac{1}{\sqrt{1-v^2}} = \gamma(v). \tag{1.25}$$

---

[6] An inertial frame is a reference system in which free particles (i.e. particles that are **not** influenced by any forces) are moving with uniform velocity. Any reference system moving with constant velocity relative to an inertial frame is another inertial frame. Note that there are infinitely many inertial frames and that the laws of physics have to have the same form in all inertial frames, i.e. the laws of physics are so-called *Lorentz covariant*.

Here v is the relative velocity of the two inertial frames. Note that, comparing Eqs. (1.22) and (1.23), it holds that

$$\cosh \xi = \gamma \quad \text{and} \quad \sinh \xi = \beta \gamma = v\gamma. \tag{1.26}$$

Thus, it follows that the rapidity is given by

$$\xi = \text{artanh}\, v \quad \Leftrightarrow \quad \tanh \xi = v. \tag{1.27}$$

The physical interpretation is as follows. A particle (or an observer K) at rest in the inertial frame S is represented by a *world-line* parallel to the time axis, i.e. the x^0-axis. Without loss of generality, let us assume that K is at rest at the origin of the three-dimensional space in S. The same K can also be viewed from another inertial frame S', which is related to S by the Lorentz transformation $\Lambda^{(01)}$. In the S'-coordinates, the world-line of K is given by

$$x'^0 = \tau \cosh \xi, \quad x'^1 = -\tau \sinh \xi, \quad x'^2 = x'^3 = 0,$$

where we have used $x^0 = \tau$ and $x^1 = x^2 = x^3 = 0$ for the S-coordinates. The velocity of K along the x'^1-axis is now

$$v' = \frac{dx'^1}{dt'} = -\tanh \xi \leq 0,$$

whereas the velocity along the other axes is zero. Thus, we can interpret this result as either (i) K (or the particle) is moving with velocity $-v'$ along the negative x'^1-axis or (ii) S' is moving with velocity $v = dx^1/dt = \tanh \xi \geq 0$ along the positive x^1-axis (see Fig. 1.1). Note that $v' = -v$.

Now, using the inner product $g(\Lambda x, \Lambda y) = \Lambda x \cdot \Lambda y = x \cdot y = g(x, y)$, where $x, y \in M$ and $\Lambda \in \mathcal{L}$, we find that

(1) $x \cdot y = x^T g y$ and
(2) $\Lambda x \cdot \Lambda y = (\Lambda x)^T g(\Lambda y) = x^T \Lambda^T g \Lambda y$.

If conditions (1) and (2) are equivalent, which they are, since the inner product is invariant under Lorentz transformations, they imply that

$$g = \Lambda^T g \Lambda, \tag{1.28}$$

which is basically the same equation as Eq. (1.20), but in matrix form. From this equation, one obtains

- $(\det \Lambda)^2 = 1 \quad \Rightarrow \quad \det \Lambda = \pm 1 \quad$ and
- $\left(\Lambda^0{}_0\right)^2 = 1 + \sum_{i=1}^3 \left(\Lambda^i{}_0\right)^2 \geq 1 \quad \Rightarrow \quad \Lambda^0{}_0 \geq 1 \quad \text{or} \quad \Lambda^0{}_0 \leq -1.$

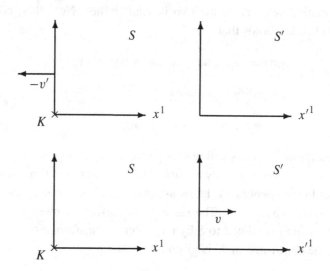

Figure 1.1 The two upper inertial frames show the first interpretation, i.e. S is moving to the left relative to S', whereas the two lower inertial frames show the second interpretation, i.e. S' is moving to the right relative to S.

The four conditions $\det \Lambda = \pm 1$, $\Lambda^0{}_0 \geq 1$, and $\Lambda^0{}_0 \leq 1$ can be used to classify the elements of the Lorentz group. Thus, we can divide the Lorentz group $\mathcal{L}$ [denoted SO(1, 3) in group theory] into the following subgroups (see Fig. 1.2):

$$\mathcal{L}_+ = \{\Lambda \in \mathcal{L} : \det \Lambda = 1\} \qquad \text{pure (or proper) Lorentz group,} \qquad (1.29)$$

$$\mathcal{L}^\uparrow = \{\Lambda \in \mathcal{L} : \Lambda^0{}_0 \geq 1\} \qquad \text{orthochronous Lorentz group,} \qquad (1.30)$$

$$\mathcal{L}^\uparrow_+ = \mathcal{L}_+ \cap \mathcal{L}^\uparrow \qquad \text{pure and orthochronous Lorentz group.} \qquad (1.31)$$

Note that the three subsets

$$\mathcal{L}^\uparrow_- = \left\{\Lambda \in \mathcal{L} : \det \Lambda = -1, \Lambda^0{}_0 \geq 1\right\}, \qquad (1.32)$$

$$\mathcal{L}^\downarrow_- = \left\{\Lambda \in \mathcal{L} : \det \Lambda = -1, \Lambda^0{}_0 \leq 1\right\}, \qquad (1.33)$$

$$\mathcal{L}^\downarrow_+ = \left\{\Lambda \in \mathcal{L} : \det \Lambda = 1, \Lambda^0{}_0 \leq 1\right\} \qquad (1.34)$$

are **not** subgroups of $\mathcal{L}$. However, $\mathcal{L}_0 = \mathcal{L}^\uparrow_+ \cup \mathcal{L}^\downarrow_-$ is a subgroup of $\mathcal{L}$. Hence, $\mathcal{L}_+$, $\mathcal{L}^\uparrow$, $\mathcal{L}^\uparrow_+$, and $\mathcal{L}_0$ are the invariant subgroups of $\mathcal{L}$. The other subsets of $\mathcal{L}$, which are not subgroups, can be connected to $\mathcal{L}^\uparrow_+$ by the three corresponding and following Lorentz transformations.

(1) *Parity* (or *space inversion*).

$$\Lambda_P = \text{diag}(1, -1, -1, -1) \in \mathcal{L}^\uparrow_- \qquad (1.35)$$

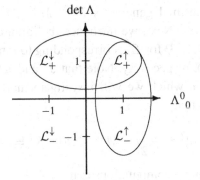

Figure 1.2 Subgroups of the Lorentz group $\mathcal{L}$. The intersection of the two ellipses corresponds to the pure and orthochronous Lorentz group $\mathcal{L}_+^\uparrow$, whereas the horizontal and vertical ellipses correspond to the subgroups $\mathcal{L}_+$ and $\mathcal{L}^\uparrow$, respectively.

(2) *Time reversal* (or *time inversion*).

$$\Lambda_T = \text{diag}(-1, 1, 1, 1) \in \mathcal{L}_-^\downarrow \tag{1.36}$$

(3) *Spacetime inversion*.

$$\Lambda_{PT} = \Lambda_P \Lambda_T = \text{diag}(-1, -1, -1, -1) \in \mathcal{L}_+^\downarrow \tag{1.37}$$

Thus, it holds that $\mathcal{L}_-^\uparrow = \Lambda_P \mathcal{L}_+^\uparrow$, $\mathcal{L}_-^\downarrow = \Lambda_T \mathcal{L}_+^\uparrow$, and $\mathcal{L}_+^\downarrow = \Lambda_{PT} \mathcal{L}_+^\uparrow$, which are so-called *cosets* of $\mathcal{L}$ with respect to $\mathcal{L}_+^\uparrow$. The four subsets (one subgroup and three cosets) $\mathcal{L}_+^\uparrow, \mathcal{L}_-^\uparrow, \mathcal{L}_-^\downarrow$, and $\mathcal{L}_+^\downarrow$ are disjoint and **not** continuously connected (see again Fig. 1.2). Finally, we obtain

$$\mathcal{L} = \mathcal{L}_+^\uparrow \cup \mathcal{L}_-^\uparrow \cup \mathcal{L}_-^\downarrow \cup \mathcal{L}_+^\downarrow = \mathcal{L}_+^\uparrow \cup \Lambda_P \mathcal{L}_+^\uparrow \cup \Lambda_T \mathcal{L}_+^\uparrow \cup \Lambda_{PT} \mathcal{L}_+^\uparrow. \tag{1.38}$$

The pure and orthochronous (or restricted) Lorentz group $\mathcal{L}_+^\uparrow$ [denoted $SO^+(1, 3)$ in group theory] is a matrix Lie group (see Appendix A.2), which means that every matrix can be written in the form $\exp\left(-\frac{i}{2}\omega_{\mu\nu} M^{\mu\nu}\right)$, where $\omega_{\mu\nu} \in \mathbb{R}$ such that $\omega_{\mu\nu} = -\omega_{\nu\mu}$ and $M^{\mu\nu}$ are the so-called *generators* of $\mathcal{L}_+^\uparrow$. The number of generators of $\mathcal{L}_+^\uparrow$ is six, since we define $M^{\mu\nu} = -M^{\nu\mu}$, which implies that $M^{00} = M^{11} = M^{22} = M^{33} = 0$. Thus, $\mathcal{L}_+^\uparrow$ is a six-parameter group. For example, the infinitesimal generator for the 'rotation' in the $x^0 x^1$-plane corresponding to the Lorentz transformation $\Lambda^{(01)}$ is given by

$$M^{01} = i \left. \frac{d\Lambda^{(01)}(\xi)}{d\xi} \right|_{\xi=0} = \begin{pmatrix} 0 & -i & 0 & 0 \\ -i & 0 & 0 & 0 \\ 0 & 0 & 0 & 0 \\ 0 & 0 & 0 & 0 \end{pmatrix}. \tag{1.39}$$

The other five infinitesimal generators M^{02}, M^{03}, M^{32}, M^{13}, and M^{21} can be derived in a similar way. Next, we introduce the infinitesimal generators J^i ($i = 1, 2, 3$) and K^i ($i = 1, 2, 3$) for $\mathcal{L}_+^\uparrow$, corresponding to rotations and boosts (or standard transformations), respectively. Note that J^i and K^i are constructed in terms of $M^{\mu\nu}$, or vice versa, which we will investigate more in Sections 1.3 and 1.4. Nevertheless, it holds that

$$J^i = -\frac{1}{2}\epsilon^{ijk}M^{jk} \quad \text{and} \quad K^i = M^{0i}. \tag{1.40}$$

Then, one finds by simple computations that

$$\left[J^i, J^j\right] = i\epsilon^{ijk}J^k, \tag{1.41}$$

$$\left[J^i, K^j\right] = i\epsilon^{ijk}K^k, \tag{1.42}$$

$$\left[K^i, K^j\right] = -i\epsilon^{ijk}J^k, \tag{1.43}$$

which form a Lie algebra (see Appendix A.3). Actually, defining the operators

$$\mathbf{j} = \frac{1}{2}(\mathbf{J} + i\mathbf{K}) \quad \text{and} \quad \mathbf{k} = \frac{1}{2}(\mathbf{J} - i\mathbf{K}), \tag{1.44}$$

i.e. linear combinations of the operators $\mathbf{J}$ and $\mathbf{K}$, one obtains the following commutation relations

$$\left[j^i, j^j\right] = i\epsilon^{ijk}j^k, \tag{1.45}$$

$$\left[j^i, k^j\right] = 0, \tag{1.46}$$

$$\left[k^i, k^j\right] = i\epsilon^{ijk}k^k, \tag{1.47}$$

which give an alternative basis for the Lie algebra. From the commutation relations in Eqs. (1.45)–(1.47), we observe that the operators $\mathbf{j}$ and $\mathbf{k}$ are decoupled. This is described by $\mathfrak{su}(2) \oplus \mathfrak{su}(2)$ called the Lorentz algebra,[7] which is the Lie algebra of the Lie group $SU(2) \otimes SU(2)$ that can be represented as $D^j \otimes D^{j'}$, where D^j ($j = 0, \frac{1}{2}, 1, \ldots$) is an irreducible representation of $SU(2)$. Note that D^j is spanned by basis vectors $|j, m\rangle$, where $m = -j, -j+1, \ldots, j$. Thus, we have the basis vectors $|j, m; j', m'\rangle = |j, m\rangle|j', m'\rangle$. In addition, we find the relations

$$\mathbf{J}^2 - \mathbf{K}^2 = 2\left(\mathbf{j}^2 + \mathbf{k}^2\right), \tag{1.48}$$

$$\mathbf{J} \cdot \mathbf{K} = -i\left(\mathbf{j}^2 - \mathbf{k}^2\right), \tag{1.49}$$

which are invariant forms (i.e. they commute with the generators $\mathbf{j}$ and $\mathbf{k}$) that are multiplets of the unit operator $\mathbb{1}$ with the eigenvalues $2[j(j+1) + j'(j'+1)]$ and $-i[j(j+1) - j'(j'+1)]$, respectively. In fact, the invariant forms can be written as

---

[7] Note that $\mathfrak{su}(2) \simeq \mathfrak{so}(3)$, but the Lie group generated by $\mathfrak{su}(2) \simeq \mathfrak{so}(3)$ is $SU(2)$ and **not** $SO(3)$.

$$\mathbf{J}^2 - \mathbf{K}^2 = \frac{1}{2} M^{\mu\nu} M_{\mu\nu}, \tag{1.50}$$

$$\mathbf{J} \cdot \mathbf{K} = \frac{1}{8} \epsilon^{\mu\nu\rho\sigma} M_{\mu\nu} M_{\rho\sigma}. \tag{1.51}$$

The irreducible representations are denoted by $D^{(j,j')} \equiv D^j \otimes D^{j'}$, where $j, j' = 0, \frac{1}{2}, 1, \ldots$ For example, an explicit representation for the generators of $D^{\left(\frac{1}{2},0\right)}$ and $D^{\left(0,\frac{1}{2}\right)}$ is given by

$$\mathbf{j} = \frac{1}{2}\boldsymbol{\sigma}, \ \mathbf{k} = 0 \quad \text{and} \quad \mathbf{j} = 0, \ \mathbf{k} = \frac{1}{2}\boldsymbol{\sigma}, \tag{1.52}$$

respectively, where $\boldsymbol{\sigma}$ is the vector of *Pauli matrices*, i.e.

$$\boldsymbol{\sigma} = (\sigma^1, \sigma^2, \sigma^3) = \left(\begin{pmatrix} 0 & 1 \\ 1 & 0 \end{pmatrix}, \begin{pmatrix} 0 & -i \\ i & 0 \end{pmatrix}, \begin{pmatrix} 1 & 0 \\ 0 & -1 \end{pmatrix} \right), \tag{1.53}$$

which satisfy the commutation relation $[\sigma^i, \sigma^j] = 2i\epsilon^{ijk}\sigma^k$. Thus, this leads to

$$D^{\left(\frac{1}{2},0\right)}: \quad \mathbf{J} = \frac{1}{2}\boldsymbol{\sigma}, \quad \mathbf{K} = -i\frac{1}{2}\boldsymbol{\sigma}, \tag{1.54}$$

$$D^{\left(0,\frac{1}{2}\right)}: \quad \mathbf{J} = \frac{1}{2}\boldsymbol{\sigma}, \quad \mathbf{K} = i\frac{1}{2}\boldsymbol{\sigma}. \tag{1.55}$$

1.3 The Poincaré group

The *Poincaré group* (or the inhomogeneous Lorentz group), which is a (linear) Lie group, is given by

$$\mathcal{P} = \left\{ (\Lambda, a) : x^\mu \mapsto x'^\mu = \Lambda^\mu{}_\nu x^\nu + a^\mu \right\}, \tag{1.56}$$

where $\Lambda \in \mathcal{L}$ is a Lorentz transformation and $a \in \mathbb{R}^4$ is a translation. Thus, the Lorentz group and the translation group are subgroups of the Poincaré group. In addition, note that one has $\mathcal{P}^\uparrow_+$ if $\Lambda \in \mathcal{L}^\uparrow_+$, i.e. the pure and orthochronous Poincaré group. The group multiplication law of the Poincaré group is

$$(\Lambda_2, a_2)(\Lambda_1, a_1) = (\Lambda_2\Lambda_1, \Lambda_2 a_1 + a_2), \tag{1.57}$$

where (Λ_1, a_1) and (Λ_2, a_2) are two elements of the Poincaré group, i.e. $\Lambda_1, \Lambda_2 \in \mathcal{L}$ and $a_1, a_2 \in \mathbb{R}^4$. In addition, the identity element is $(\mathbb{1}_4, 0)$ and the inverse is given by $(\Lambda, a)^{-1} = (\Lambda^{-1}, -\Lambda^{-1}a)$, where $(\Lambda, a) \in \mathcal{P}$. An element of the Poincaré group (Λ, a) can be represented by a 5×5 matrix in the following way:

$$(\Lambda, a) \mapsto \begin{pmatrix} \Lambda & a \\ 0 & 1 \end{pmatrix}, \tag{1.58}$$

which means that the group multiplication law (1.57) simply corresponds to ordinary matrix multiplication of 5×5 matrices. The generators of the Poincaré group are $M^{\mu\nu}$ and P^μ, which give rise to the unitary operators that represent the elements $(\Lambda, 0)$ and $(\mathbb{1}_4, a)$ of the Poincaré group, i.e.

$$U(\Lambda, 0) = \exp\left(-\tfrac{i}{2}\omega_{\mu\nu}M^{\mu\nu}\right), \tag{1.59}$$

$$U(\mathbb{1}_4, a) = \exp(ia_\mu P^\mu), \tag{1.60}$$

where again $\omega_{\mu\nu}$ and $M^{\mu\nu}$ are the parameters and generators of the Lorentz subgroup, respectively, and a_μ and P^μ are the parameters and generators of the translation subgroup, respectively. Note that if the elements of the Poincaré group are close to the identity or to first order in infinitesimal parameters of the Poincaré group, we have

$$U(\Lambda, a) \simeq \exp\left(-\tfrac{i}{2}\omega_{\mu\nu}M^{\mu\nu} + ia_\mu P^\mu\right). \tag{1.61}$$

The different generators of the Poincaré group satisfy the following commutation relations:

$$[M^{\mu\nu}, M^{\rho\sigma}] = i(g^{\mu\rho}M^{\nu\sigma} + g^{\nu\sigma}M^{\mu\rho} - g^{\nu\rho}M^{\mu\sigma} - g^{\mu\sigma}M^{\nu\rho}), \tag{1.62}$$

$$[M^{\mu\nu}, P^\sigma] = -i(g^{\nu\sigma}P^\mu - g^{\mu\sigma}P^\nu), \tag{1.63}$$

$$[P^\mu, P^\nu] = 0, \tag{1.64}$$

which are the commutation relations of the *Poincaré algebra*, i.e. the algebra that corresponds to the Poincaré group. Finally, returning to the example of the infinitesimal generator for the 'rotation' in the $x^0 x^1$-plane, i.e. M^{01} in Eq. (1.39), the corresponding generator of the Poincaré group is given by

$$M^{01} = i\left.\frac{d(\Lambda^{(01)}(\xi), 0)}{d\xi}\right|_{\xi=0} = \begin{pmatrix} 0 & -i & 0 & 0 & 0 \\ -i & 0 & 0 & 0 & 0 \\ 0 & 0 & 0 & 0 & 0 \\ 0 & 0 & 0 & 0 & 0 \\ 0 & 0 & 0 & 0 & 0 \end{pmatrix}. \tag{1.65}$$

In the case of the Lorentz group, the generator M^{01} is represented by a 4×4 matrix, whereas in the case of the Poincaré group, the generator M^{01} is represented by a 5×5 matrix.

Exercise 1.1 Verify the commutation relations $[J^1, J^2] = iJ^3$, $[K^1, K^2] = -iJ^3$, and $[J^1, K^2] = iK^3$ using Eq. (1.62) as well as the definitions $\mathbf{J} = (J^1, J^2, J^3) = (M^{32}, M^{13}, M^{21})$ and $\mathbf{K} = (K^1, K^2, K^3) = (M^{01}, M^{02}, M^{03})$.

1.4 Casimir operators

Casimir operators are constructed from the generators of a group and commute with **all** these generators and they are invariants of the given group. Note that only scalar operators can be Casimir operators of a group. For example, $\mathbf{J}^2$ is the Casimir operator of the angular momentum algebra [or the rotation group SO(3)], since $[\mathbf{J}^2, J^i] = 0$, where J^i ($i = 1, 2, 3$) are the generators of SO(3), which has the $\mathfrak{so}(3)$ algebra $[J^i, J^j] = i\epsilon^{ijk} J^k$ (see the discussion in Appendix A.5).

The Poincaré group has two Casimir operators, $P^2 = P_\mu P^\mu$ and $w^2 = w_\mu w^\mu$, where

$$w_\sigma = \frac{1}{2}\epsilon_{\sigma\mu\nu\lambda} M^{\mu\nu} P^\lambda \tag{1.66}$$

is the *Pauli–Lubanski (pseudo)vector*. The time component of the Pauli–Lubanski vector is given by $w_0 = w^0 = \mathbf{P} \cdot \mathbf{J}$, while the 3-vector part of this 4-vector can be written as $\mathbf{w} = P^0 \mathbf{J} + \mathbf{P} \times \mathbf{K}$, where $\mathbf{P} = (P^1, P^2, P^3)$, $\mathbf{J} = (J^1, J^2, J^3) = (M^{32}, M^{13}, M^{21})$, and $\mathbf{K} = (K^1, K^2, K^3) = (M^{01}, M^{02}, M^{03})$. The quantities J^1, J^2, and J^3 are the generators of the angular momentum algebra, whereas the components K^1, K^2, and K^3 are the boosts in the Cartesian coordinate directions x, y, and z, respectively. Note that w_μ is orthogonal to P^μ, since $w_\mu P^\mu = 0$. This can be shown as follows: $w_\mu P^\mu = \frac{1}{2}\epsilon_{\mu\nu\rho\sigma} M^{\nu\rho} P^\sigma P^\mu = 0$, because $\epsilon_{\mu\nu\rho\sigma}$ is a totally antisymmetric tensor and $P^\sigma P^\mu$ is a symmetric tensor with respect to the indices σ and μ. In addition, the Pauli–Lubanski vector satisfies the following commutation relations with the generators of the Poincaré group and itself:

$$[M_{\mu\nu}, w_\sigma] = -i\left(g_{\nu\sigma} w_\mu - g_{\mu\sigma} w_\nu\right), \tag{1.67}$$

$$[P_\mu, w_\nu] = 0, \tag{1.68}$$

$$[w_\mu, w_\nu] = i\epsilon_{\mu\nu\rho\sigma} w^\rho P^\sigma. \tag{1.69}$$

Now, there is also another way of writing the Pauli–Lubanski vector. Introducing the quantity

$$v^{\mu\nu\rho} = P^\mu M^{\nu\rho} + P^\nu M^{\rho\mu} + P^\rho M^{\mu\nu}, \tag{1.70}$$

the Pauli–Lubanski vector can be written as

$$w = (w^0, w^1, w^2, w^3) = (v^{321}, v^{320}, v^{130}, v^{210}). \tag{1.71}$$

Thus, the second Casimir operator is given by

$$w^2 = -\frac{1}{6}v^{\mu\nu\rho} v_{\mu\nu\rho}. \tag{1.72}$$

Next, one can show that the scalar operators $P^2 = P_\mu P^\mu$ and $w^2 = w_\mu w^\mu$ generate **all** invariants of the Poincaré group $\mathcal{P}_+^\uparrow$. Note that P^2 and w^2 are sometimes referred to as the *first* and *second Casimir invariants* of the Poincaré group, respectively. Actually, since the Poincaré group has rank 2, there are only two Casimir operators or invariants. Now, using the definition of the Pauli–Lubanski vector (1.66), the contraction in Eq. (1.18), and the commutation relation in Eq. (1.63), the second Casimir invariant can be written as

$$w^2 = -\frac{1}{2} M^{\mu\nu} M_{\mu\nu} P^2 + M^{\mu\sigma} M_{\nu\sigma} P_\mu P^\nu, \tag{1.73}$$

where the first term is proportional to the first Casimir invariant. Of course, the commutation relations between the Casimir operators and the generators of the Poincaré group are equal to zero, i.e.

$$[P^2, M^{\mu\nu}] = 0, \tag{1.74}$$

$$[P^2, P^\mu] = 0, \tag{1.75}$$

$$[w^2, M^{\mu\nu}] = 0, \tag{1.76}$$

$$[w^2, P^\mu] = 0. \tag{1.77}$$

In what follows, we will use the Casimir operators, or actually their eigenvalues, in order to classify the so-called *irreducible representations* of the Poincaré group.

Exercise 1.2   Verify the commutation relations in Eqs. (1.67)–(1.69).

Exercise 1.3   Show that

$$\frac{1}{6} v^{\mu\nu\rho} v_{\mu\nu\rho} = \frac{1}{2} M^{\mu\nu} M_{\mu\nu} P^2 - M^{\mu\rho} M_{\nu\rho} P_\mu P^\nu$$

by direct computation.

1.5 General description of relativistic states

A general problem when describing relativistic states is to find the meaning of the time t and the position vector $\mathbf{x}$ in a wave function $\Psi(t, \mathbf{x})$. Therefore, it is better to choose more 'safe' variables to describe the states, which could be the 3-momentum vector $\mathbf{p}$ and the spin ζ. Next, we want to look at relativistic transformations of the variables $\mathbf{p}$ and ζ in order to observe how they behave during such transformations. The group of relativistic transformations is the Poincaré group.

Now, let us introduce the concept of a *representation* of a group (see also the discussion in Appendix A.2). Assume that $\mathcal{G}$ is a group. It follows that the operator $U(g)$, where $g \in \mathcal{G}$, is a representation of $\mathcal{G}$ if $U(g_1 g_2) = U(g_1)U(g_2)$, where $g_1, g_2 \in \mathcal{G}$, and $U(g^{-1}) = U^{-1}(g)$. Note that $U(g)$ acts on a given Hilbert space. In addition, the representation is unitary if $U(g)$ is unitary.

Definition A unitary representation $U(g)$ is *reducible* if it may be written as

$$U(g) = \begin{pmatrix} U_1(g) & 0 \\ 0 & U_2(g) \end{pmatrix}, \tag{1.78}$$

otherwise it is *irreducible*.

Lemma 1.1 (Schur's lemma) *If $U(g)$ is irreducible, and the operator A commutes with all the $U(g)$, then A is a multiple of the unit operator, i.e. $A = \alpha \mathbb{1}$, where α is a constant.*

The generators of the Poincaré group may be described as generators of rotations (three generators), boosts (three generators), and translations (four generators). Thus, in total, there are ten generators for the Poincaré group.

Now, consider any representation of the Poincaré group, then the unitary operator $U(\Lambda, a)$ can be written as follows.

- For rotations:

$$U(R(\theta), 0) \simeq 1 - i\boldsymbol{\theta} \cdot \mathbf{J}. \tag{1.79}$$

- For boosts:

$$U(L(\xi), 0) \simeq 1 - i\boldsymbol{\xi} \cdot \mathbf{K}. \tag{1.80}$$

- For translations:

$$U(\mathbb{1}_4, a) \simeq 1 + ia \cdot P. \tag{1.81}$$

Commutation relations for J^i, K^i, and P^μ are implicitly given by Eqs. (1.62)–(1.64). However, note that the forms of J^i, K^i, and P^μ are **not** unique, but differ for different representations.

Next, consider a quantum system represented by the state $|\psi\rangle$. A Poincaré transformation g will carry it over to a new state $|\psi_g\rangle$. We have that

$$U(g) = U(\Lambda, a) : \quad |\psi_g\rangle = U(g)|\psi\rangle = U(\Lambda, a)|\psi\rangle. \tag{1.82}$$

Then, consider a unitary representation of the Poincaré group. Since a reducible representation can be decomposed into orthogonal irreducible representations, we will consider only irreducible representations, which we will identify as those describing elementary systems called *particles*.

1.6 Irreducible representations of the Poincaré group

Let us now look for unitary irreducible representations of the Poincaré group $\mathcal{P}_+^\uparrow$, i.e. $U(\mathcal{P}_+^\uparrow)$. In order to investigate these representations, we can study the eigenvalues of the Casimir operators, i.e. we want to determine λ and λ' in the relations

$P^2 = \lambda \mathbb{1}$ and $w^2 = \lambda' \mathbb{1}$. Using the *correspondence principle*, the eigenvalues of the 4-momentum operator $P = (P^\mu)$ are $p = (p^\mu) = (E, \mathbf{p})$, where E is the energy and $\mathbf{p}$ is the 3-momentum. Thus, $\lambda = E^2 - \mathbf{p}^2$. In addition, according to the basic postulate of special theory of relativity, i.e. the *principle of relativity*, we have that $\lambda = m^2$, where m is the mass of a particle. In order to determine λ', we can study the effect of the Pauli–Lubanski vector on an arbitrary state $|m, p^\mu, \gamma\rangle$, where γ refers to some other quantum numbers, describing a particle

$$w^\mu |m, p^\mu, \gamma\rangle = \frac{1}{2}\epsilon^{\mu\nu\rho\sigma} M_{\nu\rho} P_\sigma |m, p^\mu, \gamma\rangle. \tag{1.83}$$

Since w^2 should also be a multiple of the unit operator $\mathbb{1}$, it is sufficient to look for its value for one vector in Hilbert space. Therefore, we can choose $p' = (m, \mathbf{0})$, which describes a particle in its rest frame, and thus we obtain

$$w^0 |m, (m, \mathbf{0}), \gamma\rangle = \frac{1}{2}\epsilon^{0\nu\rho\sigma} M_{\nu\rho} p'_\sigma |m, (m, \mathbf{0}), \gamma\rangle$$

$$= \frac{1}{2}\epsilon^{0\nu\rho 0} M_{\nu\rho} m |m, (m, \mathbf{0}), \gamma\rangle = 0, \tag{1.84}$$

$$w^i |m, (m, \mathbf{0}), \gamma\rangle = \frac{1}{2}\epsilon^{ijk0} M_{jk} m |m, (m, \mathbf{0}), \gamma\rangle = m J^i |m, (m, \mathbf{0}), \gamma\rangle. \tag{1.85}$$

Thus, in the particle's rest frame, by using Eqs. (1.84) and (1.85), w is space-like, i.e. $w = (0, \mathbf{w}) = (0, m\mathbf{J})$, which means that $\mathbf{w}$ is equal to the mass of the particle times the total angular momentum.[8] Hence, we have

$$w^2 |m, (m, \mathbf{0}), \gamma\rangle = \left(w^{0^2} - \mathbf{w}^2\right) |m, (m, \mathbf{0}), \gamma\rangle = -m^2 \mathbf{J}^2 |m, (m, \mathbf{0}), \gamma\rangle$$

$$= -m^2 s(s+1) |m, (m, \mathbf{0}), \gamma\rangle, \tag{1.86}$$

where $s(s+1)$ are the eigenvalues of the operator $\mathbf{J}^2$. Thus, in order to label the different unitary irreducible representations of the Poincaré group, we can use the two quantum numbers m and s, where m is the mass and s is the spin of a particle, both of which are intrinsic properties of particles and should therefore be described by quantities that are invariant under Poincaré transformations. A complete set of commuting observables for characterizing the eigenvectors that describe particles is therefore given by P^2, $\mathbf{P}$, w^2, and $\mathbf{w}$, where the last operator is an operator that gives the projection of the spin along a specific axis, e.g. the z-axis in the rest frame of the particle or the direction of the 3-momentum $\mathbf{p}$. Note that the projection of the spin along the 3-momentum is usually called the *helicity*. In conclusion, the generalized eigenvectors of the unitary irreducible representations of the Poincaré group can be written as $|m, \mathbf{p}; s, m_s\rangle$ or $|m, \mathbf{p}; s, h\rangle$, where h is the helicity, i.e. $h = \mathbf{J} \cdot \mathbf{P}/|\mathbf{p}|$.

[8] Note that this is only true for massive particles, since rest frames do not exist for massless particles.

The irreducible representations of the Poincaré group are classified according to the eigenvalues of the Casimir operators P^2 and w^2. These irreducible representations can be thought of as describing particles. The classification was developed by Wigner in 1939 and it is known as *Wigner's classification*.[9] The classification of the eigenvalues for P^2 and P^0 can be divided into four different main classes, where two classes contain two sub-classes each, and is given as follows.

(1) (1) $p^2 = m^2 > 0,\quad p^0 > 0$
    (2) $p^2 = m^2 > 0,\quad p^0 < 0$
(2) (1) $p^2 = 0,\quad p \neq 0,\quad p^0 > 0$
    (2) $p^2 = 0,\quad p \neq 0,\quad p^0 < 0$
(3) $p^2 = 0,\quad p^0 = 0 \quad\Leftrightarrow\quad p = (p^\mu) = (0,0,0,0) = 0$
(4) $p^2 = m^2 < 0$

Note that Wigner originally denoted the two sub-classes of class 1 by P_+ and P_-, respectively, whereas he denoted the two sub-classes of class 2 by 0_+ and 0_-, respectively. In addition, he denoted class 3 by 0_0, and he attached no symbol to class 4. Obviously, since P^2 is a Casimir operator, all final states obtained by performing Lorentz transformations on an initial state with the eigenvalue $p^2 = m^2$ will have the same value of p^2, and in fact, the sign of p^0 will always be unchanged by the application of a Lorentz transformation. Thus, the two sub-classes with $p^0 < 0$, i.e. classes 12 and 22, are 'unphysical', since the corresponding particles would have negative energy. However, mathematically, there is no problem with particles having negative energy, and since they exist, we cannot simply ignore them. The correct interpretation of the case $\text{sgn}(p^0) = -1$ will become clearer in quantum field theory. Physically, the quantity p^0 is definitely an energy when $\text{sgn}(p^0) = 1$, but it is not appropriate to name it an energy when $\text{sgn}(p^0) = -1$. In an abstract language, we will call the states with $\text{sgn}(p^0) = \pm 1$, positive- and negative-energy states, respectively. Note that sometimes the positive- and negative-energy states are called positive- and negative-frequency states, respectively. The negative-energy states do **not** describe real particles. However, in quantum field theory, they will play an important role describing so-called *antiparticles*, whereas the positive-energy states will describe particles.

Class 1 (or actually class 11) describes massive ($m \neq 0$) particles. We can perform a Lorentz transformation to a frame in which the particles are at rest, i.e. $\mathbf{p} = \mathbf{0}$. In this rest frame, the eigenvalues of $P = (P^\mu)$ are $p = (p^\mu) = (m, \mathbf{0})$, which means that $p^2 = p_\mu p^\mu = m^2$ and $-w^2 = -w^{02} + \mathbf{w}^2 = \mathbf{w}^2 = p^{02} \mathbf{J}^2 = m^2 s(s+1)$, since the eigenvalues of $\mathbf{J}^2$ (in the rest frame) are the values of the

[9] For the interested reader, please see: E. Wigner, On unitary representations of the inhomogeneous Lorentz group, *Ann. Math.* **40**, 149–204 (1939). This classic article has also been reprinted in *Nucl. Phys. B (Proc. Suppl.)* **6**, 9–64 (1989).

total (intrinsic) angular momentum and spin of the particles. Thus, massive particles may be classified according to their mass and spin, which uniquely determine the corresponding unitary irreducible representations of the Poincaré group. In this book, we will mainly focus our attention on $s = 0$, $s = 1/2$, and $s = 1$, but we will also mention some of the other possibilities such as $s = 3/2$.

Class 2 (or actually class 21) describes massless ($m = 0$) particles. In this case, $p^2 = 0$. However, there is **no** rest frame for these particles, since massless particles cannot be at rest. In fact, class 2 can be further divided into massless particles with *discrete* spin (integer multiples of 1/2) and massless particles with *continuous* spin. However, particles with continuous spin have not been observed in Nature, and are therefore assumed to be unphysical. In this book, we will consider $s = 0$, $s = 1/2$, and $s = 1$, but also $s = 2$.

Class 3 corresponds to the vacuum. The only finite-dimensional unitary representation is the trivial representation called the vacuum.

Finally, class 4 describes *virtual particles* (which can have space-like 4-momenta), but it also corresponds to so-called *tachyons*, i.e. 'particles' with imaginary mass that travel faster than the speed of light. In this book, tachyons will not be investigated any further.

1.7 One-particle relativistic states

In this section, we want to obtain the transformation properties of one-particle relativistic states in the plane-wave basis under Poincaré transformations. We will use *Wigner's method*, which means that for a state with 4-momentum p, the action of a Lorentz transformation is to change p, but to leave the inner product $p^2 = p \cdot p$ unchanged.

Unitary operators that represent translations are denoted by

$$U(a) \equiv U(\mathbb{1}_4, a), \tag{1.87}$$

which, in exponential form, reads

$$U(a) = \exp(ia \cdot P), \tag{1.88}$$

where $P = (P^0, \mathbf{P})$ is the 4-momentum operator. Here the generator of time-translations P^0 is the Hamiltonian and the generator of space-translations $\mathbf{P}$ is the 3-momentum operator. Since $P^2 = P \cdot P$ commutes with all generators of the Poincaré group, it implies that P^2 commutes with the unitary operator $U(\Lambda, a)$. Using Schur's lemma, we find that $P^2 = \text{const}$. This constant is m^2. Thus, we obtain that $P^2 = m^2$.

For free particles, we can express the Hamiltonian as

$$P^0 = +\sqrt{m^2 + \mathbf{P}^2},$$ (1.89)

which also means that we have $p^0 = \sqrt{m^2 + \mathbf{p}^2}$. We introduce the plane-wave basis $|p, \zeta\rangle$ as the eigenstates of the 4-momentum operator P^μ such that

$$P^\mu |p, \zeta\rangle = p^\mu |p, \zeta\rangle,$$ (1.90)

where ζ denotes all other quantum numbers (except from p) that describe the states. Exponentiating and using Eq. (1.88), we obtain

$$U(a)|p, \zeta\rangle = e^{ia \cdot P}|p, \zeta\rangle = e^{ia \cdot p}|p, \zeta\rangle.$$ (1.91)

Now, let p' be a fixed 4-momentum such that $p' \cdot p' = m^2$, where $p'^0 > 0$. Then, we can write $p = \Lambda(p)p'$ for any 4-momentum p. Next, we can define a new basis $|\Lambda(p), \zeta\rangle = U(\Lambda(p))|p', \zeta\rangle$. Let us show that $|\Lambda(p), \zeta\rangle$ corresponds to the 4-momentum p. Acting with the unitary operator $U(a)$ on the basis, we find that

$$U(a)|\Lambda(p), \zeta\rangle = U(a)U(\Lambda(p))|p', \zeta\rangle.$$ (1.92)

Using the fact that $U(a)U(\Lambda(p)) = U(\Lambda(p), a) = U(\Lambda(p))U(\Lambda(p)^{-1}a)$, we obtain

$$U(a)|\Lambda(p), \zeta\rangle = U(\Lambda(p))U(\Lambda(p)^{-1}a)|p', \zeta\rangle.$$ (1.93)

Now, $U(a)|p', \zeta\rangle = e^{ia \cdot p'}|p', \zeta\rangle$ and $[\Lambda(p)^{-1}a] \cdot p' = a \cdot \Lambda(p)p' = a \cdot p$ imply that

$$U(a)|\Lambda(p), \zeta\rangle = U[\Lambda(p)]e^{i[\Lambda(p)^{-1}a] \cdot p'}|p', \zeta\rangle = e^{ia \cdot p}U(\Lambda(p))|p', \zeta\rangle$$
$$= e^{ia \cdot p}|\Lambda(p), \zeta\rangle.$$ (1.94)

Thus, in addition, we have

$$P^\mu |\Lambda(p), \zeta\rangle = p^\mu |\Lambda(p), \zeta\rangle.$$ (1.95)

What happens for Lorentz transformations? In order to investigate what happens for Lorentz transformations, we need to introduce the so-called 'little group' (or *stabilizer*) of p', which is a subgroup of the Poincaré group that leaves p' invariant. The little group of p' was first defined by Wigner and it is therefore denoted by $W(p')$. Consider $\Gamma \in W(p')$, i.e. Γ leaves p' invariant, which means that we can write

$$U(\Gamma)|p', \zeta\rangle = \sum_{\zeta'} D_{\zeta'\zeta}(\Gamma)|p', \zeta'\rangle,$$ (1.96)

where $D_{\zeta'\zeta}$ are some coefficients. The matrix $D = (D_{\zeta\zeta'})$ builds up a unitary representation of the little group $W(p')$. If $U(\Lambda, a)$ is irreducible, then the representation D must also be irreducible. We will study the specific form of D for

(1) massive particles (i.e. class 11),
(2) massless particles (i.e. class 21).

1.7.1 Massive particles

In the case of massive particles, the method is comparably simple. Consider the inertial frame with $p'_0 = m$ and $p'_i = 0$, i.e. a particle at rest. In addition, let ζ be the third component of spin, and use λ instead of ζ. Thus, we have the states $|p', \lambda\rangle$. The little group of p' is the group consisting of three-dimensional rotations, which means that we can write the representation as

$$D(R) = D^{(s)}(R(\boldsymbol{\theta})) = \exp\left(-i\boldsymbol{\theta} \cdot \mathbf{s}\right), \tag{1.97}$$

where s is the spin. Specifically, for $s = 1/2$, we have

$$D^{(1/2)}(R(\boldsymbol{\theta})) = \exp\left(-i\boldsymbol{\theta} \cdot \frac{\boldsymbol{\sigma}}{2}\right), \tag{1.98}$$

whereas for arbitrary s, using Eq. (1.96), we write

$$U(R)|p', \lambda\rangle = \sum_{\lambda'} D^{(s)}_{\lambda'\lambda}(R)|p', \lambda'\rangle. \tag{1.99}$$

In order to change from the rest frame to an arbitrary inertial frame with 4-momentum p, we perform without loss of generality a boost $L(p)$ on the states such that $L(p)p' = p$. Thus, we find the new states

$$|L(p), \lambda\rangle = U(L(p))|p', \lambda\rangle, \tag{1.100}$$

where the normalization is given by $\langle L(p), \lambda|L(p'), \lambda'\rangle = p^0\delta(\mathbf{p} - \mathbf{p}')\delta_{\lambda\lambda'}$ for $s = 0$ and $\langle L(p), \lambda|L(p'), \lambda'\rangle = \delta(\mathbf{p} - \mathbf{p}')\delta_{\lambda\lambda'}$ for $s = 1/2$. Now, we want to investigate the effect of an arbitrary Lorentz transformation Λ on the new states $|L(p), \lambda\rangle$. In addition, the Lorentz transformation Λ will change p to Λp. Therefore, we have to perform the following steps.

- Decelerating by the re-boost $L(p)^{-1}$, which goes from the arbitrary inertial frame to the rest frame.
- In the rest frame, we know how the 'old' states transform.
- Accelerating by the boost $L(\Lambda p)$, which goes from the rest frame back to the arbitrary inertial frame.

In other words, performing a Lorentz transformation Λ on the states $|L(p), \lambda\rangle$ and using the fact that $U(L(\Lambda p))U(L(\Lambda p))^{-1} = \mathbb{1}$ as well as the 'representation law' $U(AB) = U(A)U(B)$, we obtain

$$U(\Lambda)|L(p), \lambda\rangle = U(\Lambda)U(L(p))|p', \lambda\rangle$$
$$= U(L(\Lambda p))U(L(\Lambda p))^{-1}U(\Lambda)U(L(p))|p', \lambda\rangle$$
$$= U(L(\Lambda p))U(R(p, \Lambda))|p', \lambda\rangle, \tag{1.101}$$

where $R(p, \Lambda) \equiv L(\Lambda p)^{-1}\Lambda L(p)$ is the so-called *Wigner rotation* and $L(p)$ changes p' into p, Λ changes p into Λp, and finally, $L(\Lambda p)^{-1}$ changes Λp into p'. Thus, operating with $R(p, \Lambda)$ on p' gives p' back, which means that $R(p, \Lambda)$ describes indeed a rotation. Finally, we obtain

$$U(\Lambda)|L(p), \lambda\rangle = U(L(\Lambda p))U(R(p, \Lambda))|p', \lambda\rangle$$
$$= U(L(\Lambda p)) \sum_{\lambda'} D_{\lambda'\lambda}^{(s)}(R(p, \Lambda))|p', \lambda'\rangle$$
$$= \sum_{\lambda'} D_{\lambda'\lambda}^{(s)}(R(p, \Lambda))|L(\Lambda p), \lambda'\rangle, \tag{1.102}$$

where the states $|L(p), \lambda\rangle$ span the covariant spin basis. Another basis is the helicity basis, for example.

Thus, in the case of massive particles, the conclusion of Wigner's method is that, for the knowledge on the representations of the Lorentz group, we need only the knowledge of the representations of the rotation group. Therefore, spin, and in fact any other label that states may have which is affected by Lorentz transformations, is given by the rotation group. This is true for all massive particles. Hence, in the case of massive particles with spin s, the helicity λ can take on any integer value between $-s$ and s, i.e. $\lambda \in \{s, s-1, s-2, \ldots, -s\}$, which means that massive particles have $2s + 1$ states.

1.7.2 Massless particles

Since a particle without mass cannot be at rest, we therefore need a different approach for massless particles compared with that in the case of massive particles. In the case of massless particles, we have $p'^2 = 0$. Consider that p' points along the z-axis, i.e. $p'_1 = p'_2 = 0$ and $p'_3 = p'_0$. Without loss of generality, we can choose $p'_0 = 1$. Assume that $\Gamma = \Lambda_t R_z(\theta)$ is an element of the little group of p', where

$$\Lambda_t = \begin{pmatrix} 1 & 0 & 0 \\ 0 & 1 & 0 \\ \lambda_{31} & \lambda_{32} & 1 \end{pmatrix} \tag{1.103}$$

with λ_{31} and λ_{32} being arbitrary. One can show that in order for the particle to have **discrete** spin, then the representation D must have the form

$$D(\Gamma) = D(R_z(\theta)), \tag{1.104}$$

i.e. we have $D(\Lambda_t) \equiv 1$. In addition, $D(R_z(\theta))$ must be at most double-valued, $D(R_z(2\pi)) = \pm 1$. Note that, in principle, there is no reason to exclude particles with continuous spin, but all observed particles in Nature have discrete spin.

The irreducible representations of $R_z(\theta)$ are trivial. Since $R_z(\theta)$ belongs to an Abelian group, Schur's lemma implies that the representations must be one-dimensional. This means that λ can take on only **one** value. Therefore, the numbers $D_{\lambda'\lambda}(\Gamma)$ are given by

$$D_{\lambda'\lambda}(\Gamma) = \delta_{\lambda\lambda'}d_\lambda(\theta), \tag{1.105}$$

where $d_\lambda(\theta) = \mathrm{e}^{-\mathrm{i}\lambda\theta}$. Since $D(R_z(2\pi)) = \pm 1$, it implies that λ is an integer or a half-integer.[10] This in turn means that the unitary operator can be written as

$$U(R_z(\theta)) = \mathrm{e}^{-\mathrm{i}\theta S_z}. \tag{1.106}$$

In the case of massless particles, λ is the spin along the z-axis and is called the helicity. There is only one value for λ, which means that the helicity is (relativistically) invariant for massless particles. In general, if parity is included, then massless particles with spin λ have only two *independent* helicity states $\pm\lambda$ (if $\lambda = 0$, there is only one helicity state). No other states are allowed. This result is a property of any massless particle and one of the consequences of Wigner's method. For example, photons have spin 1, but only two helicity states ± 1 are allowed. The absence of the helicity state 0 is due to the transversal nature of the field, which in turn is due to the absence of a photon rest mass. Other examples are gluons, and plausibly gravitons, which also exist in two helicity states only. Gluons, like photons, have helicity states ± 1, whereas gravitons would have helicity states ± 2. Finally, neutrinos are particles with spin 1/2 and have historically been assumed to be massless. In fact, there are now strong evidences that neutrinos are massive particles, albeit their masses are indeed extremely small. Nevertheless, if we assume neutrinos to be massless, then neutrinos would have only helicity $-1/2$, whereas antineutrinos would always have helicity $1/2$.

Let us discuss the transformation properties of the states $|p', \lambda\rangle$. Acting by a unitary operator of an element in the little group of p' on these states gives

$$U(\Gamma)|p', \lambda\rangle = \mathrm{e}^{-\mathrm{i}\lambda\theta(\Gamma)}|p', \lambda\rangle. \tag{1.107}$$

What about $\Lambda(p)$? Here $\Lambda(p)$ is the Lorentz transformation which transforms p' to p. We choose $p_0' = 1$ and $\Lambda(p) = R(\mathbf{z} \to \mathbf{p})L(p^z)$, where $L(p^z)$ is a pure boost along the z-axis such that $L(p^z)p' = p^z$, $p^{z0} = p^0$, $p^{z1} = p^{z2} = 0$, and $p^{z3} = p^0$ and $R(\mathbf{z} \to \mathbf{p})$ is a rotation around the axis $\mathbf{z} \times \mathbf{p}$. Since $|p, \lambda\rangle \equiv U(\Lambda(p))|p', \lambda\rangle$, we have

---

[10] Integer values correspond to single-valued representations, whereas half-integer values correspond to double-valued representations.

$$U(\Lambda)|p, \lambda\rangle = e^{-i\lambda\theta(p,\Lambda)}|\Lambda p, \lambda\rangle, \tag{1.108}$$

where $\theta(p, \Lambda)$ is the angle of the rotation around the z-axis in $\Gamma(p, \Lambda) = H(\Lambda p)^{-1}$ $\Lambda H(p)$, where $H(p) = R(\mathbf{z} \rightarrow \mathbf{p})L(p^z)$, which becomes $\Gamma(p, \Lambda) = \Lambda_t R_z$ $(\theta(p, \Lambda))$. Here we have used the normalization $\langle p, \lambda|p', \lambda'\rangle = p^0\delta(\mathbf{p} - \mathbf{p}')\delta_{\lambda\lambda'}$.

Problems

(1) Given an infinitesimal Lorentz transformation

$$\Lambda^{\mu}{}_{\nu} = \delta^{\mu}_{\nu} + \omega^{\mu}{}_{\nu},$$

show that the infinitesimal parameters $\omega_{\mu\nu}$ are antisymmetric.

(2) Construct explicitly the antisymmetric matrix $M = (M^{\mu\nu})$, where $M^{\mu\nu}$ are the generators of the Lorentz transformations, in terms of the three components of the angular momentum and the boosts in the three Cartesian directions.
 Hint: Be careful with covariant and contravariant indices.

(3) Verify that P^2 and w^2 are Casimir operators, i.e. that they commute with P^{μ} and $M^{\mu\nu}$.

Guide to additional recommended reading

The following books (see the indicated pages) and their authors have similar treatments of the content in the present chapter.

- A. Z. Capri, *Relativistic Quantum Mechanics and Introduction to Quantum Field Theory*, World Scientific (2002), pp. 3–6.
- F. Gross, *Relativistic Quantum Mechanics and Field Theory*, Wiley (1993), pp. 55–56, 593–597.
- J. Mickelsson, T. Ohlsson, and H. Snellman, *Relativity Theory*, KTH (2005), pp. 1–17.
- Y. Ohnuki, *Unitary Representations of the Poincaré Group and Relativistic Wave Equations*, World Scientific (1988), pp. 1–208.
- L. H. Ryder, *Quantum Field Theory*, 2nd edn., Cambridge (1996), pp. 55–64.
- F. Schwabl, *Advanced Quantum Mechanics*, Springer (1999), pp. 115–116, 131–134.
- S. S. Schweber, *An Introduction to Relativistic Quantum Field Theory*, Dover (2005), pp. 18–53.
- H. Snellman, *Elementary Particle Physics*, KTH (2004), pp. 21–25, 29–34.
- F. J. Ynduráin, *Relativistic Quantum Mechanics and Introduction to Field Theory*, Springer (1996), pp. 1–22, 109–123.
- For the interested reader: E. Wigner, On unitary representations of the inhomogeneous Lorentz group, *Ann. Math.* **40**, 149–204 (1939).

2

The Klein–Gordon equation

W. Pauli[1] said that the Klein–Gordon equation is 'the equation with many fathers'. Some of the many fathers are E. Schrödinger, W. Gordon, O. Klein, V. Fock, J. Kudar, T. de Donder, and H. van Dungen.[2] The Klein–Gordon equation is an equation for spin-0 particles, i.e. a relativistic quantum mechanical wave equation for particles with no internal degrees of freedom (i.e. **no** spin). In order to be more precise, the Klein–Gordon equation is a partial differential equation, which is **both** second order in time and space derivatives.

Let us find an equation for a free spin-0 (or spinless) particle with mass m in both the non-relativistic and relativistic cases.

(1) In the non-relativistic case, for a free particle, the energy is given by

$$E = \frac{\mathbf{p}^2}{2m}. \tag{2.1}$$

Using the correspondence principle, we replace the 4-momentum $p = (E, \mathbf{p})$ by the corresponding quantum mechanical operators, and we obtain

$$p = (p^\mu) = (i\partial^\mu) = (i\partial_t, -i\nabla), \tag{2.2}$$

which means that we have the following equation from the energy

$$i\frac{\partial}{\partial t}\Psi = -\frac{\nabla^2}{2m}\Psi \equiv H\Psi. \tag{2.3}$$

This is the well-known Schrödinger equation in ordinary quantum mechanics. However, if the Schrödinger wave function $\Psi(x)$ is a solution to Eq. (2.3), then,

---

[1] In 1945, Pauli was awarded the Nobel Prize in physics 'for the discovery of the Exclusion Principle, also called the Pauli Principle'.

[2] For the interested reader, please see: E. Schrödinger, Quantisierung als Eigenwertproblem IV, *Ann. Phys.* **81**, 109–139 (1926); W. Gordon, Der Comptoneffekt nach der Schrödingerschen Theorie, *Z. Phys.* **40**, 117–133 (1926); and O. Klein, Elektrodynamik und Wellenmechanik vom Standpunkt des Korrespondenzprinzips, *Z. Phys.* **41**, 407–442 (1927).

in general, $\tilde{\Psi}(x) = \Psi(x') = \Psi(\Lambda x)$, where Λ is a Lorentz transformation, is **not** a solution to Eq. (2.3). Thus, the Schrödinger equation is **not** Lorentz covariant.[3]

(2) In the relativistic case, for a free particle, the energy is given by

$$E^2 = m^2 + \mathbf{p}^2. \tag{2.4}$$

Again, using the correspondence principle, we have the equation

$$-\frac{\partial^2}{\partial t^2}\phi = (m^2 - \nabla^2)\phi, \tag{2.5}$$

which can be written as

$$\frac{\partial^2}{\partial t^2}\phi - \nabla^2\phi + m^2\phi = 0. \tag{2.6}$$

Since $c = 1$, it holds that $t = x^0$, which means that we can rewrite the above equation as

$$\frac{\partial^2}{\partial x^{0^2}}\phi - \nabla^2\phi + m^2\phi = 0. \tag{2.7}$$

Now, introducing the derivatives

$$(\partial^\mu) \equiv \left(\frac{\partial}{\partial x_\mu}\right) = (\partial^0, -\nabla) \quad \text{and} \quad (\partial_\mu) \equiv \left(\frac{\partial}{\partial x^\mu}\right) = (\partial_0, \nabla), \tag{2.8}$$

where $\partial_0 = \partial^0$ (since $x^0 = x_0$), we can construct the operator

$$\Box \equiv \partial_\mu \partial^\mu = \partial_0^2 - \nabla^2, \tag{2.9}$$

which is the *d'Alembert operator* (or *d'Alembertian*). Using this operator, we can write Eq. (2.7) as

$$(\Box + m^2)\phi = 0, \tag{2.10}$$

which is the *Klein–Gordon equation*. In general, the Klein–Gordon wave function $\phi = \phi(x)$ is a complex-valued function, i.e. a complex scalar. In order to describe charged particles, the wave function needs to be a complex scalar, but for neutral particles, it is enough if the wave function is a real scalar. Note that Eq. (2.10) is Lorentz covariant, which follows from the fact that both $\Box$ and m^2 behave as Lorentz scalars, i.e. $\Box' = \partial'_\mu \partial'^\mu = \partial_\mu \partial^\mu$, since the inner product of two 4-vectors is invariant under Lorentz transformations, and m^2 is a constant, as well as $\phi = \phi(x)$ is assumed to be a scalar. Thus, performing a

---

[3] Note that an equation is Lorentz covariant if the equation consists of Lorentz covariant quantities, which are quantities that transform as tensors under Lorentz transformations. Especially, a (Lorentz) scalar does not change under Lorentz transformations and is a so-called *Lorentz invariant*.

Lorentz transformation Λ from the old coordinate system described by x to a new coordinate system described by x', we obtain in the new inertial frame

$$(\Box' + m^2)\phi'(x') = 0, \qquad (2.11)$$

which has the same form as the Klein–Gordon equation in the old inertial frame.

In the special case when $m = 0$, i.e. the particle is massless, the Klein–Gordon equation (2.10) reduces to $\Box\phi = 0$, which is the relativistic wave equation.

Finally, what is the physical interpretation of the Klein–Gordon equation? As we will see later in this chapter, it provides, for example, relativistic corrections to bound states of atoms and has antiparticle degrees of freedom, but no spin degrees of freedom.

2.1 Transformation properties

The wave function ϕ transforms as a scalar under a proper (no parity change) and orthochronous (no time reversal) Lorentz transformation and a spacetime translation $x \mapsto x' = \Lambda x + a$, i.e. $(\Lambda, a) \in \mathcal{P}_+^\uparrow$. Now, there are two interpretations of the wave function ϕ (and later, the field ϕ). First, one has the *passive interpretation*, which says that $\phi'(x') = \phi(x)$ and means that the coordinate system is transformed. However, this interpretation has problems with parity and time reversal transformations. Second, one has the *active interpretation*, which says that $\phi'(x) = (U(\Lambda, a)\phi)(x) = \phi(\Lambda^{-1}(x - a))$ and means that the wave function is transformed. Thus, if $\phi(x)$ is a solution to the Klein–Gordon equation, then so is also the actively transformed wave function $\phi'(x)$. Consequently, the Klein–Gordon equation is relativistically invariant (see Exercise 2.1). Therefore, one usually uses the active interpretation.

Next, using Eq. (1.35), the *parity transformation* of a 4-position vector $x = (x^0, \mathbf{x})$ is given by $\Lambda_P(x^0, \mathbf{x}) = (x^0, -\mathbf{x})$. Note that one defines *scalar* and *pseudoscalar* wave functions as $\phi_s(x^0, -\mathbf{x}) = \phi_s(x^0, \mathbf{x})$ and $\phi_{ps}(x^0, -\mathbf{x}) = -\phi_{ps}(x^0, \mathbf{x})$, respectively. Assuming the active interpretation and applying the parity transformation (actually, the unitary representation of the parity transformation) to a scalar wave function $\phi_s(x)$, we obtain

$$\phi_s'(x) = \phi_s'(x^0, \mathbf{x}) = (U(\Lambda_P, 0)\phi_s)(x^0, \mathbf{x}) = \phi_s\left(\Lambda_P^{-1}(x^0, \mathbf{x})\right) = \phi_s(x^0, -\mathbf{x})$$
$$= \phi_s(x^0, \mathbf{x}) = \phi_s(x), \qquad (2.12)$$

i.e. scalar wave functions are invariant under parity transformations. On the other hand, for a pseudoscalar wave function $\phi_{ps}(x)$, we have

$$\phi'_{\text{ps}}(x) = \phi'_{\text{ps}}(x^0, \mathbf{x}) = (U(\Lambda_P, 0)\phi_{\text{ps}})(x^0, \mathbf{x}) = \phi_{\text{ps}}\left(\Lambda_P^{-1}(x^0, \mathbf{x})\right) = \phi_{\text{ps}}(x^0, -\mathbf{x})$$
$$= -\phi_{\text{ps}}(x^0, \mathbf{x}) = -\phi_{\text{ps}}(x). \tag{2.13}$$

For example, pions (or π mesons) are represented by pseudoscalar wave functions.

Exercise 2.1 Prove that the Klein–Gordon equation is relativistically invariant.

2.2 The current

The Klein–Gordon equation (wave function or field) has a conserved current. The density and the current for the Klein–Gordon equation are given by

$$\rho = \frac{i}{2m}\left[\phi^* \partial_0 \phi - (\partial_0 \phi^*)\phi\right], \tag{2.14}$$

$$\mathbf{j} = \frac{1}{2im}\left[\phi^* \nabla \phi - (\nabla \phi^*)\phi\right], \tag{2.15}$$

respectively. Note that the current for the Klein–Gordon equation is the same as for the Schrödinger equation. However, the density for the Klein–Gordon equation is **not** the same as for the Schrödinger equation. The density for the Schrödinger equation is the well-known expression $\rho_S = \Psi^*\Psi = |\Psi|^2$ that describes a positive (semi)definite probability density to detect a particle with wave function $\Psi = \Psi(t, \mathbf{x})$ at time t and position $\mathbf{x}$. Nevertheless, using the Klein–Gordon equation together with the density (2.14) and the current (2.15), it follows that the continuity equation is fulfilled

$$\partial_0 \rho + \nabla \cdot \mathbf{j} = 0. \tag{2.16}$$

Next, we introduce the 4-vector current

$$j = (j^\mu) = (\rho, \mathbf{j}), \tag{2.17}$$

which means that we can write the continuity equation for j as

$$\partial_\mu j^\mu = 0. \tag{2.18}$$

Exercise 2.2 Using the Klein–Gordon equation, construct the 4-current.

Then, we investigate the density of the Klein–Gordon equation, which is performed by inserting the stationary states $i\partial_0 \phi = \epsilon \phi$ and $-i\partial_0 \phi^* = \epsilon \phi^*$ into the expression for the density, and we obtain

$$\rho = \frac{i}{2m}\left[\phi^* \frac{\epsilon}{i} \phi - \left(-\frac{\epsilon}{i}\phi^*\right)\phi\right] = \frac{\epsilon}{m}\phi^*\phi. \tag{2.19}$$

Thus, in the non-relativistic limit ($\epsilon \simeq m$), we find that

$$\rho \simeq \phi^*\phi, \tag{2.20}$$

which is the (correct) non-relativistic probability density, whereas in the relativistic limit, the relativistic energy–momentum relation $\epsilon^2 = m^2 + \mathbf{p}^2$ has **two** roots $\epsilon = \pm\sqrt{m^2 + \mathbf{p}^2}$, which means that ρ may be **both** negative and positive. Thus, ρ cannot be consistently interpreted as a probability density, since it is not positive (semi)definite. The solution to this problem is to reinterpret the Klein–Gordon equation as a field equation satisfied by an operator $\phi(x)$, which is **not** a wave function, but a quantum field, whose excitations can be an arbitrary number of particles. See the discussion on the Klein paradox in Section 2.5. Hence, since the number of particles can change, there is no reason to have a single-particle equation. In such a quantum field theory framework, $\rho(x)$ becomes an operator that is called the *charge density operator*, and is not an operator of particle probability. See the discussion on quantization of the Klein–Gordon field in Chapter 6. Thus, the Klein–Gordon equation is the proper equation for spinless (scalar) fields. For example, Pauli and V. Weisskopf showed that the Klein–Gordon equation can describe spinless mesons, e.g. pions.[4]

Historically, the indefiniteness of the density for the Klein–Gordon equation led physicists to reject the Klein–Gordon equation in order to search for a Lorentz covariant quantum mechanical wave equation. However, today, the Klein–Gordon equation is undoubtedly considered to be a suitable equation for spinless particles, such as pions, described by spinless scalar fields.

2.3 Solutions to the Klein–Gordon equation

For free particles, the solutions to the Klein–Gordon equation are plane-wave solutions, which are given by

$$\phi_p(x) = N e^{-ip\cdot x}, \qquad (2.21)$$

where N is an arbitrary normalization constant, with the requirement that $p^2 \equiv p^{0^2} - \mathbf{p}^2 = m^2$, which means that $p^0 = \pm E = \pm\sqrt{m^2 + \mathbf{p}^2}$. Thus, we have a potential problem, since there are negative-energy solutions, which are unphysical. In principle, the negative-energy solutions can be discarded, but as we will see, they will return as antiparticles. In fact, the problem with the negative energies of the Klein–Gordon equation is due to the time derivative in Eq. (2.6) that is of second order. In the Schrödinger equation, the time derivative is of first order, but second order for the space derivative as in the case of the Klein–Gordon equation. However, since we are considering free particles, i.e. no interactions are involved, then there are no problems.[5] The reason is that if at some time the solution has

---

[4] For the interested reader, please see: W. Pauli and V. Weisskopf, Über die Quantisierung der skalaren relativistischen Wellengleichung, *Helv. Phys. Acta* **7**, 709–731 (1934).

[5] In the case that there are interactions included, the solution will be given as a linear combination of **both** positive- and negative-energy plane-wave solutions.

$E > 0$, then the solution maintains this condition. In this case, it is also possible to replace the Klein–Gordon equation by

$$i\partial_0\phi = \sqrt{m^2 - \nabla^2}\phi, \tag{2.22}$$

where the square-root is defined (by expansion in Fourier space) as

$$\sqrt{m^2 - \nabla^2}\phi = \int \sqrt{m^2 + \mathbf{k}^2}\chi(x^0, \mathbf{k})e^{i\mathbf{k}\cdot\mathbf{x}}\, d^3k \tag{2.23}$$

with

$$\phi(x^0, \mathbf{x}) = \int \chi(x^0, \mathbf{k})e^{i\mathbf{k}\cdot\mathbf{x}}\, d^3k. \tag{2.24}$$

The function $\chi(x^0, \mathbf{k})$ satisfies the equation

$$i\partial_0\chi(x^0, \mathbf{k}) = \omega(\mathbf{k})\chi(x^0, \mathbf{k}), \tag{2.25}$$

where $\omega(\mathbf{k}) = \sqrt{m^2 + \mathbf{k}^2}$.

The spectrum of the Klein–Gordon equation is a continuum of positive energies $E \geq m$ and negative energies $-E \leq -m$. The proper interpretation of the two parts of the spectrum is that the Klein–Gordon equation describes both particles (with energy $E > 0$ and charge density $\rho > 0$) and 'unphysical' particles (with energy $-E < 0$ and charge density $\rho < 0$), which are, however, reinterpreted as antiparticles (with energy $E > 0$ and charge density $\rho < 0$) (cf. Section 1.6). Thus, particles and antiparticles have the same mass m, the same positive energy E, opposite charge, opposite current, and, of course, in this case, no spin. For example, positive pions π^+ and negative pions π^- are such particles and antiparticles that can be described by the positive $[\phi_+(x) = e^{-iEx^0}\Phi_+(\mathbf{x})]$ and negative $[\phi_-(x) = e^{iEx^0}\Phi_-(\mathbf{x})]$ energy solutions to the Klein–Gordon equation.

For free particles at rest, i.e. $\nabla\phi = 0$, we have the simple Klein–Gordon equation

$$\left(\partial_0^2 + m^2\right)\phi(x^0, \mathbf{x}) = 0, \tag{2.26}$$

which has the solutions

$$\phi_\pm(x^0, \mathbf{0}) = e^{\mp imx^0}\phi_\pm(0, \mathbf{0}) \tag{2.27}$$

that are the *stationary solutions*. These two solutions are independent and have opposite signs for the rest energy of the particles, i.e. $p^0 = \pm m$. In fact, for moving particles, using that $mx^0 = Ex^0 - \mathbf{p}\cdot\mathbf{x} = p\cdot x$ is Lorentz invariant (cf. the discussion in Section 1.6), Eq. (2.27) can be generalized to

$$\phi_\pm(x^0, \mathbf{x}) = e^{\mp i(Ex^0 - \mathbf{p}\cdot\mathbf{x})} = e^{\mp ip\cdot x} = \phi_\pm(x). \tag{2.28}$$

In the time-independent case, i.e. $\phi(x) = \Phi(\mathbf{x})$, the Klein–Gordon equation becomes

$$\left(\nabla^2 - m^2\right) \Phi(\mathbf{x}) = 0, \tag{2.29}$$

which is the homogeneous screened Poisson equation.

What about solving the Klein–Gordon equation with initial and boundary conditions? The Klein–Gordon equation is a second-order partial differential equation in both time and space derivatives. Therefore, we need two initial conditions and one or more boundary conditions to uniquely determine its solutions without any unknowns, i.e. we should have a 'well-posed' problem. In general, assuming the solution to the Klein–Gordon equation to be of the form $\phi = \phi(x)$, where $x \in M$, which can be done without loss of generality, the initial conditions consist of given functions of space $\alpha(\mathbf{x})$ and $\beta(\mathbf{x})$ for the solution itself and its first-order time derivative at time $x^0 = 0$, i.e. $\phi(0, \mathbf{x}) = \alpha(\mathbf{x})$ and $\partial_0 \phi(0, \mathbf{x}) = \beta(\mathbf{x})$, whereas the boundary conditions consist of given functions of time $A(x^0)$ and/or $B(x^0)$ for the solution itself and/or its first-order space derivative on the boundary $\partial\Omega$ of the investigated space domain $\Omega \subseteq \mathbb{R}^3$, i.e. $\phi(x^0, \mathbf{x})|_{\mathbf{x}\in\partial\Omega} = A(x^0)$ and/or $\mathbf{n} \cdot \nabla\phi(x^0, \mathbf{x})|_{\mathbf{x}\in\partial\Omega} = B(x^0)$, where $\mathbf{n}$ is the normal to the boundary $\partial\Omega$. In particular, if the solution region under consideration consists of different parts, i.e. $\Omega = \cup_i \Omega_i$, then the solution itself as well as its first-order space derivative must be continuous at the common borders of the different parts.

2.4 Charged particles

Next, we discuss the coupling of a charged Klein–Gordon particle to an external electromagnetic field. We consider only so-called *minimal coupling*.[6] If we introduce the electromagnetic 4-vector potential $A = (A^\mu) = (A^0, \mathbf{A})$, where A^0 is the electric scalar potential and $\mathbf{A}$ is the magnetic 3-vector potential such that

$$\mathbf{E} = -\nabla A^0 - \partial_0 \mathbf{A}, \quad \mathbf{B} = \nabla \times \mathbf{A}, \tag{2.30}$$

then we can define the *minimal coupling* as the replacement

$$p^\mu \mapsto p^\mu - q A^\mu, \tag{2.31}$$

where q is the charge of the particle described by the charged Klein–Gordon wave function. Using the correspondence principle, we obtain

$$i\partial^\mu \mapsto i\partial^\mu - q A^\mu. \tag{2.32}$$

Performing this replacement in the Klein–Gordon equation leads to

$$[i\partial_0 - q A^0(x)]^2 \phi(x) - [-i\nabla - q\mathbf{A}(x)]^2 \phi(x) = m^2 \phi(x). \tag{2.33}$$

---

[6] It is assumed that the concept of minimal coupling is known from classical electromagnetism.

In this case, the conserved 4-current is given by [cf. Eqs. (2.14)–(2.15)]

$$\rho = \frac{i}{2m} \left[\phi^* \partial_0 \phi - (\partial_0 \phi^*) \phi \right] - \frac{q}{m} A^0 \phi^* \phi, \tag{2.34}$$

$$\mathbf{j} = \frac{1}{2im} \left[\phi^* \nabla \phi - (\nabla \phi^*) \phi \right] - \frac{q}{m} \mathbf{A} \phi^* \phi, \tag{2.35}$$

and no simple interpretation for ρ is possible. For example, in the static case, i.e. $A^0(x) = A^0(\mathbf{x})$ and $\mathbf{A} = 0$, we have the stationary solutions $\phi(x^0, \mathbf{x}) = e^{-i\epsilon x^0} u(\mathbf{x})$, which implies the density

$$\rho(\mathbf{x}) = \frac{\epsilon - q A^0(\mathbf{x})}{m} u^*(\mathbf{x}) u(\mathbf{x}) = \frac{\epsilon - q A^0(\mathbf{x})}{m} |u(\mathbf{x})|^2, \tag{2.36}$$

which can become negative even for $\epsilon > 0$. Thus, again, a one-particle theory is **not** possible for the Klein–Gordon equation, since the density can be **both** positive and negative in space. In particular, the Klein–Gordon equation does not lead to the concept of a probability density for the position of a particle.

Now, inserting the positive- and negative-energy solutions (2.28) into Eq. (2.33), we obtain

$$[E \mp q A^0(x)]^2 \phi_\pm(x) - [\mathbf{p} \mp q\mathbf{A}(x)]^2 \phi_\pm(x) = m^2 \phi_\pm(x). \tag{2.37}$$

Thus, the positive- and negative-energy solutions correspond to particles and antiparticles, respectively. Instead of using positive- and negative-energy solutions, we could use the solution ϕ and its complex conjugate ϕ^*. In this case, ϕ describes a particle, whereas ϕ^* describes an antiparticle.

What happens with the Klein–Gordon equation for a charged particle moving in an electromagnetic field in the non-relativistic limit? In this limit, we assume that the mass of the particle m is much larger than all the other energy terms in the equation, i.e.

$$\left| \frac{i \partial_0 \phi}{\phi} \right| \ll m \quad \text{and} \quad |q A^0| \ll m.$$

In addition, we separate the Klein–Gordon wave function as follows

$$\phi(x) = e^{-imt} \Psi(t, \mathbf{x}), \tag{2.38}$$

where the factor e^{-imt} describes oscillations of the wave funtion due to the parameter m. Thus, we want to find an equation for the part $\Psi(t, \mathbf{x})$ of the wave function in Eq. (2.38). Now, using the above assumptions and separation, the first term in Eq. (2.33) can be approximated by

$$[i\partial_0 - qA^0(x)]^2\phi(x) = (i\partial_t - qA^0)^2 e^{-imt}\Psi(t, \mathbf{x})$$

$$= (i\partial_t - qA^0)(me^{-imt}\Psi + ie^{-imt}\partial_t\Psi - qA^0 e^{-imt}\Psi)$$

$$= m^2 e^{-imt}\Psi + 2ime^{-imt}\partial_t\Psi - e^{-imt}\partial_t^2\Psi$$

$$- 2mqA^0 e^{-imt}\Psi - 2iqA^0 e^{-imt}\partial_t\Psi + q^2(A^0)^2 e^{-imt}\Psi$$

$$\simeq m^2 e^{-imt}\Psi + 2ime^{-imt}\partial_t\Psi - 2mqA^0 e^{-imt}\Psi, \qquad (2.39)$$

where all terms that do not contain m (except for the exponential factors) have been neglected. Thus, inserting Eq. (2.39) into Eq. (2.33) and simplifying, we obtain the Klein–Gordon equation in the non-relativistic limit

$$i\partial_t \Psi(t, \mathbf{x}) = \left[\frac{(-i\nabla - q\mathbf{A})^2}{2m} + qA^0\right]\Psi(t, \mathbf{x}), \qquad (2.40)$$

but this equation is exactly the Schrödinger equation for a charged particle moving in an electromagnetic field.

2.5 The Klein paradox

Originally, Klein investigated the *Klein paradox* by using the Dirac equation.[7] (See Problem 5 in Chapter 3.) However, for simplicity, we discuss this paradox by using the Klein–Gordon equation instead.

Consider scattering of a Klein–Gordon particle with energy $E = \sqrt{m^2 + \mathbf{k}^2}$ (i.e. imposing positive energy for the incoming beam) off a step-function potential

$$A = (A^0, \mathbf{A}) = (V(z), 0), \qquad (2.41)$$

where

$$V(z) = \begin{cases} 0, & z < 0 \\ V, & z > 0 \end{cases} \qquad (2.42)$$

with $V > 0$, which is illustrated in Fig. 2.1. This implies that the Klein–Gordon equation (2.33) takes the form

$$\begin{cases} -\partial_t^2\phi + \nabla^2\phi - m^2\phi = 0, & z < 0 \\ (i\partial_t - V)^2\phi + \nabla^2\phi - m^2\phi = 0, & z > 0 \end{cases} \qquad (2.43)$$

For $z < 0$, the incoming beam is given by $\exp[-i(Et - kz)]$, which gives the following form for the solution $\phi_<(t, z) = e^{-iEt}\left(e^{ikz} + Re^{-ikz}\right)$, $z < 0$ and $\phi_>(t, z) = Te^{-iEt}e^{ik'z}$, $z > 0$, where R and T are constants. Next, we need to determine the

[7] For the interested reader, please see: O. Klein, Die Reflexion von Elektronen an einem Potentialsprung nach der relativistischen Dynamik von Dirac, *Z. Phys.* **53**, 157–165 (1929).

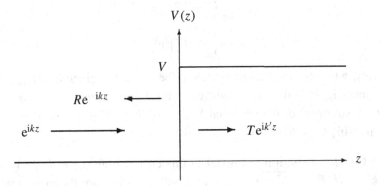

Figure 2.1 The setup of the potential for the Klein paradox. To the left ($z < 0$), there are an incoming beam and a reflected beam, whereas to the right ($z > 0$), there is only a transmitted beam. The height of the step-function potential is V.

energy–momentum dispersion relations as well as the constants R and T. Inserting the assumed solution into Eq. (2.43), we find that

$$\begin{cases} E^2 - k^2 - m^2 = 0, & z < 0 \\ (E - V)^2 - k'^2 - m^2 = 0, & z > 0 \end{cases}, \qquad (2.44)$$

which means that the momenta in the two different regions are given by $k = \pm\sqrt{E^2 - m^2}, z < 0$ and $k' = \pm\sqrt{(E - V)^2 - m^2}, z > 0$, where we have to choose the solutions with plus signs in order to reproduce the situation that is shown in Fig. 2.1. Note that the two momenta can also be called *local wave numbers*. Since the solution to the Klein–Gordon equation ϕ as well as its spatial derivative $\partial_z \phi$ have to be continuous at the border $z = 0$ between the two regions $z < 0$ and $z > 0$, i.e. $\phi_<(t, 0) = \phi_>(t, 0)$ and $\partial_z \phi_<(t, 0) = \partial_z \phi_>(t, 0)$, we have the system of equations

$$\begin{cases} 1 + R = T \\ k(1 - R) = k'T \end{cases}, \qquad (2.45)$$

which has the solution

$$R = \frac{k - k'}{k + k'}, \qquad (2.46)$$

$$T = \frac{2k}{k + k'}. \qquad (2.47)$$

Then, inserting the solution to the Klein–Gordon equation into the definition of the current (2.15) and using the fact that k is real, the current for $z < 0$ and $z > 0$ is

$$j_< = \frac{k}{m} \left(1 - |R|^2\right),$$

(2.48)

$$j_> = \frac{k' + k'^*}{2m} |T|^2 e^{i(k' - k'^*)z},$$

(2.49)

respectively, while for the incoming beam, the current is given by $j_{\text{incident}} = k/m$. Now, we investigate in detail three different interesting situations for the energy E (or different strengths of the potential V), i.e. (i) $E > V + m$, (ii) $V - m < E < V + m$, and (iii) $V > E + m$.

- **Above the step-function potential** In this energy region, i.e. $E > V + m$, we have $k' = \sqrt{(E - V)^2 - m^2} < k$ and both k and k' are therefore real, which implies, using Eq. (2.14), that the density for $z < 0$ and $z > 0$ is

$$\rho_< = \frac{E}{m} |\phi_<|^2 > 0,$$

(2.50)

$$\rho_> = \frac{E - V}{m} |\phi_>|^2 > 0,$$

(2.51)

respectively. Furthermore, using Eqs. (2.46) and (2.47), we find that $R < 1$, which means that there is less reflected beam than incoming beam. In addition, using Eq. (2.49) and the fact that k' is real, we have that

$$j_> = \frac{k'}{m} |T|^2,$$

(2.52)

so that the transmission and reflection coefficients are given by

$$\mathcal{T} = \frac{j_>}{j_{\text{incident}}} = \frac{4kk'}{(k + k')^2},$$

(2.53)

$$\mathcal{R} = -\frac{j_< - j_{\text{incident}}}{j_{\text{incident}}} = \left(\frac{k - k'}{k + k'}\right)^2,$$

(2.54)

respectively, which means that we obtain the expected incoming, reflected, and transmitted beams for non-relativistic scattering using the Schrödinger equation. Note that it holds that $\mathcal{T} + \mathcal{R} = 1$, as it must be. In addition, the constants R and T should not be confused with the reflection and transmission coefficients $\mathcal{R}$ and $\mathcal{T}$.

- **In the step-function potential** In this energy region, $V - m < E < V + m$, which means that the incoming kinetic energy is less than the height of the step-function potential and k' is purely imaginary. Therefore, we can write $k' = i\sqrt{m^2 - (E - V)^2} = i\kappa$. Thus, using Eq. (2.46), we obtain

$$R = \frac{k - i\kappa}{k + i\kappa},$$

(2.55)

which implies that $|R| = 1$. Inserting this into Eq. (2.48) yields $j_< = 0$, which means that we have total internal reflection, i.e. $|j_{\text{incident}}| = |j_{\text{reflected}}|$, since the beam for $z < 0$ is composed of both the incoming and reflected beams. In addition, since k' is purely imaginary, using Eq. (2.49), we find that $j_> = 0$. Thus, in this case, the transmission and reflection coefficients are always $\mathcal{T} = 0$ and $\mathcal{R} = 1$, which is again in agreement with non-relativistic scattering. Obviously, it holds that $\mathcal{T} + \mathcal{R} = 1$. Using the definition of the density (2.14) for $z > 0$ and the fact that k' is purely imaginary, i.e. $k' = i\kappa$, we obtain

$$\rho_> = \frac{E - V}{m}|\phi_>|^2 = \frac{E - V}{m}|T|^2 e^{-2\kappa z} = \frac{E - V}{m}\frac{4k^2}{k^2 + \kappa^2}e^{-2\kappa z}, \qquad (2.56)$$

which means that the density decays exponentially as it tunnels through the step-function potential. In this case, the density can be positive or negative. When $E > V$, we have $\rho_> > 0$, whereas when $E < V$, we have $\rho_> < 0$. Thus, we observe that when $E < V$ particle–antiparticle pairs are created. The particles are repelled by the step-function potential, whereas the antiparticles are attracted by the same, since they 'feel' V with the opposite sign compared with the particles. However, these particle–antiparticle pairs are virtual, since there is no net flow of the antiparticle current.

- **Extremely strong potential** In this energy region, i.e. $V > E + m$, we have $k' = \pm\sqrt{(E - V)^2 - m^2}$, which means that $k'^2 > k^2$ and both k and k' are real. Again, the density and the current for $z > 0$ are given by

$$\rho_> = \frac{E - V}{m}|T|^2, \qquad (2.57)$$

$$j_> = \frac{k'}{m}|T|^2, \qquad (2.58)$$

respectively, where $\rho_> < 0$, since $V > E$. Using $k'^2 = (E - V)^2 - m^2$, the group velocity for $z > 0$ is given by

$$v_g = \frac{\partial E}{\partial k'} = \frac{k'}{E - V}. \qquad (2.59)$$

If a particle moves to the right in the step-function potential, i.e. $v_g > 0$, then it implies that $k' < 0$, i.e. k' is negative. Thus, in this case, we find

$$\mathcal{T} = -\frac{4k|k'|}{(k - |k'|)^2} < 0, \qquad (2.60)$$

$$\mathcal{R} = \left(\frac{k + |k'|}{k - |k'|}\right)^2 > 1. \qquad (2.61)$$

Note that $\mathcal{T} + \mathcal{R} = 1$ is obeyed, but a negative transmission probability is totally unphysical.

All cases seem to agree with non-relativistic scattering using the Schrödinger equation, except from the extremely strong potential $V > E + m$. This is the *Klein paradox*. The solution to this paradox is that particle–antiparticle pairs are created from vacuum quantum fluctuations and the particles add to the reflected beam, which means that $\mathcal{R} > 1$. However, note that there is a theoretical problem in addition to the paradox. Initially, we had only a single-particle theory, and now, we have both particles and antiparticles. Thus, we would need a many-particle theory and a description that can handle both creation and annihilation of particles and antiparticles. The solution to this problem is to use quantum field theory (see again the discussion on quantization of the Klein–Gordon field in Chapter 6).

2.6 The pionic atom

To use the Klein–Gordon equation for a physical system, we consider the so-called *pionic atom*. What is a pionic atom and how is such an atom created? A pionic atom is an atom in which one or more electrons have been replaced by π^- mesons. Since pions are spinless particles, we can use the Klein–Gordon equation. To create pionic atoms, we first produce π^- mesons through inelastic proton–proton scattering, i.e.

$$p + p \rightarrow p + p + \pi^- + \pi^+.$$

Second, the π^- mesons are slowed down by successive electromagnetic interactions with electrons and nuclei, filtered out of the beam and, in principle, stopped in matter. Third, exotic atoms are formed when the π^- mesons are captured by the atoms of ordinary elements, ejecting bound electrons from their Bohr orbits. This happens when the π^- mesons reach the typical velocity of atomic electrons. Note that a typical principal quantum number n of the π^- mesons can be estimated as $n \sim \sqrt{m_\pi/m_e} \simeq 17$. The process of π^- meson capture includes the so-called *Auger effect*, which means that the transition of an electron (here: a π^- meson) in an atom filling in an inner-shell vacancy causes the emission of another electron, which is called the Auger electron that carries off the excess energy. Thus, subsequent stepwise deexcitation, mostly by Auger and X-ray emission, occurs until the captured π^- mesons are in low-lying bound states or even in the ground states. Note that Auger emission dominates the deexcitation from higher n-states, whereas X-ray emission dominates the deexcitations from lower n-states. The emitted X-rays are photons with energy approximately 270 times the energy of hydrogen photons, since the mass of a π^- meson is approximately 270 times the mass of an electron. For simplicity, we can write the π^- meson capture as follows

$$\pi^- + \text{atom} \rightarrow (\text{atom} - e^- + \pi^-) + e^-.$$

Thus, the physical system becomes a hydrogen-like pionic atom. Note that for hydrogen, i.e. when a π^- meson is captured, replacing the atomic electron, the physical system is called pionic hydrogen.

Now, we investigate a stationary state of a pionic atom having a nucleus with charge Ze and a π^- meson with charge $-e$. In order to find the energy levels of the pionic atom, we insert the Ansatz

$$\phi(x) = T(t)\Phi(\mathbf{x}), \tag{2.62}$$

where $T(t) = e^{-i\epsilon t}$ describes a stationary state, into the Klein–Gordon equation (2.33), which separates the solution into a product of functions for the space and time dependencies, respectively. In addition, we use the Coulomb potential for the 4-vector potential, i.e.

$$A = (A^0, \mathbf{A}), \quad \text{where} \quad qA^0(t, \mathbf{x}) = -\frac{Ze^2}{4\pi r} = -\frac{Z\alpha}{r} \quad \text{and} \quad \mathbf{A}(t, \mathbf{x}) = 0, \tag{2.63}$$

where $\alpha = e^2/(4\pi)$. Note that the mass of the pion π^- is m_π and its charge is $-e$. Thus, separating space and time, we obtain the time-independent Klein–Gordon equation for the pionic atom

$$\left[\left(\epsilon + \frac{Z\alpha}{r}\right)^2 + \nabla^2 - m_\pi^2\right]\Phi(\mathbf{x}) = 0, \tag{2.64}$$

cf. Eq. (2.29). Now, the radial symmetry of the Coulomb potential suggests that a good choice of coordinates would be spherical coordinates described by the three variables r, θ, and φ. Therefore, using the Laplacian in spherical coordinates

$$\nabla^2\Phi = \frac{1}{r}\partial_r^2(r\Phi) + \frac{1}{r^2\sin\theta}\partial_\theta(\sin\theta\,\partial_\theta\,\Phi) + \frac{1}{r^2\sin^2\theta}\partial_\varphi^2\Phi = \frac{1}{r}\partial_r^2(r\Phi) - \frac{1}{r^2}\mathbf{L}^2\Phi, \tag{2.65}$$

we find that Eq. (2.64) can be rewritten as

$$\frac{1}{r}\left[\partial_r^2 - \frac{\mathbf{L}^2 - (Z\alpha)^2}{r^2} + \frac{2\epsilon Z\alpha}{r} + \epsilon^2 - m_\pi^2\right]r\Phi(\mathbf{x}) = 0. \tag{2.66}$$

Next, the operator $\mathbf{L}^2$ in the Laplacian means that the spatial solution can be written as

$$\Phi(\mathbf{x}) = \frac{R_\ell(r)}{r}Y_{\ell m}(\theta, \varphi), \tag{2.67}$$

where $Y_{\ell m}(\theta, \varphi)$ are the so-called *spherical harmonics*, which are eigenfunctions to the operator $\mathbf{L}^2$, i.e.

$$\mathbf{L}^2 Y_{\ell m}(\theta, \varphi) = \ell(\ell + 1)Y_{\ell m}(\theta, \varphi), \tag{2.68}$$

where ℓ is the *orbital angular momentum quantum number*. Here $\ell = 0, 1, \ldots$
and $m = -\ell, -\ell + 1, \ldots, \ell - 1, \ell$. Then, inserting Eqs. (2.67) and (2.68) into
Eq. (2.66), we have

$$\left[\partial_r^2 - \frac{\ell(\ell + 1) - (Z\alpha)^2}{r^2} + \frac{2\epsilon Z\alpha}{r} + \epsilon^2 - m_\pi^2 \right] R_\ell(r) = 0. \tag{2.69}$$

Instead, using the Schrödinger equation, we would obtain

$$\left[\partial_r^2 - \frac{\ell(\ell + 1)}{r^2} + \frac{2m_\pi Z\alpha}{r} + 2m_\pi E \right] R_\ell(r) = 0, \tag{2.70}$$

which has the well-known spectrum (that is due to N. Bohr[8])

$$E_n = -\frac{m_\pi (Z\alpha)^2}{2n^2}, \quad \text{where} \quad n = 1, 2, \ldots; \quad \ell = 0, 1, \ldots, n - 1. \tag{2.71}$$

Note that this spectrum is independent of **both** ℓ and m, which means that there are
degeneracies in **both** ℓ and m for the spectrum. Here, the quantity $n' := n - \ell - 1$
gives the number of nodes of the wave function, which has to be an integer. Identi-
fying Eqs. (2.69) and (2.70), i.e. the structure of the Schrödinger equation and the
Klein–Gordon equation, we find the relation between orbital angular momentum
quantum numbers in the two equations as

$$\lambda(\ell) [\lambda(\ell) + 1] = \ell(\ell + 1) - (Z\alpha)^2, \tag{2.72}$$

which has the solution

$$\lambda(\ell) = -\frac{1}{2} + \sqrt{\left(\ell + \frac{1}{2} \right)^2 - (Z\alpha)^2}. \tag{2.73}$$

In order to be able to use the spectrum for the Schrödinger equation to find the spectrum for the Klein–Gordon equation, we have to perform the following replacements

$$E \rightarrow \frac{\epsilon^2 - m_\pi^2}{2m_\pi}, \quad \ell \rightarrow \lambda(\ell), \quad \alpha \rightarrow \alpha \frac{\epsilon}{m_\pi},$$

which also come from comparing Eqs. (2.69) and (2.70) with each other. Hence,
inserting these replacements into the formula for the spectrum (2.71), we find the
equation

$$\frac{\epsilon^2 - m_\pi^2}{2m_\pi} = -\frac{m_\pi (Z\alpha)^2 \epsilon^2}{2 [n' + \lambda(\ell) + 1]^2 m_\pi^2}, \tag{2.74}$$

[8] In 1922, Bohr was awarded the Nobel Prize in physics 'for his services in the investigation of the structure of
atoms and of the radiation emanating from them'.

which has the solution (choosing the root that gives $0 < \epsilon \leq m_\pi$, i.e. which corresponds to bound states)

$$\epsilon = \frac{m_\pi}{\sqrt{1 + \dfrac{(Z\alpha)^2}{\left[n' + \lambda(\ell) + 1\right]^2}}}, \quad \text{where} \quad n' = 0, 1, \ldots; \quad \ell = 0, 1, \ldots \quad (2.75)$$

Thus, using the Klein–Gordon equation, the energy levels of the pionic atom are given by

$$E_{\text{KG}}(n, \ell, m) = \frac{m_\pi}{\sqrt{1 + \dfrac{(Z\alpha)^2}{\left[n - \ell - \frac{1}{2} + \sqrt{\left(\ell + \frac{1}{2}\right)^2 - (Z\alpha)^2}\right]^2}}},$$

$$\text{where} \quad n = 1, 2, \ldots; \quad \ell = 0, 1, \ldots, n - 1. \quad (2.76)$$

We observe that Eq. (2.76) is dependent on ℓ, but independent of m. Hence, the degeneracy in ℓ for the energy levels is removed using the Klein–Gordon equation instead of the Schrödinger equation, which is due to the replacement $\ell(\ell + 1) \rightarrow \ell(\ell + 1) - (Z\alpha)^2$. Since $\alpha \lesssim 1/137$, we have in general that $Z\alpha \ll 1$, which means that we can use $Z\alpha$ as an expansion parameter. Performing an expansion of Eq. (2.76), we obtain

$$E_{\text{KG}}(n, \ell, m) = m_\pi - \frac{m_\pi (Z\alpha)^2}{2n^2} - \frac{m_\pi (Z\alpha)^4}{2n^3} \left(\frac{1}{\ell + \frac{1}{2}} - \frac{3}{4n}\right) + \cdots, \quad (2.77)$$

where the first term m_π is the rest energy of the pion (i.e. the mass of the pion), the second term is the non-relativistic energy proportional to $(Z\alpha)^2$, which is the same as the spectrum for the Schrödinger equation (i.e. a so-called *Bohr term*), and finally, the third term is the leading-order relativistic correction proportional to $(Z\alpha)^4$ that gives rise to a *fine structure*, which is in agreement with experiments. Therefore, the relativistic correction term shows that the Klein–Gordon equation gives a more accurate solution than the Schrödinger equation. However, the relativistic correction term is **not** in agreement with observations of the hydrogen spectrum, which has, for example, a splitting between the six states with $n = 2$ and $\ell = 1$ that forms groups of two and four degenerate states (see the discussion in Section 3.11). Thus, in order to describe the hydrogen spectrum, i.e. the spectrum of the electron in the hydrogen atom, we must have a relativistic quantum mechanical wave equation for spin-1/2 particles, which will be investigated in great detail in the next chapter.

Problems

(1) Calculate the transmission coefficient of a Klein–Gordon particle with mass m and charge q that is incident on a potential barrier of the form

$$V(z) = \begin{cases} 0 & z < 0, z > a \\ V_0 & 0 < z < a \end{cases},$$

where V_0 and a are positive constants. In addition, find the energy of the particle for which the transmission coefficient is equal to one.

(2) Solve the Klein–Gordon equation for a square-well potential of the form

$$V(r) = \begin{cases} -V_0 & r \le R \\ 0 & r > R \end{cases},$$

where $V_0 > 0$ and $R > 0$.

(3) *Klein–Gordon hydrogen atom.* A spinless electron is bound by the Coulomb potential

$$V(r) = -\frac{Ze^2}{4\pi r}$$

in a stationary state of total energy $E \le m$.

(a) Derive the time-independent Klein–Gordon equation for this potential.

(b) Assume that the radial and angular parts of the wave function $\psi(\mathbf{r})$ separate, and verify that this yields the radial Klein–Gordon equation

$$\frac{d^2 R_\ell(r)}{dr^2} + \left[\frac{2EZ\alpha}{r} - (m^2 - E^2) - \frac{\ell(\ell+1) - (Z\alpha)^2}{r^2} \right] R_\ell(r) = 0,$$

where $R_\ell(r)$ is the radial wave function and $\alpha = e^2/(4\pi)$ is the Sommerfeld fine structure constant, i.e. $\alpha \simeq 1/137$, which characterizes the range of the potential.

(c) Show that this equation can be written in the dimensionless form

$$\frac{d^2 R_\ell(\rho)}{d\rho^2} + \left[\frac{2ZE\alpha}{\gamma\rho} - \frac{1}{4} - \frac{\ell(\ell+1) - (Z\alpha)^2}{\rho^2} \right] R_\ell(\rho) = 0,$$

where $\rho = \gamma r$ and $\gamma^2 = 4(m^2 - E^2)$.

(d) Assume that this equation has a solution of the usual form of a power series times the $\rho \to 0$ and $\rho \to \infty$ solutions:

$$R_\ell(\rho) = \rho^k (1 + c_1\rho + c_2\rho^2 + c_3\rho^3 + \cdots)e^{-\rho/2}.$$

Show that

$$k = k_\pm = \frac{1}{2} \pm \sqrt{\left(\ell + \frac{1}{2}\right)^2 - (Z\alpha)^2}.$$

(4) Solve the Klein–Gordon equation for a potential of an homogeneously charged sphere with radius R of the form

$$V(r) = -\frac{Z\alpha}{2R}\left(3 - \frac{r^2}{R^2}\right), \quad \text{for } r \le R,$$

where Z is the coupling strength parameter and α is the Sommerfeld fine structure constant, i.e. $\alpha \simeq 1/137$, which characterizes the range of the potential. Note that this potential gives a better model for a realistic pionic atom than the ordinary Coulomb potential.

(5) Solve the Klein–Gordon equation for an exponential potential of the form $V(r) = -Z\alpha e^{-r/a}$, where Z is the coupling strength parameter, α is the Sommerfeld fine structure constant, i.e. $\alpha \simeq 1/137$, which characterizes the range of the potential, and a is the range parameter. You need only to consider the case of s states, i.e. $\ell = 0$.

(6) Assume that a pion is bound by a scalar potential of the form

$$V(r) = -V_0\,\delta^{(3)}(r),$$

where V_0 is positive constant. Solve the Klein–Gordon equation for the special case when the solution is static, i.e. independent of time. Discuss the significance of your result.

Guide to additional recommended reading

The following books (see the indicated pages) and their authors have similar treatments of the content in the present chapter.

- H. A. Bethe and R. Jackiw, *Intermediate Quantum Mechanics*, 3rd edn., Westview Press (1997), pp. 341–348.
- A. Z. Capri, *Relativistic Quantum Mechanics and Introduction to Quantum Field Theory*, World Scientific (2002), pp. 8–22.
- F. Gross, *Relativistic Quantum Mechanics and Field Theory*, Wiley (1993), pp. 89–118.
- R. H. Landau, *Quantum Mechanics II – A Second Course in Quantum Theory*, 2nd edn., Wiley-VCH (2004), pp. 204–216.
- F. Schwabl, *Advanced Quantum Mechanics*, Springer (1999), pp. 116–120.
- S. S. Schweber, *An Introduction to Relativistic Quantum Field Theory*, Dover (2005), pp. 54–64.
- F. J. Ynduráin, *Relativistic Quantum Mechanics and Introduction to Field Theory*, Springer (1996), pp. 23–34.
- For the interested reader: W. Gordon, Der Comptoneffekt nach der Schrödingerschen Theorie, *Z. Phys.* **40**, 117–133 (1926); O. Klein, Elektrodynamik und Wellenmechanik vom Standpunkt des Korrespondenzprinzips, *Z. Phys.* **41**, 407–442 (1927); O. Klein, Die Reflexion von Elektronen an einem Potentialsprung nach der relativistischen Dynamik von Dirac, *Z. Phys.* **53**, 157–165 (1929); W. Pauli and V. Weisskopf, Über die Quantisierung der skalaren relativistischen Wellengleichung, *Helv. Phys. Acta* **7**, 709–731 (1934); and E. Schrödinger, Quantisierung als Eigenwertproblem IV, *Ann. Phys.* **81**, 109–139 (1926).

3

The Dirac equation

In this chapter, we investigate the Dirac equation, which is named after P. A. M. Dirac, who is one of the fathers of quantum field theory.[1] The Dirac equation is a relativistic quantum mechanical wave equation for spin-1/2 particles (e.g. electrons), which was derived by Dirac in 1928.[2] The difficulties in finding a consistent single-particle theory from the Klein–Gordon equation led Dirac to search for an equation that

- had a positive-definite conserved probability density and
- was first order **both** in time and space.

One can show that these two conditions imply that a matrix equation is required. The reason why the Klein–Gordon equation did not yield a positive-definite probability density is connected with the second-order time derivative in this equation, which arises because the Klein–Gordon equation is related to the relativistic energy–momentum relation $E^2 = m^2 + \mathbf{p}^2$ via the correspondence principle that includes a term E^2. Thus, a 'better' Lorentz covariant wave equation with a positive-definite probability density should have a first-order time derivative only. However, the equivalence of time and space coordinates in Minkowski space requires that such an equation also have only first-order space derivatives.

Now, heuristically constructing the Dirac equation, consider the equation

$$\frac{\partial \psi}{\partial t} + (\boldsymbol{\alpha} \cdot \nabla)\psi + im\beta\psi = 0, \tag{3.1}$$

or equivalently,

$$\frac{\partial \psi_i}{\partial t} + \sum_j (\boldsymbol{\alpha} \cdot \nabla)_{ij}\psi_j + im \sum_j \beta_{ij}\psi_j = 0, \tag{3.2}$$

[1] In 1933, Dirac shared the Nobel Prize in physics together with Schrödinger 'for the discovery of new productive forms of atomic theory'.

[2] For the interested reader, please see: P. A. M. Dirac, The quantum theory of the electron, *Proc. R. Soc. (London)* A **117**, 610–624 (1928); The quantum theory of the electron. Part II, *ibid.* **118**, 351–361 (1928).

where $\boldsymbol{\alpha} = (\alpha^1, \alpha^2, \alpha^3)$ and β are some constant $N \times N$ matrices to be determined. Note that $\psi = (\psi_i)$ is an $N \times 1$ column matrix, i.e. a column vector with N components, where the idea is that each component ψ_i ($i = 1, 2, \ldots, N$) should satisfy the Klein–Gordon equation, i.e. $(\Box + m^2) \psi_i = 0$. Furthermore, we want to have

$$\rho = \psi^\dagger \psi = \sum_{i=1}^{N} \psi_i^* \psi_i = \sum_{i=1}^{N} |\psi_i|^2 \geq 0, \tag{3.3}$$

$$\mathbf{j} = \psi^\dagger \boldsymbol{\alpha} \psi, \tag{3.4}$$

where ρ and $\mathbf{j}$ are the density and the current for the Dirac equation, respectively, which should imply that the continuity equation

$$\frac{\partial \rho}{\partial t} + \nabla \cdot \mathbf{j} = 0 \tag{3.5}$$

is fulfilled. In addition, the conjugate (or Hermitian adjoint) $\psi^\dagger = (\psi^*)^T$, which is a row vector with N components, satisfies the equation

$$\frac{\partial \psi^\dagger}{\partial t} + \nabla \psi^\dagger \cdot \boldsymbol{\alpha}^\dagger - im \psi^\dagger \beta^\dagger = 0. \tag{3.6}$$

Thus, multiplying Eq. (3.1) by $\psi^\dagger$ from the left, Eq. (3.6) by ψ from the right, and adding the results together, we obtain

$$\frac{\partial}{\partial t} \left(\psi^\dagger \psi \right) + \left[\psi^\dagger (\boldsymbol{\alpha} \cdot \nabla) \psi + \nabla \psi^\dagger \cdot \boldsymbol{\alpha}^\dagger \psi \right] + im \left(\psi^\dagger \beta \psi - \psi^\dagger \beta^\dagger \psi \right) = 0. \tag{3.7}$$

Then, imposing $\boldsymbol{\alpha} = \boldsymbol{\alpha}^\dagger$ and $\beta = \beta^\dagger$, i.e. $\boldsymbol{\alpha}$ and β are Hermitian, the continuity equation will be fulfilled. Thus, the density and the current become

$$\rho = \psi^\dagger \psi \quad \text{and} \quad \mathbf{j} = \psi^\dagger \boldsymbol{\alpha} \psi, \tag{3.8}$$

respectively. In order to 'square' Eq. (3.1), we act on it with the 'conjugate' $\partial/(\partial t) - \boldsymbol{\alpha} \cdot \nabla - im\beta$ from the left, and we find

$$\frac{\partial^2 \psi}{\partial t^2} = \sum_{i,j} \frac{1}{2} \left(\alpha^i \alpha^j + \alpha^j \alpha^i \right) \frac{\partial^2}{\partial x^i \partial x^j} \psi - m^2 \beta^2 \psi + im \sum_i \left(\alpha^i \beta + \beta \alpha^i \right) \frac{\partial}{\partial x^i} \psi. \tag{3.9}$$

The right-hand side of Eq. (3.9) will be identical to $(\nabla^2 - m^2) \psi$ in the case that we choose $\boldsymbol{\alpha}$ and β to satisfy the following relations

$$\frac{1}{2} \left(\alpha^i \alpha^j + \alpha^j \alpha^i \right) = \delta^{ij} \mathbb{1}_N, \tag{3.10}$$

$$\alpha^i \beta + \beta \alpha^i = 0, \tag{3.11}$$

$$\beta^2 = \mathbb{1}_N, \tag{3.12}$$

where $\mathbb{1}_N$ is the $N \times N$ identity matrix. Note that Eq. (3.10) implies that $(\alpha^i)^2 = \mathbb{1}_N$ for all i. Since the square of any of the matrices α^i and β is equal to the identity matrix, the eigenvalues have to be ± 1. Using Eq. (3.11), we can rewrite this equation as

$$\alpha^i \beta = -\mathbb{1}_N \beta \alpha^i. \tag{3.13}$$

Taking the determinant of both sides of this equation, we obtain

$$\det(\alpha^i \beta) = (-1)^N \det(\beta \alpha^i). \tag{3.14}$$

This implies that

$$(-1)^N = 1, \tag{3.15}$$

which only holds if N is even. In addition, one can show that $\text{tr}(\alpha^i) = 0$ and $\text{tr}(\beta) = 0$, using $\alpha^i = -\beta \alpha^i \beta$ and $\beta = -\alpha^i \beta \alpha^i$ as well as the cyclic property for the trace of a product of matrices. Note that the traces of the matrices α^i and β can only be zero if N is even, since a trace is the sum of the eigenvalues, which for α^i and β are only ± 1. In the case of $N = 2$, the Pauli matrices would nearly work, since they satisfy the anticommutation relation $\{\sigma^i, \sigma^j\} = 2\delta^{ij}\mathbb{1}_2$, where σ^i could serve as α^i, but the Pauli matrices span the space of 2×2 matrices together with the identity matrix $\mathbb{1}_2$, which cannot serve as β, since $\text{tr}(\mathbb{1}_2) = 2 \neq 0$ and $\mathbb{1}_2$ commutes with all matrices in contradiction with Eq. (3.11). Thus, we must have $N \geq 4$. Actually, for irreducible representations of α^i and β, we must have that $N = 4$. For example, an explicit representation of α^i and β is given by

$$\alpha^i = \begin{pmatrix} 0 & \sigma^i \\ \sigma^i & 0 \end{pmatrix} \quad \Leftrightarrow \quad \alpha = \begin{pmatrix} 0 & \sigma \\ \sigma & 0 \end{pmatrix} \quad \text{and} \quad \beta = \begin{pmatrix} \mathbb{1}_2 & 0 \\ 0 & -\mathbb{1}_2 \end{pmatrix}, \tag{3.16}$$

where $\beta^2 = \mathbb{1}_4$, that were found by Dirac using trial-and-error. Furthermore, we introduce the notations $\gamma^0 \equiv \beta$ and $\gamma^i \equiv \beta\alpha^i$, where γ^0 is Hermitian [i.e. $(\gamma^0)^\dagger = \gamma^0$] and γ^i is anti-Hermitian [i.e. $(\gamma^i)^\dagger = -\gamma^i$].

Multiplying Eq. (3.1) by β from the left, we obtain

$$(\beta \partial_0 + \beta \alpha^i \partial_i)\psi + im\psi = 0, \tag{3.17}$$

which is equivalent to

$$(i\gamma^\mu \partial_\mu - m\mathbb{1}_4)\psi = 0, \tag{3.18}$$

using the facts that $\beta = \gamma^0$ and $\beta\alpha^i = \gamma^i$. This equation is the *Dirac equation*. In addition, the quantity ψ is the Dirac wave function (sometimes known as the Dirac spinor).

The algebra of the γ^μs (or the *Dirac algebra*) is given by

$$\gamma^\mu \gamma^\nu + \gamma^\nu \gamma^\mu = 2g^{\mu\nu}\mathbb{1}_4, \tag{3.19}$$

which can also be written using the notation for the anticommutator as

$$\{\gamma^\mu, \gamma^\nu\} = 2g^{\mu\nu}\mathbb{1}_4, \tag{3.20}$$

defining what is known as a *Clifford algebra* (see Section 3.3 for a detailed discussion on the properties of the gamma matrices). In addition, the gamma matrices satisfy the Hermiticity condition $(\gamma^\mu)^\dagger = \gamma^0\gamma^\mu\gamma^0$. Another useful notation is the *Feynman-slash symbol*, which is the following

$$\slashed{a} = \gamma^\mu a_\mu = \gamma \cdot a = \gamma^0 a^0 - \boldsymbol{\gamma} \cdot \mathbf{a}, \tag{3.21}$$

where $\boldsymbol{\gamma} \equiv (\gamma^1, \gamma^2, \gamma^3)$ and $a = (a^0, \mathbf{a})$ is a 4-vector. Thus, we can write the Dirac equation as $(i\slashed{\partial} - m\mathbb{1}_4)\psi = 0$.

Now, we can write the probability and current densities as

$$\rho = j^0 = \psi^\dagger\psi = \psi^\dagger\gamma^0\gamma^0\psi, \tag{3.22}$$
$$j^i = \psi^\dagger\alpha^i\psi = \psi^\dagger\gamma^0\gamma^i\psi, \tag{3.23}$$

respectively. Finally, introducing the *Dirac adjoint* $\bar{\psi} \equiv \psi^\dagger\gamma^0$, we are able to write the 4-current density as

$$j^\mu = \bar{\psi}\gamma^\mu\psi, \tag{3.24}$$

which fulfils the continuity equation $\partial_\mu j^\mu = 0$.

In general, note that **any** solution to the Dirac equation (3.18) automatically solves the Klein–Gordon equation (2.10), since the Klein–Gordon equation can be decomposed as

$$-(\Box + m^2)\psi = (i\gamma^\mu\partial_\mu + m\mathbb{1}_4)(i\gamma^\mu\partial_\mu - m\mathbb{1}_4)\psi = 0. \tag{3.25}$$

However, the converse does not hold, i.e. **not** all solutions to the Klein–Gordon equation solve the Dirac equation.

In fact, for spin-3/2 particles, there exists a similar relativistic wave equation to the Dirac equation for spin-1/2 particles. This equation, which is called the *Rarita–Schwinger equation*,[3] is given by

$$(i\slashed{\partial} - m\mathbb{1}_4)\psi^\mu(x) = 0, \tag{3.26}$$

where m is the mass and ψ^μ is a vector-valued Dirac spinor that can be considered to be the direct product of a 4-vector A^μ and a Dirac spinor ψ, i.e. $\psi^\mu = A^\mu \otimes \psi$, which means that ψ^μ has $4 \cdot 4 = 16$ components.[4] In addition, ψ^μ should fulfil either of the two conditions

---

[3] For the interested reader, please see: W. Rarita and J. Schwinger, On a theory of particles with half-integral spin, *Phys. Rev.* **60**, 61 (1941) and S. Kusaka, β-Decay with neutrino of spin $\frac{3}{2}$, *ibid.* **60**, 61–62 (1941).

[4] Note that the Rarita–Schwinger equation can be rewritten as $\epsilon^{\mu\nu\rho\sigma}\gamma^5\gamma_\nu\partial_\rho\psi_\sigma + m\psi^\mu = 0$.

$$\partial_\mu \psi^\mu(x) = 0, \tag{3.27}$$

$$\gamma_\mu \psi^\mu(x) = 0, \tag{3.28}$$

which are equivalent for vector-valued Dirac spinors that satisfy Eq. (3.26). The Rarita–Schwinger equation can be used to describe composite particles like Δ baryons or proposed elementary particles such as the gravitino, which is the supersymmetric partner of the graviton.[5] However, no fundamental particle with spin-3/2 has yet been found experimentally. It turns out that Eq. (3.26) together with one of the two equivalent conditions, i.e. Eq. (3.27) or (3.28), can be generalized to describe particles with spin $n + 1/2$. This generalization is given by

$$\left(i\partial\!\!\!/ - m\mathbb{1}_4\right) \psi^{\{\mu_1,\mu_2,\dots,\mu_n\}}(x) = 0, \tag{3.29}$$

$$\partial_{\mu_i} \psi^{\{\mu_1,\mu_2,\dots,\mu_n\}}(x) = 0, \quad i = 1, 2, \dots, n, \tag{3.30}$$

$$\gamma_{\mu_i} \psi^{\{\mu_1,\mu_2,\dots,\mu_n\}}(x) = 0, \quad i = 1, 2, \dots, n, \tag{3.31}$$

where the symbol $\{\mu_1, \mu_2, \dots, \mu_n\}$ is the symmetrization of the n Lorentz indices $\mu_1, \mu_2, \dots, \mu_n$ and the quantity $\psi^{\{\mu_1,\mu_2,\dots,\mu_n\}}$ is the wave function that describes a particle with mass m and spin $s = n + 1/2$. In addition, one can describe particles with spin-$n/2$, which is performed by the so-called *Bargmann–Wigner equations*,[6] i.e.

$$\left(i\partial\!\!\!/ - m\mathbb{1}_4\right)_{\alpha\alpha'} \psi_{\{\alpha',\beta,\dots,\gamma\}}(x) = 0, \tag{3.32}$$

$$\left(i\partial\!\!\!/ - m\mathbb{1}_4\right)_{\beta\beta'} \psi_{\{\alpha,\beta',\dots,\gamma\}}(x) = 0, \tag{3.33}$$

$$\vdots$$

$$\left(i\partial\!\!\!/ - m\mathbb{1}_4\right)_{\gamma\gamma'} \psi_{\{\alpha,\beta,\dots,\gamma'\}}(x) = 0, \tag{3.34}$$

where the symbol $\{\alpha, \beta, \dots, \gamma\}$ now consists of n Dirac spinor indices (and not Lorentz indices) and the quantity $\psi_{\{\alpha,\beta,\dots,\gamma\}}$ is a Bargmann–Wigner multispinor that can be regarded as the wave function for a particle with mass m and spin $s = n/2$, which is composed of n identical elementary Dirac (spin-1/2) spinors. For example, in the case of spin-3/2, the Bargmann–Wigner multispinor has three Dirac spinor indices, and we obtain the three equations

$$\left(i\partial\!\!\!/ - m\mathbb{1}_4\right)_{\alpha\alpha'} \psi_{\{\alpha',\beta,\gamma\}}(x) = 0, \tag{3.35}$$

$$\left(i\partial\!\!\!/ - m\mathbb{1}_4\right)_{\beta\beta'} \psi_{\{\alpha,\beta',\gamma\}}(x) = 0, \tag{3.36}$$

$$\left(i\partial\!\!\!/ - m\mathbb{1}_4\right)_{\gamma\gamma'} \psi_{\{\alpha,\beta,\gamma'\}}(x) = 0. \tag{3.37}$$

---

[5] Supersymmetry is a symmetry where the so-called *supersymmetric partner* of a particle has spin that differs by one-half. This means that for every boson there exists a corresponding fermion, and vice versa.

[6] For the interested reader, please see: V. Bargmann and E. P. Wigner, Group theoretical discussion of relativistic wave equations, *Proc. Nat. Acad. Sci. (USA)* **34**, 211–223 (1948).

As a matter of fact, one can even show that the generalized Rarita–Schwinger equations (3.29)–(3.31) and the Bargmann–Wigner equations (3.32)–(3.34) are equivalent for the same value of the spin. Especially, it holds that the Rarita–Schwinger equation (3.26) and the three Bargmann–Wigner equations (3.35)–(3.37) give an equivalent description for spin-3/2 particles.

3.1 Free particle solutions to the Dirac equation

For free particles, the solutions to the Dirac equation $(i\gamma^\mu \partial_\mu - m\mathbb{1}_4)\psi = 0$ are plane-wave solutions of the forms

$$\psi_p(x) = e^{-ip \cdot x} u(p) \quad \text{and} \quad \psi_p(x) = e^{ip \cdot x} v(p), \tag{3.38}$$

for positive- and negative-energy states, respectively, where $u(p)$ and $v(p)$ are so-called *spinors*, i.e. 4×1 column matrices, which are independent from the 4-position vector x. Note that the plane-wave solutions are linearly independent solutions to the Dirac equation. Inserting these plane-wave solutions into the Dirac equation, we obtain

$$(\not{p} - m\mathbb{1}_4)u(p) = (\gamma \cdot p - m\mathbb{1}_4)u(p) = 0 \quad \text{and}$$
$$(-\not{p} - m\mathbb{1}_4)v(p) = (-\gamma \cdot p - m\mathbb{1}_4)v(p) = 0, \tag{3.39}$$

which are the corresponding equations for the spinors in momentum space. Multiplying from the left with the 'conjugates' $\gamma \cdot p + m\mathbb{1}_4$ and $\gamma \cdot p - m\mathbb{1}_4$, respectively, we find that

$$(\gamma \cdot p + m\mathbb{1}_4)(\gamma \cdot p - m\mathbb{1}_4)u(p) = 0 \quad \text{and} \quad (\gamma \cdot p - m\mathbb{1}_4)(-\gamma \cdot p - m\mathbb{1}_4)v(p) = 0, \tag{3.40}$$

which are equivalent to

$$(\gamma^\mu \gamma^\nu p_\mu p_\nu - m^2\mathbb{1}_4)u(p) = 0 \quad \text{and} \quad (-\gamma^\mu \gamma^\nu p_\mu p_\nu + m^2\mathbb{1}_4)v(p) = 0, \tag{3.41}$$

respectively. We now compute the quantity $\gamma^\mu \gamma^\nu p_\mu p_\nu$. We have that

$$\gamma^\mu \gamma^\nu p_\mu p_\nu = \gamma^\nu \gamma^\mu p_\nu p_\mu = \frac{1}{2}(\gamma^\mu \gamma^\nu + \gamma^\nu \gamma^\mu) p_\mu p_\nu = g^{\mu\nu}\mathbb{1}_4 p_\mu p_\nu$$
$$= p_\mu p^\mu \mathbb{1}_4 = p^2 \mathbb{1}_4, \tag{3.42}$$

which means that

$$(p^2 - m^2) u(p) = 0 \quad \text{and} \quad (p^2 - m^2) v(p) = 0. \tag{3.43}$$

Thus, the 4-momentum vector p_μ must satisfy the Klein–Gordon equation (or more accurately the relativistic energy–momentum relation) with $p^2 = m^2$. From this condition, it follows that $p_0^2 - \mathbf{p}^2 = m^2$, which can be written as $p_0^2 = m^2 + \mathbf{p}^2$.

This equation has the two solutions $p_0 = \pm\sqrt{m^2 + \mathbf{p}^2}$. Since $p^2 = m^2 > 0$, this means that p is time-like. Thus, a rest frame exists, which is given by $\mathbf{p} = \mathbf{0}$ in which $p_0 = \pm m$. Therefore, we have $p = (\pm m, \mathbf{0})$.

Next, in addition to the ordinary gamma matrices, i.e. the γ^μs, we introduce the following useful matrix[7]

$$\gamma^5 \equiv i\gamma^0\gamma^1\gamma^2\gamma^3 = -\frac{i}{4!}\epsilon^{\mu\nu\rho\sigma}\gamma_\mu\gamma_\nu\gamma_\rho\gamma_\sigma = -\frac{i}{4!}\epsilon_{\mu\nu\rho\sigma}\gamma^\mu\gamma^\nu\gamma^\rho\gamma^\sigma. \qquad (3.44)$$

Note that this matrix is Hermitian, i.e. $(\gamma^5)^\dagger = \gamma^5$, and it also holds that $(\gamma^5)^2 = \mathbb{1}_4$ and $\gamma_5 = -i\gamma_0\gamma_1\gamma_2\gamma_3 = i\gamma^0\gamma^1\gamma^2\gamma^3 = \gamma^5$. Furthermore, the matrix anticommutes with all the other gamma matrices, i.e. $\{\gamma^5, \gamma^\mu\} = 0$. Then, we present an explicit representation of the gamma matrices. Using Eq. (3.16), it is given by

$$\gamma^0 = \begin{pmatrix} \mathbb{1}_2 & 0 \\ 0 & -\mathbb{1}_2 \end{pmatrix}, \quad \gamma^i = \begin{pmatrix} 0 & \sigma^i \\ -\sigma^i & 0 \end{pmatrix} \quad \Leftrightarrow \quad \gamma = \begin{pmatrix} 0 & \sigma \\ -\sigma & 0 \end{pmatrix}, \quad \text{and}$$

$$\gamma^5 = \begin{pmatrix} 0 & \mathbb{1}_2 \\ \mathbb{1}_2 & 0 \end{pmatrix} \qquad (3.45)$$

and it is called the *Dirac* (or '*standard*') *representation*.

In general, the Hamiltonian for a free Dirac particle is given by

$$H = \beta m + \boldsymbol{\alpha} \cdot \mathbf{p}. \qquad (3.46)$$

Thus, in particular, we effectively have the Hamiltonian

$$H = \beta m \qquad (3.47)$$

in the rest frame of the particle, i.e. when $\mathbf{p} = \mathbf{0}$. Especially, note that the operator

$$\Sigma = \begin{pmatrix} \sigma & 0 \\ 0 & \sigma \end{pmatrix}, \qquad (3.48)$$

which has to do with the spin and will later be interpreted as twice the spin operator (see Section 3.4), commutes with this Hamiltonian, i.e. $[\Sigma, H] = 0$. The Dirac equation for a free particle (in the rest frame) has four linearly independent spinor solutions, which are labelled as follows,

$$u_+ = \begin{pmatrix} 1 \\ 0 \\ 0 \\ 0 \end{pmatrix}, \quad u_- = \begin{pmatrix} 0 \\ 1 \\ 0 \\ 0 \end{pmatrix}, \quad v_+ = \begin{pmatrix} 0 \\ 0 \\ 1 \\ 0 \end{pmatrix}, \quad v_- = \begin{pmatrix} 0 \\ 0 \\ 0 \\ 1 \end{pmatrix}, \qquad (3.49)$$

---

[7] Note that we define $\gamma_5 \equiv \gamma^5$. Although γ^5 uses the Greek letter γ, it is **not** one of the gamma matrices, since the label 5 is only a relic of an old notation in which γ^0 was called 'γ^4'.

corresponding to the eigenvalues m, m, $-m$, and $-m$, respectively, of the Hamiltonian (3.47). The two solutions $u_\pm$ are the positive-energy solutions with spin eigenvalues $\pm 1/2$, whereas the two solutions $v_\pm$ are the negative-energy solutions with spin eigenvalues $\pm 1/2$. Note that the spin eigenvalues are the eigenvalues of the 'spin operator' $\Sigma^3/2$, which commutes with the Hamiltonian, and therefore, can be simultaneously diagonalized. In addition, the positive-energy solutions are related to particles, while the negative-energy solutions are related to antiparticles, cf. the discussion on the solutions to the Klein–Gordon equation in Section 2.3.

In a general Lorentz system, i.e. $\mathbf{p} \neq \mathbf{0}$, we have the Hamiltonian (3.46) and, in addition, the eigenvalues of the Hamiltonian H are $p_0 = \pm E(\mathbf{p})$, where $E(\mathbf{p}) = \sqrt{m^2 + \mathbf{p}^2}$. The four linearly independent four-component spinor solutions can also be written in terms of two-component spinors. These spinors will be denoted u_1, u_2, v_1, and v_2. Using the two-component spinors, we can write the four-component spinors as

$$u_\pm = \begin{pmatrix} u_1 \\ u_2 \end{pmatrix} \quad \text{and} \quad v_\pm = \begin{pmatrix} v_1 \\ v_2 \end{pmatrix}. \tag{3.50}$$

Now, using the two-component spinors for the equation $Hu_\pm = E(\mathbf{p})u_\pm$ with the Hamiltonian (3.46), we arrive at the following coupled system of equations for the positive-energy states of the Dirac equation

$$\boldsymbol{\sigma} \cdot \mathbf{p} u_2 + m u_1 = E(\mathbf{p}) u_1, \tag{3.51}$$
$$\boldsymbol{\sigma} \cdot \mathbf{p} u_1 - m u_2 = E(\mathbf{p}) u_2. \tag{3.52}$$

Note that Eqs. (3.51) and (3.52) are linearly dependent, because the coefficient matrix of their corresponding homogeneous system of equations has determinant equal to zero, which equivalently means that $E(\mathbf{p})^2 = m^2 + \mathbf{p}^2$. Since it holds for the positive-energy solutions that $E(\mathbf{p}) + m \geq 2m \neq 0$, we can use Eq. (3.52) and solve for the spinor u_2, which gives

$$u_2 = \frac{\boldsymbol{\sigma} \cdot \mathbf{p}}{E(\mathbf{p}) + m} u_1, \tag{3.53}$$

where u_1 is arbitrary. Thus, we have the positive-energy solutions with spin up, i.e. spin $+1/2$,

$$u_+ = u_+(E(\mathbf{p}), \mathbf{p}) = \begin{pmatrix} u_{1,+} \\ \frac{\boldsymbol{\sigma} \cdot \mathbf{p}}{E(\mathbf{p}) + m} u_{1,+} \end{pmatrix}, \tag{3.54}$$

where $u_{1,+} = (1 \quad 0)^T$, and with spin down, i.e. spin $-1/2$,

$$u_- = u_-(E(\mathbf{p}), \mathbf{p}) = \begin{pmatrix} u_{1,-} \\ \frac{\boldsymbol{\sigma} \cdot \mathbf{p}}{E(\mathbf{p}) + m} u_{1,-} \end{pmatrix}, \tag{3.55}$$

where $u_{1,-} = (0 \quad 1)^T$. Note that the two-component spinors $u_{1,+}$ and $u_{1,-}$ must be linearly independent from each other, i.e. $u_{1,+} \perp u_{1,-}$. In addition, we have that $u_1^\dagger u_1 = 1$. Using Eq. (3.39) and the definition of the Dirac adjoint as well as the identity $\gamma^\mu = \gamma^0 (\gamma^\mu)^\dagger \gamma^0$, which can easily be verified (see Exercise 3.3), we obtain for the conjugated Dirac spinors (or actually the Dirac adjoint spinors) $\bar{u}$ and $\bar{v}$ the following equations

$$\bar{u}(p)(\not{p} - m\mathbb{1}_4) = 0 \quad \text{and} \quad \bar{v}(p)(\not{p} + m\mathbb{1}_4) = 0. \tag{3.56}$$

Finally, using the normalization condition[8]

$$\bar{u}u = u^\dagger \gamma^0 u = 1 \quad \text{or} \quad u^\dagger u = \frac{E(\mathbf{p})}{m} \neq 1, \tag{3.57}$$

since

$$\bar{u}u = \frac{m}{E(\mathbf{p})} u^\dagger u, \tag{3.58}$$

instead of

$$u^\dagger u = 1, \tag{3.59}$$

we obtain the positive-energy solutions $[p_0 = E(\mathbf{p})]$

$$u_\pm = u_\pm(E(\mathbf{p}), \mathbf{p}) = \sqrt{\frac{E(\mathbf{p}) + m}{2m}} \begin{pmatrix} u_1 \\ \frac{\sigma \cdot \mathbf{p}}{E(\mathbf{p}) + m} u_1 \end{pmatrix}, \tag{3.60}$$

where $u_1 = u_{1,+}$ for spin up and $u_1 = u_{1,-}$ for spin down, respectively. In addition, $u_\pm$ constitute two linearly independent orthogonal solutions to the Dirac equation with energy $E(\mathbf{p})$ and 3-momentum $\mathbf{p}$. Similarly, using the two component spinors for the equation $Hu_\pm = -E(\mathbf{p})u_\pm$, we find that

$$u_1 = -\frac{\sigma \cdot \mathbf{p}}{E(\mathbf{p}) + m} u_2 = \frac{\sigma \cdot (-\mathbf{p})}{E(\mathbf{p}) + m} u_2, \tag{3.61}$$

where u_2 is arbitrary. Thus, using the normalization condition $\bar{u}u = 1$, we obtain the 'negative-energy solutions' $[p_0 = -E(\mathbf{p})]$

$$u_\pm(-E(\mathbf{p}), \mathbf{p}) = \sqrt{-\frac{E(\mathbf{p}) + m}{2m}} \begin{pmatrix} \frac{\sigma \cdot (-\mathbf{p})}{E(\mathbf{p}) + m} u_2 \\ u_2 \end{pmatrix}. \tag{3.62}$$

---

[8] This normalization may cause confusion for a massless particle, since the spinors will become infinite in the limit $m \to 0$. However, the mass m in the denominator will always cancel in physical quantities such as cross-sections and decay rates. In principle, there is an infinite number of ways to choose a normalization for the solutions to the Dirac equation, but in practice, there is only a limited amount that is convenient. See, for example, the normalizations used by Mandl & Shaw ($\bar{u}u = 1$, which is the same as is used here) and Peskin & Schroeder ($\bar{u}u = 2m$).

However, for the 'negative-energy solutions' the normalization factor is imaginary. Therefore, it is convenient to use the spinors $v_\pm$ for the negative-energy solutions, which are defined as

$$v_\pm(E(\mathbf{p}), \mathbf{p}) = u_\pm(-E(\mathbf{p}), -\mathbf{p}).\qquad(3.63)$$

In addition, $v_\pm$ are normalized as $\bar{v}v = -1$ and constitute two linearly independent orthogonal solutions to the Dirac equation corresponding to energy $-E(\mathbf{p})$ and 3-momentum $-\mathbf{p}$. Note the replacement $\mathbf{p} \mapsto -\mathbf{p}$ in the relation between the 'negative-energy solutions' $u_\pm$ and the 'real' negative-energy solutions $v_\pm$. Thus, using the normalization condition $\bar{v}v = -1$, we obtain the negative-energy solutions $[p_0 = -E(\mathbf{p})]$

$$v_\pm = v_\pm(E(\mathbf{p}), \mathbf{p}) = \sqrt{\frac{E(\mathbf{p}) + m}{2m}} \begin{pmatrix} \frac{\boldsymbol{\sigma}\cdot\mathbf{p}}{E(\mathbf{p})+m} v_2 \\ v_2 \end{pmatrix},\qquad(3.64)$$

where $v_2 = v_{2,+}$ for spin up and $v_2 = v_{2,-}$ for spin down, respectively. In conclusion, we introduce the shorter and more useful notation for the positive- and negative-energy solutions, i.e.

$$u(\mathbf{p}, s) = \begin{cases} u_+(E(\mathbf{p}), \mathbf{p}), & s = +1/2 \\ u_-(E(\mathbf{p}), \mathbf{p}), & s = -1/2 \end{cases}\qquad(3.65)$$

and

$$v(\mathbf{p}, s) = \begin{cases} v_+(E(\mathbf{p}), \mathbf{p}), & s = +1/2 \\ v_-(E(\mathbf{p}), \mathbf{p}), & s = -1/2 \end{cases},\qquad(3.66)$$

respectively. This notation is not as pedagogical as the longer notation, but it is certainly less cumbersome.

To this end, let us consider an example of the positive- and negative-energy solutions. The two positive-energy solutions can describe a free spin-1/2 (fermion) particle such as an electron, whereas the two negative-energy solutions describe its antiparticle. In the case of the electron, the antiparticle is the positron.[9] Indeed, a positron with energy E and 3-momentum $\mathbf{p}$ is described by one of the negative-energy electron solutions with energy $-E$ and 3-momentum $-\mathbf{p}$. See Section 3.2 for a further discussion.

Exercise 3.1 Show that the spinors $u(\mathbf{p}, s)$ and $v(\mathbf{p}, s)$ are orthogonal solutions to the Dirac equation [see Eqs. (3.128)–(3.130)].

---

[9] In 1936, C. Anderson was awarded the Nobel Prize in physics 'for his discovery of the positron' in 1932, which had been postulated in 1928 by Dirac.

3.2 Problems with the Dirac equation: the hole theory and the Dirac sea

Although the Dirac equation solves some of the problems that the Klein–Gordon equation has (e.g. no positive-definite probability density), there are still some other problems with the Dirac equation. In addition, note that both the Klein–Gordon equation and the Dirac equation have the problem connected with the Klein paradox.

As we have observed, the Dirac equation has negative-energy solutions. Now, how should we interpret them? For some time, their physical interpretation was questioned and caused a lot of difficulty and confusion. In 1930, Dirac 'resolved' this problem by introducing the so-called *hole theory*.[10] He postulated the existence of a '*sea*' of electrons occupying all the negative-energy states,[11] or to use his own words: 'Assume that nearly all the negative energy states are occupied with one electron in each state in accordance with the exclusion principle of Pauli.' Since electrons must obey the Pauli exclusion principle, it is impossible for the positive-energy electrons to perform transitions into this sea containing negative-energy states by radiating photons with energies larger than $2m$. Thus, a catastrophic instability of the hole theory is avoided. In the picture of the Dirac sea (see Fig. 3.1), all the negative-energy electron states are occupied. However, if one of the negative-energy electrons absorbs a photon with energy larger than $2m$, then it is lifted to an unoccupied positive-energy electron state, leaving a hole in the Dirac sea (see the illustration to the left in Fig. 3.1). Indeed, if the sea electron had energy $-E$, 3-momentum $-\mathbf{p}$, and charge $-e$, then the new sea state, i.e. the hole, has energy E, 3-momentum $\mathbf{p}$, and charge e. Therefore, the hole behaves as a positive-energy state with positive charge, i.e. it describes a positron. Thus, the physical interpretation of this process is production of an electron–positron pair from the vacuum (i.e. the Dirac sea). In the case that there exists a hole in the Dirac sea (see the illustration to the right in Fig. 3.1), a positive-energy electron can fall into that hole under emission of a photon, which is equivalent to annihilation of an electron–positron pair. Note that annihilation of an electron–positron pair is the inverse process to production of an electron–positron pair, and vice versa.

In conclusion, a one-particle theory is only applicable for an isolated free particle, i.e. when there are no interactions present. However, there are no real situations without interactions. Therefore, in the case of interactions, we must use a many-particle theory, in which the number of particles is not conserved. Quantum field theory is such a theory, which we will investigate in Chapters 4–13, but not in Chapter 9.

---

[10] For the interested reader, please see: P. A. M. Dirac, A theory of electrons and protons, *Proc. R. Soc.* (*London*) A **126**, 360–365 (1930).

[11] The Dirac sea is a theoretical construction of the vacuum as an infinite sea of particles with negative energy.

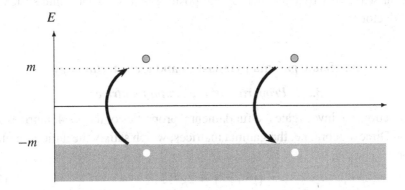

Figure 3.1 The energy spectrum of the free Dirac equation. Note that only solutions with $E > m$ (positive-energy solutions) and $E < m$ (negative-energy solutions) are possible. Therefore, there is a 'mass gap' between $E = -m$ and $E = m$ of size $2m$. The Dirac sea is marked in light grey. The process shown to the left in the picture is the production of an electron–positron pair, whereas the process shown to the right in the picture is the annihilation of an electron–positron pair.

In addition, the theoretical construction of the Dirac sea has at least one large aesthetical problem, since the existence of such a sea means that there is an infinite negative electric charge occupying all of space. Therefore, one has to assume that the 'bare vacuum', sometimes known as the 'jellium background', must have an infinite positive electric charge, which exactly cancels the Dirac sea. Thus, in the revised version of the hole theory, this problem has been circumvented and there is a complete symmetry between positive and negative charges. However, in quantum field theory, the Dirac equation describes the positron as a **real** particle (**not** as the absence of a particle) and the vacuum is the state, where there are **no** particles (**not** an infinite sea of particles). In fact, transforming an occupied negative-energy electron state into an unoccupied positive-energy positron state and an unoccupied negative-energy electron state into an occupied positive-energy positron state is known as a so-called *Bogoliubov transformation* and gives us the possibility not to use the Dirac sea. Anyway, note that the concept of the Dirac sea cannot be used for the Klein–Gordon equation, since spinless particles do not obey the Pauli exclusion principle.

Finally, in some applications of condensed matter physics, the concepts of the hole theory and the Dirac sea are still used. For example, the sea of conduction electrons in an electrical conductor is called the *Fermi sea* and consists of such electrons with energy up to the chemical potential of the conductor. Thus, an unoccupied state in the Fermi sea behaves as an electron with positive charge, i.e. a positron, but it is normally referred to as a hole. In this case, the negative charge of the Fermi sea is exactly cancelled by the positive charge of the ionic structure of the conductor.

3.3 Some gamma gymnastics and trace technology

3.3.1 Properties of the gamma matrices

In this section, we investigate the fundamental properties of the 4×4 matrices spanning the Dirac algebra, i.e. the gamma matrices, which satisfy the anticommutation relation

$$\{\gamma^\mu, \gamma^\nu\} = 2g^{\mu\nu}\mathbb{1}_4. \tag{3.67}$$

For a detailed study of the properties of the gamma matrices, please see R. H. Good, Jr., Properties of the Dirac matrices, *Rev. Mod. Phys.* **27**, 187–211 (1955). In analogy to the Pauli matrices together with the identity matrix $\mathbb{1}_2$ that span the space of the 2×2 complex matrices, there are 16 matrices that span the space of 4×4 complex matrices.[12] The matrices that can be constructed from the Dirac algebra (and the identity matrix $\mathbb{1}_4$) are the following:

$$\Gamma_1 = \mathbb{1}_4,$$
$$\Gamma_2 = \gamma^0, \quad \Gamma_3 = i\gamma^1, \quad \Gamma_4 = i\gamma^2, \quad \Gamma_5 = i\gamma^3,$$
$$\Gamma_6 = i\gamma^2\gamma^3, \quad \Gamma_7 = i\gamma^3\gamma^1, \quad \Gamma_8 = i\gamma^1\gamma^2, \quad \Gamma_9 = \gamma^0\gamma^1, \quad \Gamma_{10} = \gamma^0\gamma^2,$$
$$\Gamma_{11} = \gamma^0\gamma^3,$$
$$\Gamma_{12} = i\gamma^0\gamma^2\gamma^3, \quad \Gamma_{13} = i\gamma^0\gamma^3\gamma^1, \quad \Gamma_{14} = i\gamma^0\gamma^1\gamma^2, \quad \Gamma_{15} = \gamma^1\gamma^2\gamma^3,$$
$$\Gamma_{16} = i\gamma^0\gamma^1\gamma^2\gamma^3 = \gamma^5.$$

All the Γ_is fulfil the normalization condition

$$(\Gamma_i)^2 = \mathbb{1}_4, \quad i = 1, 2, \ldots, 16, \tag{3.68}$$

since factors of i have been included in the definitions for some of the Γ_is. Furthermore, using the Dirac algebra, note that all other products of gamma matrices can be reduced to one of the 16 Γ_is. Using the Γ_is, the following four properties can be derived:

- $\Gamma_i\Gamma_j = \mathbb{1}_4 \quad \Leftrightarrow \quad i = j,$
- $\Gamma_i\Gamma_j = \pm\Gamma_j\Gamma_i,$
- $\Gamma_i\Gamma_j = c_{ij}\Gamma_k,$ where $c_{ij} \in \{\pm 1, \pm i\},$
- If $\Gamma_i \neq \mathbb{1}_4,$ then there exists a Γ_j such that $\Gamma_j\Gamma_i\Gamma_j = -\Gamma_i.$

---

[12] Note that since different gamma matrices anticommute [cf. Eq. (3.20) or Eq. (3.67)], we need only to consider products of different gamma matrices and the order of them is not important. Thus, there are $2^4 - 1 = 15$ combinations of the gamma matrices. Thus, we have 16 different matrices including the identity matrix $\mathbb{1}_4$.

Table 3.1 *Products of two Γ_is. By inspection, we note that all coefficients in front of the Γ_is in the right-hand sides belong to the set $\{\pm 1, \pm i\}$, according to the third property.*

$\Gamma_1\Gamma_i = \Gamma_i$	$\Gamma_3\Gamma_{13} = \Gamma_{11}$	$\Gamma_5\Gamma_{14} = -i\Gamma_{16}$	$\Gamma_8\Gamma_{11} = \Gamma_{16}$	$\Gamma_{12}\Gamma_{13} = i\Gamma_8$
$\Gamma_2\Gamma_3 = i\Gamma_9$	$\Gamma_3\Gamma_{14} = -\Gamma_{10}$	$\Gamma_5\Gamma_{15} = -\Gamma_8$	$\Gamma_8\Gamma_{12} = i\Gamma_{13}$	$\Gamma_{12}\Gamma_{14} = -i\Gamma_7$
$\Gamma_2\Gamma_4 = i\Gamma_{10}$	$\Gamma_3\Gamma_{15} = -\Gamma_6$	$\Gamma_5\Gamma_{16} = i\Gamma_{14}$	$\Gamma_8\Gamma_{13} = -i\Gamma_{12}$	$\Gamma_{12}\Gamma_{15} = -i\Gamma_9$
$\Gamma_2\Gamma_5 = i\Gamma_{11}$	$\Gamma_3\Gamma_{16} = i\Gamma_{12}$	$\Gamma_6\Gamma_7 = i\Gamma_8$	$\Gamma_8\Gamma_{14} = \Gamma_2$	$\Gamma_{12}\Gamma_{16} = -i\Gamma_3$
$\Gamma_2\Gamma_6 = \Gamma_{12}$	$\Gamma_4\Gamma_5 = i\Gamma_6$	$\Gamma_6\Gamma_8 = -i\Gamma_7$	$\Gamma_8\Gamma_{15} = -\Gamma_5$	$\Gamma_{13}\Gamma_{14} = i\Gamma_6$
$\Gamma_2\Gamma_7 = \Gamma_{13}$	$\Gamma_4\Gamma_6 = -i\Gamma_5$	$\Gamma_6\Gamma_9 = \Gamma_{16}$	$\Gamma_8\Gamma_{16} = \Gamma_{11}$	$\Gamma_{13}\Gamma_{15} = -i\Gamma_{10}$
$\Gamma_2\Gamma_8 = \Gamma_{14}$	$\Gamma_4\Gamma_7 = -\Gamma_{15}$	$\Gamma_6\Gamma_{10} = i\Gamma_{11}$	$\Gamma_9\Gamma_{10} = i\Gamma_8$	$\Gamma_{13}\Gamma_{16} = -i\Gamma_4$
$\Gamma_2\Gamma_9 = -i\Gamma_3$	$\Gamma_4\Gamma_8 = i\Gamma_3$	$\Gamma_6\Gamma_{11} = -i\Gamma_{10}$	$\Gamma_9\Gamma_{11} = -i\Gamma_7$	$\Gamma_{14}\Gamma_{15} = -i\Gamma_{11}$
$\Gamma_2\Gamma_{10} = -i\Gamma_4$	$\Gamma_4\Gamma_9 = \Gamma_{14}$	$\Gamma_6\Gamma_{12} = \Gamma_2$	$\Gamma_9\Gamma_{12} = -i\Gamma_{15}$	$\Gamma_{14}\Gamma_{16} = -i\Gamma_5$
$\Gamma_2\Gamma_{11} = -i\Gamma_5$	$\Gamma_4\Gamma_{10} = i\Gamma_2$	$\Gamma_6\Gamma_{13} = i\Gamma_{14}$	$\Gamma_9\Gamma_{13} = -\Gamma_5$	$\Gamma_{15}\Gamma_{16} = -i\Gamma_2$
$\Gamma_2\Gamma_{12} = \Gamma_6$	$\Gamma_4\Gamma_{11} = -\Gamma_{12}$	$\Gamma_6\Gamma_{14} = -i\Gamma_{13}$	$\Gamma_9\Gamma_{14} = \Gamma_4$	
$\Gamma_2\Gamma_{13} = \Gamma_7$	$\Gamma_4\Gamma_{12} = -\Gamma_{11}$	$\Gamma_6\Gamma_{15} = -\Gamma_3$	$\Gamma_9\Gamma_{15} = i\Gamma_{12}$	
$\Gamma_2\Gamma_{14} = \Gamma_8$	$\Gamma_4\Gamma_{13} = -i\Gamma_{16}$	$\Gamma_6\Gamma_{16} = \Gamma_9$	$\Gamma_9\Gamma_{16} = \Gamma_6$	
$\Gamma_2\Gamma_{15} = -i\Gamma_{16}$	$\Gamma_4\Gamma_{14} = \Gamma_9$	$\Gamma_7\Gamma_8 = i\Gamma_6$	$\Gamma_{10}\Gamma_{11} = i\Gamma_6$	
$\Gamma_2\Gamma_{16} = i\Gamma_{15}$	$\Gamma_4\Gamma_{15} = -\Gamma_7$	$\Gamma_7\Gamma_9 = -i\Gamma_{11}$	$\Gamma_{10}\Gamma_{12} = \Gamma_5$	
$\Gamma_3\Gamma_4 = i\Gamma_8$	$\Gamma_4\Gamma_{16} = i\Gamma_{13}$	$\Gamma_7\Gamma_{10} = \Gamma_{16}$	$\Gamma_{10}\Gamma_{13} = -i\Gamma_{15}$	
$\Gamma_3\Gamma_5 = -i\Gamma_7$	$\Gamma_5\Gamma_6 = i\Gamma_4$	$\Gamma_7\Gamma_{11} = i\Gamma_9$	$\Gamma_{10}\Gamma_{14} = -\Gamma_3$	
$\Gamma_3\Gamma_6 = -\Gamma_{15}$	$\Gamma_5\Gamma_7 = -i\Gamma_3$	$\Gamma_7\Gamma_{12} = -i\Gamma_{14}$	$\Gamma_{10}\Gamma_{15} = i\Gamma_{13}$	
$\Gamma_3\Gamma_7 = i\Gamma_5$	$\Gamma_5\Gamma_8 = -\Gamma_{15}$	$\Gamma_7\Gamma_{13} = \Gamma_2$	$\Gamma_{10}\Gamma_{16} = \Gamma_7$	
$\Gamma_3\Gamma_8 = -i\Gamma_4$	$\Gamma_5\Gamma_9 = -\Gamma_{13}$	$\Gamma_7\Gamma_{14} = i\Gamma_{12}$	$\Gamma_{11}\Gamma_{12} = -\Gamma_4$	
$\Gamma_3\Gamma_9 = i\Gamma_2$	$\Gamma_5\Gamma_{10} = \Gamma_{12}$	$\Gamma_7\Gamma_{15} = -\Gamma_4$	$\Gamma_{11}\Gamma_{13} = \Gamma_3$	
$\Gamma_3\Gamma_{10} = -\Gamma_{14}$	$\Gamma_5\Gamma_{11} = i\Gamma_2$	$\Gamma_7\Gamma_{16} = \Gamma_{10}$	$\Gamma_{11}\Gamma_{14} = -i\Gamma_{15}$	
$\Gamma_3\Gamma_{11} = \Gamma_{13}$	$\Gamma_5\Gamma_{12} = \Gamma_{10}$	$\Gamma_8\Gamma_9 = i\Gamma_{10}$	$\Gamma_{11}\Gamma_{15} = i\Gamma_{14}$	
$\Gamma_3\Gamma_{12} = -i\Gamma_{16}$	$\Gamma_5\Gamma_{13} = -\Gamma_9$	$\Gamma_8\Gamma_{10} = -i\Gamma_9$	$\Gamma_{11}\Gamma_{16} = \Gamma_8$	

The first property is an extension of the normalization condition (3.68) and the second property gives commutation and anticommutation relations for different Γ_is. Actually, the third property can be derived explicitly and the result is presented in Table 3.1. In addition, for the fourth property, using the products in Table 3.1, we can find a Γ_j for each Γ_i, where $j \neq i$ and $i \neq 1$, that fulfils the relation $\Gamma_j\Gamma_i\Gamma_j = -\Gamma_i$. This result is displayed in Table 3.2.

In order to show that the Γ_is are linearly independent matrices, we use the equation

$$\sum_{i=1}^{16} c_i\Gamma_i = 0 \tag{3.69}$$

and prove that it holds that $c_i = 0$ for all $i = 1, 2, \ldots, 16$. In fact, it holds that the trace of all Γ_is, except from Γ_1, is equal to zero, i.e. $\mathrm{tr}(\Gamma_i) = 0$ for

Table 3.2 *Values of i and j that fulfil the relation $\Gamma_j \Gamma_i \Gamma_j = -\Gamma_i$. Note that we have chosen to display the lowest possible values of j only.*

i	2	3	4	5	6	7	8	9	10	11	12	13	14	15	16
j	3	2	2	2	4	3	3	2	2	2	3	4	5	2	2

$i = 2, 3, \ldots, 16$,[13] but $\mathrm{tr}(\Gamma_1) = N = 4$. Therefore, taking the trace of Eq. (3.69), we obtain $c_1 = 0$. Now, the sum in Eq. (3.69) runs instead from 2 to 16. Next, multiplying the sum by a matrix Γ_j, where $j \neq 1$, and again taking the trace of this result, we find that

$$4c_j + \sum_{i \neq j} c_i \mathrm{tr}(\Gamma_i \Gamma_j) = 0, \quad j = 2, 3, \ldots, 16. \tag{3.70}$$

Then, using the definitions of the Γ_is, one can show that any product of two Γ_is can always be reduced to another such matrix, i.e. $\Gamma_i \Gamma_j = c\Gamma_k$, where $c \in \{\pm 1, \pm i\}$. In the case that $i \neq j$, we have $\Gamma_k \neq \mathbb{1}_4$ [only in the case that $i = j$, we have $\Gamma_k = \mathbb{1}_4$, cf. Eq. (3.68)], and therefore, the trace in Eq. (3.70) is equal to zero for any $j \in \{2, 3, \ldots, 16\}$. Consequently, $c_j = 0$ for $j = 2, 3, \ldots, 16$, and we have already shown that $c_1 = 0$. In conclusion, $c_i = 0$ for $i = 1, 2, \ldots, 16$. Thus, we have shown that the 16 matrices Γ_i ($i = 1, 2, \ldots, 16$) are linearly independent, and hence, it means that a representation of the Dirac algebra must consist of at least 4×4 matrices, since one cannot construct 16 linearly independent matrices using matrices of lower dimensionality than four. On the other hand, since there are exactly 16 linearly independent 4×4 matrices, the gamma matrices can be represented by 4×4 matrices, and this representation is irreducible.

We conclude with the following important theorem.

Theorem 3.1 (Pauli's fundamental theorem) *If $\{\gamma^\mu\}$ and $\{\gamma'^\mu\}$ are two sets of gamma matrices satisfying the Dirac algebra, i.e.*

$$\{\gamma^\mu, \gamma^\nu\} = 2g^{\mu\nu} \mathbb{1}_4, \tag{3.71}$$

$$\{\gamma'^\mu, \gamma'^\nu\} = 2g^{\mu\nu} \mathbb{1}_4, \tag{3.72}$$

then there exists a non-singular matrix S, such that

$$\gamma'^\mu = S\gamma^\mu S^{-1}, \tag{3.73}$$

---

[13] Taking the trace of the relation $\Gamma_j \Gamma_i \Gamma_j = -\Gamma_i$, using the cyclic property of the trace, and finally using the property $\Gamma_j^2 = \mathbb{1}_4$, it is straightforward to show that $\mathrm{tr}(\Gamma_i) = 0$ for $i = 2, 3, \ldots, 16$.

which is a so-called similarity transformation. Especially, if $\gamma^{0\dagger} = \gamma^0$, $\gamma^{i\dagger} = -\gamma^i$, $\gamma'^{0\dagger} = \gamma'^0$, and $\gamma'^{i\dagger} = -\gamma'^i$, then S can be chosen to be unitary.

This theorem implies that any irreducible representation of the gamma matrices is only unique up to a similarity transformation. In principle, there is an infinite number of representations for the gamma matrices due to the similarity transformation, but there is only a limited amount that is useful. Such representations for the gamma matrices are, for example, the Dirac representation (or the standard representation) (3.45), the Weyl representation (or chiral representation) (3.324), and the Majorana representation. The Dirac representation is most useful for physical systems with small kinetic energies (e.g. atomic physics), whereas the Weyl representation is more convenient for physical systems with ultra-relativistic energies. The Majorana representation is used for situations when particles are their own antiparticles (see the discussion in Section 7.6). In addition, note that the similarity transformation needs to be unitary if the transformed gamma matrices should satisfy the Hermiticity condition $(\gamma'^\mu)^\dagger = \gamma'^0 \gamma'^\mu \gamma'^0$.

In general, using the linear independence of the Γ_is, every 4×4 matrix A can be written as

$$A = \sum_{i=1}^{16} a_i \Gamma_i, \tag{3.74}$$

where the coefficients a_i $(i = 1, 2, \ldots, 16)$ are given by

$$a_i = \frac{1}{4} \text{tr}(A\Gamma_i). \tag{3.75}$$

In addition, Schur's lemma implies that any 4×4 matrix A that commutes with each of the gamma matrices, $[A, \gamma^\mu] = 0$ for all $\mu = 0, 1, 2, 3$, and therefore also with all the Γ_is, is a constant multiple of the unit operator, since the gamma matrices are irreducible.

3.3.2 Gamma matrix identities

Using the algebra of the Dirac gamma matrices, we can derive the following useful identities (which are valid for all representations of the gamma matrices):

$$\gamma^\mu \gamma_\mu = 4\mathbb{1}_4, \tag{3.76}$$
$$\gamma^\mu \gamma^\nu \gamma_\mu = -2\gamma^\nu, \tag{3.77}$$
$$\gamma^\mu \gamma^\nu \gamma^\rho \gamma_\mu = 4g^{\nu\rho}\mathbb{1}_4, \tag{3.78}$$
$$\gamma^\mu \gamma^\nu \gamma^\rho \gamma^\sigma \gamma_\mu = -2\gamma^\sigma \gamma^\rho \gamma^\nu. \tag{3.79}$$

Note that these identities are valid in four dimensions only.

3.3.3 Trace identities

Furthermore, by using the properties of the gamma matrices and traces, we find the following, for example.

- The trace of an odd number of γs is equal to zero.
  Since γ^5 anticommutes with γ^μ, we have that $\gamma^5 \gamma^{\mu_1} \ldots \gamma^{\mu_n} \gamma^5 = (-1)^n \gamma^{\mu_1} \ldots \gamma^{\mu_n}$. Now, taking the trace, we obtain $\mathrm{tr}(\gamma^{\mu_1} \ldots \gamma^{\mu_n}) = (-1)^n \mathrm{tr}(\gamma^{\mu_1} \ldots \gamma^{\mu_n})$, which implies that $\mathrm{tr}(\gamma^{\mu_1} \ldots \gamma^{\mu_n}) = 0$ when n is odd.

- If n is even, then one can use $\gamma^\mu \gamma^\nu + \gamma^\nu \gamma^\mu = 2g^{\mu\nu} \mathbb{1}_4$ to reduce 'n' to '$n-2$'.
  Since $\mathrm{tr}(\gamma^\mu \gamma^\nu) = \mathrm{tr}(\gamma^\nu \gamma^\mu) = \frac{1}{2}\mathrm{tr}(\gamma^\mu \gamma^\nu + \gamma^\nu \gamma^\mu) = g^{\mu\nu}\mathrm{tr}(\mathbb{1}_4) = 4g^{\mu\nu}$, we find that $\mathrm{tr}(\gamma^\mu \gamma^\nu) = 4g^{\mu\nu}$.

- The trace of γ^5 is equal to zero.
  We have that $\mathrm{tr}(\gamma^5) = \mathrm{tr}[\gamma^5 \gamma^0 (\gamma^0)^{-1}] = -\mathrm{tr}[\gamma^0 \gamma^5 (\gamma^0)^{-1}] = -\mathrm{tr}[\gamma^5 (\gamma^0)^{-1}\gamma^0] = -\mathrm{tr}(\gamma^5)$, where it has been used that $\gamma^5 \gamma^0 = -\gamma^0 \gamma^5$ and that the trace is cyclic in its arguments. Thus, it immediately follows that $\mathrm{tr}(\gamma^5) = 0$. Note that since $(\gamma^0)^2 = \mathbb{1}_4$, we could as well have used that $(\gamma^0)^{-1} = \gamma^0$.

Thus, in conclusion, we can list the following useful trace identities:

$$\mathrm{tr}(\mathbb{1}_4) = 4, \tag{3.80}$$

$$\mathrm{tr}(\gamma^{\mu_1}\gamma^{\mu_2}\ldots\gamma^{\mu_n}) = 0, \quad n \text{ odd}, \tag{3.81}$$

$$\mathrm{tr}(\gamma^\mu \gamma^\nu) = 4g^{\mu\nu}, \tag{3.82}$$

$$\mathrm{tr}(\gamma^\mu \gamma^\nu \gamma^\rho \gamma^\sigma) = 4(g^{\mu\nu}g^{\rho\sigma} - g^{\mu\rho}g^{\nu\sigma} + g^{\mu\sigma}g^{\nu\rho}), \tag{3.83}$$

$$\mathrm{tr}(\gamma^5) = 0, \tag{3.84}$$

$$\mathrm{tr}(\gamma^\mu \gamma^5) = 0, \tag{3.85}$$

$$\mathrm{tr}(\gamma^\mu \gamma^\nu \gamma^5) = 0, \tag{3.86}$$

$$\mathrm{tr}(\gamma^\mu \gamma^\nu \gamma^\rho \gamma^5) = 0, \tag{3.87}$$

$$\mathrm{tr}(\gamma^\mu \gamma^\nu \gamma^\rho \gamma^\sigma \gamma^5) = -4i\epsilon^{\mu\nu\rho\sigma}, \quad \epsilon^{0123} = 1. \tag{3.88}$$

In addition, we have

$$\mathrm{tr}(\gamma^{\mu_1}\gamma^{\mu_2}\ldots\gamma^{\mu_n}) = \mathrm{tr}(\gamma^{\mu_n}\ldots\gamma^{\mu_2}\gamma^{\mu_1}), \quad n \text{ even}. \tag{3.89}$$

Exercise 3.2 Show that $\mathrm{tr}(\gamma^{\mu_1}\gamma^{\mu_2}\ldots\gamma^{\mu_n}\gamma^5) = 0$, where n is odd.

3.3.4 Identities with Feynman-slash notation

Finally, we present some useful identities with the Feynman-slash notation:

$$\slashed{a}\slashed{b} = a \cdot b\mathbb{1}_4 - ia_\mu \sigma^{\mu\nu} b_\nu, \tag{3.90}$$

$$\slashed{a}\slashed{a} = a^2 \mathbb{1}_4, \tag{3.91}$$

$$\mathrm{tr}(\slashed{a}\slashed{b}) = 4a \cdot b, \tag{3.92}$$

$$\mathrm{tr}(\slashed{a}\slashed{b}\slashed{c}\slashed{d}) = 4\left[(a \cdot b)(c \cdot d) - (a \cdot c)(b \cdot d) + (a \cdot d)(b \cdot c)\right], \tag{3.93}$$

$$\mathrm{tr}(\gamma^5 \slashed{a}\slashed{b}\slashed{c}\slashed{d}) = 4i\epsilon_{\mu\nu\rho\sigma} a^\mu b^\nu c^\rho d^\sigma, \tag{3.94}$$

$$\gamma_\mu \slashed{a}\gamma^\mu = -2\slashed{a}, \tag{3.95}$$

$$\gamma_\mu \slashed{a}\slashed{b}\gamma^\mu = 4a \cdot b\mathbb{1}_4, \tag{3.96}$$

$$\gamma_\mu \slashed{a}\slashed{b}\slashed{c}\gamma^\mu = -2\slashed{c}\slashed{b}\slashed{a}, \tag{3.97}$$

where a, b, c, and d are 4-vectors and

$$\sigma^{\mu\nu} = \frac{i}{2}\left[\gamma^\mu, \gamma^\nu\right]. \tag{3.98}$$

Note that the $\sigma^{\mu\nu}$s satisfy a corresponding Hermiticity condition as the γ^μs, i.e. $(\sigma^{\mu\nu})^\dagger = \gamma^0 \sigma^{\mu\nu}\gamma^0$.

Exercise 3.3   For any representation of the gamma matrices that is obtained by a unitary similarity transformation from the standard representation, verify the identities $\gamma^{\mu\dagger} = \gamma^0 \gamma^\mu \gamma^0$, $(\gamma^5 \gamma^\mu)^\dagger = \gamma^0 \gamma^5 \gamma^\mu \gamma^0$, and $\sigma^{\mu\nu\dagger} = \gamma^0 \sigma^{\mu\nu}\gamma^0$.

Exercise 3.4   Show that $[\gamma^5, \sigma^{\mu\nu}] = 0$ and $\mathrm{tr}(\gamma^5 \slashed{a}\slashed{b}) = 0$.

3.4 Spin operators

In analogy to the Pauli matrices, which are 2×2 matrices and that are used in the treatment of spin in ordinary quantum mechanics, we should be able to find corresponding 4×4 matrices for the Dirac equation and the treatment of spin in relativistic quantum mechanics. The operator

$$\Sigma^i = \frac{1}{2}\epsilon^{ijk}\sigma^{jk} = \frac{i}{2}\epsilon^{ijk}\gamma^j \gamma^k \tag{3.99}$$

has the same commutation relations as the Pauli matrices, i.e. $[\Sigma^i, \Sigma^j] = 2i\epsilon^{ijk}\Sigma^k$, and thus, the operator $S^i = \Sigma^i/2$ fulfils the commutation relations for the angular momentum algebra. Therefore, S^i is interpreted as the *spin operator*. Using Eq. (3.99) and the definition of γ^5 as well as the properties of the gamma matrices, we find that Σ^i can be expressed as

$$\Sigma^i = \gamma^5 \gamma^0 \gamma^i, \tag{3.100}$$

and in addition, in the standard representation, Σ^i can be written as [cf. Eq. (3.48)]

$$\Sigma^i = \begin{pmatrix} \sigma^i & 0 \\ 0 & \sigma^i \end{pmatrix}. \tag{3.101}$$

Note that Eq. (3.99) [or Eq. (3.100)] defines the spatial part of the 'spin operator' only. In the rest frame of a particle, the covariant form of the spin operator is given by

$$(\Sigma^\mu) = (0, \Sigma^1, \Sigma^2, \Sigma^3). \tag{3.102}$$

Now, in the rest frame, we have the 4-momentum $(p^\mu) = (m, 0, 0, 0)$, where m is the mass of the particle. Thus, for massive particles, we observe that the covariant spin operator is given by

$$\Sigma^\mu = \frac{1}{2m} \epsilon^{\mu\nu\lambda\omega} p_\nu \sigma_{\lambda\omega}. \tag{3.103}$$

For example, evaluating the right-hand side of Eq. (3.103) for $\mu = 3$ using Eq. (3.99), we have

$$\frac{1}{2m} \epsilon^{3\nu\lambda\omega} p_\nu \sigma_{\lambda\omega} = \frac{1}{2m} \left(\epsilon^{3012} p_0 \sigma_{12} + \epsilon^{3021} p_0 \sigma_{21} \right)$$

$$= \frac{1}{2m} [1 \cdot m \cdot \sigma_{12} + (-1) \cdot m \cdot (-\sigma_{12})]$$

$$= \sigma_{12} = \sigma^{12} = \Sigma^3. \tag{3.104}$$

Next, introducing the normalized 4-vector $s = (s^\mu)$, which is called the *polarization vector* and is equal to $(0, \hat{\mathbf{n}})$ in the rest frame, we can obtain the covariant spin operator Σ^μ in an arbitrary direction $\hat{\mathbf{n}}$. Note that $\not{s} = -\gamma \cdot \hat{\mathbf{n}}$ and $s^2 = -1$. Thus, in general, the covariant operator $-s_\mu \Sigma^\mu$ gives the direction of the spin in any inertial frame. In particular, choosing $(s^\mu) = (s_z^\mu) \equiv (0, 0, 0, 1)$ and using Eq. (3.100), one finds

$$-s_{z\mu} \Sigma^\mu = \Sigma^3 = \gamma^5 \gamma^0 \gamma^3 = -\gamma^5 \gamma^0 \not{s}_z = \gamma^5 \not{s}_z \gamma^0, \tag{3.105}$$

since $\not{s}_z = \gamma^\mu s_{z\mu} = \gamma^3 \cdot (-1) = -\gamma^3$. However, in the rest frame, we have

$$\frac{\not{p}}{m} = \frac{\gamma^0 m}{m} = \gamma^0, \tag{3.106}$$

and in addition, the matrix γ^0 acting on the free Dirac spinors (3.49) will give ± 1, i.e. $\gamma^0 u_\pm = u_\pm$ and $\gamma^0 v_\pm = -v_\pm$. Therefore, we find that

$$-s_\mu \Sigma^\mu u_\pm = \gamma^5 \not{s}_z \gamma^0 u_\pm = \gamma^5 \not{s}_z u_\pm = \pm u_\pm, \tag{3.107}$$

$$-s_\mu \Sigma^\mu v_\pm = \gamma^5 \not{s}_z \gamma^0 v_\pm = -\gamma^5 \not{s}_z v_\pm = \pm v_\pm, \tag{3.108}$$

which means that the operator $-s_\mu \Sigma^\mu$ acts on all spinors in the same way and that the operators

$$S_\pm = \frac{1}{2} \left[\mathbb{1}_4 \pm \left(-s_\mu \Sigma^\mu \right) \right] = \frac{1}{2} \left(\mathbb{1}_4 \mp s_\mu \Sigma^\mu \right) \qquad (3.109)$$

are *spin projection operators*[14] for spinors, i.e.

$$S_\pm u_\pm = \frac{1}{2} \left(\mathbb{1}_4 \pm \gamma^5 \slashed{s}_z \right) u_\pm = u_\pm \quad \text{and} \quad S_\pm v_\pm = \frac{1}{2} \left(\mathbb{1}_4 \mp \gamma^5 \slashed{s}_z \right) v_\pm = v_\pm. \qquad (3.110)$$

These operators correspond to the two-component spin-up and spin-down projection operators in the z-direction in the non-relativistic case, i.e.

$$S_\pm^z = \frac{1}{2} \left(\mathbb{1}_2 \pm \sigma_z \right) = \frac{1}{2} \left(\mathbb{1}_2 \pm \sigma^3 \right). \qquad (3.111)$$

In fact, there is an alternative way of expressing and interpreting the operator Σ^μ and an operator similar to the operator $-s_\mu \Sigma^\mu$ in terms of the Pauli–Lubanski vector [cf. Eq. (1.66)]. Using the correspondence principle, one can write the generators of the Poincaré group as

$$M^{\mu\nu} = -i \left(x^\mu \partial^\nu - x^\nu \partial^\mu \right) - \frac{1}{2} \sigma^{\mu\nu}, \qquad (3.112)$$

$$P^\mu = i \partial^\mu, \qquad (3.113)$$

where, in the standard representation, it holds that

$$\sigma^{ij} = \epsilon^{ijk} \Sigma^k = \epsilon^{ijk} \begin{pmatrix} \sigma^k & 0 \\ 0 & \sigma^k \end{pmatrix}, \qquad (3.114)$$

or explicitly, we can write

$$\sigma^{23} = \begin{pmatrix} \sigma^1 & 0 \\ 0 & \sigma^1 \end{pmatrix}, \quad \sigma^{31} = \begin{pmatrix} \sigma^2 & 0 \\ 0 & \sigma^2 \end{pmatrix}, \quad \text{and} \quad \sigma^{12} = \begin{pmatrix} \sigma^3 & 0 \\ 0 & \sigma^3 \end{pmatrix}, \qquad (3.115)$$

i.e. the 4×4 matrices $\sigma^{\mu\nu}$ defined in Eq. (3.98) can effectively be written as the 2×2 Pauli matrices, which means that we have, for example,

$$J^3 = M^{21} = -i \left(x^2 \partial^1 - x^1 \partial^2 \right) - \frac{1}{2} \sigma^{21} = -i \left(x^1 \partial_2 - x^2 \partial_1 \right) + \frac{1}{2} \sigma^{12} = L^3 + \frac{1}{2} \sigma^3. \qquad (3.116)$$

---

[14] In general, note that projection operators (or projectors) $P_\pm$ fulfil the following relations: $P_\pm^2 = P_\pm$, $P_\pm P_\mp = 0$, and $P_\pm + P_\mp = \mathbb{1}$.

Therefore, inserting Eqs. (3.112) and (3.113) into the definition of the Pauli–Lubanski vector (1.66) and using the fact that a contraction of a symmetric tensor and an antisymmetric tensor is equal to zero, we find that

$$-w^\mu = \frac{i}{4}\epsilon^{\mu\nu\lambda\omega}\sigma_{\nu\lambda}\partial_\omega = \left\{\epsilon^{\mu\nu\lambda\omega} = -\epsilon^{\mu\nu\omega\lambda} = \epsilon^{\mu\omega\nu\lambda}\right\}$$

$$= \frac{i}{4}\epsilon^{\mu\omega\nu\lambda}\sigma_{\nu\lambda}\partial_\omega = \frac{i}{4}\epsilon^{\mu\nu\lambda\omega}\sigma_{\lambda\omega}\partial_\nu, \tag{3.117}$$

where we have renamed the summation indices in the last step. First, comparing Eqs. (3.103) and (3.117) as well as again using the correspondence principle, we obtain the following relation between the Pauli–Lubanski vector and the covariant spin operator

$$-w^\mu = \frac{m}{2}\Sigma^\mu. \tag{3.118}$$

Second, investigating the inner product of the Pauli–Lubanski vector and the polarization vector, i.e. using Eq. (3.117), we have

$$-w^\mu s_\mu = \frac{i}{4}\epsilon^{\mu\nu\lambda\omega}\sigma_{\nu\lambda}\partial_\omega s_\mu. \tag{3.119}$$

Then, using the identity

$$\gamma^5\sigma^{\mu\nu} = \frac{i}{2}\epsilon^{\mu\nu\lambda\omega}\sigma_{\lambda\omega}, \tag{3.120}$$

which is quite tedious to prove, and the fact that $\epsilon^{\mu\nu\lambda\omega} = -\epsilon^{\nu\mu\lambda\omega} = \epsilon^{\nu\lambda\mu\omega} = -\epsilon^{\nu\lambda\omega\mu}$, we find that Eq. (3.119) can be rewritten as

$$-w^\mu s_\mu = \frac{1}{2}\gamma^5\sigma^{\mu\omega}\partial_\omega s_\mu. \tag{3.121}$$

Next, in addition, using the identity $\sigma^{\mu\nu} = i\left(\gamma^\mu\gamma^\nu - g^{\mu\nu}\mathbb{1}_4\right)$, which follows from the definition (3.98), and the Dirac algebra, we obtain

$$-w^\mu s_\mu = \frac{i}{2}\gamma^5\left(\gamma^\mu\gamma^\nu - g^{\mu\nu}\mathbb{1}_4\right)\partial_\omega s_\mu = \frac{i}{2}\gamma^5\gamma^\mu\gamma^\nu\partial_\omega s_\mu = \frac{i}{2}\gamma^5 \slashed{s}\slashed{\partial}$$

$$= \left\{\slashed{p} = i\slashed{\partial}\right\} = \frac{1}{2}\gamma^5\slashed{s}\slashed{p}, \tag{3.122}$$

which is the covariant operator that projects the spin operator onto the polarization vector.

Finally, in the rest frame of a particle, we again have $s = (0, \hat{\mathbf{n}})$ and $p = (m, 0)$, which imply that $\slashed{s} = -\hat{\mathbf{n}} \cdot \boldsymbol{\gamma}$ and $\slashed{p}/m = \gamma^0$. Therefore, by using Eq. (3.122) as well as Eq. (3.100), we obtain

$$-\frac{1}{m}w^{\mu}s_{\mu} = \frac{1}{2}\gamma^{5}\displaystyle{\not\!s}\frac{\displaystyle{\not\!p}}{m} = \frac{1}{2}\gamma^{5}\left(-\hat{\mathbf{n}}\cdot\boldsymbol{\gamma}\right)\gamma^{0} = -\frac{1}{2}\gamma^{5}\gamma^{i}\gamma^{0}n^{i} = \frac{1}{2}\gamma^{5}\gamma^{0}\gamma^{i}n^{i}$$

$$= \frac{1}{2}\Sigma^{i}n^{i} = \frac{1}{2}\boldsymbol{\Sigma}\cdot\hat{\mathbf{n}} = \mathbf{S}\cdot\hat{\mathbf{n}}. \tag{3.123}$$

Thus, in the rest frame, the inner product of the Pauli–Lubanski vector and the polarization vector is equal to the projection of the spin operator $\mathbf{S}$ in the direction $\hat{\mathbf{n}} = \mathbf{p}/|\mathbf{p}|$, which is, in principle, the helicity operator (see the detailed discussion in Section 3.13). In the case that the polarization vector is given by $s = (0, \hat{\mathbf{n}})$, applying the boost for a general 3-momentum $\mathbf{p}$, i.e.

$$\Lambda = \begin{pmatrix} E/m & p^{j}/m \\ p^{i}/m & \delta^{ij} + p^{i}p^{j}/[m(E+m)] \end{pmatrix}, \tag{3.124}$$

we would obtain

$$s' = \Lambda s = \left(\frac{\mathbf{p}\cdot\hat{\mathbf{n}}}{m}, \hat{\mathbf{n}} + \frac{\mathbf{p}\cdot\hat{\mathbf{n}}}{m(E+m)}\mathbf{p}\right). \tag{3.125}$$

In addition, if $\hat{\mathbf{n}} = \mathbf{p}/|\mathbf{p}|$, then we find that

$$s' = \left(\frac{|\mathbf{p}|}{m}, \frac{E}{m}\hat{\mathbf{n}}\right), \tag{3.126}$$

which in the ultra-relativistic limit $m \ll E$ becomes

$$s' \simeq \left(\frac{E}{m}, \frac{1}{m}\mathbf{p}\right) = \frac{1}{m}p. \tag{3.127}$$

Exercise 3.5 In the rest frame of a Dirac particle, the spin operator is defined by $\mathbf{S} = \boldsymbol{\Sigma}/2$. Show that $\boldsymbol{\Sigma} = \gamma^{5}\gamma^{0}\boldsymbol{\gamma}$, $[S^{i}, S^{j}] = i\epsilon^{ijk}S^{k}$, and $\mathbf{S}^{2} = 3\mathbb{1}_{4}/4$.

3.5 Orthogonality conditions and energy projection operators

For the Dirac spinors, we have the orthogonality conditions

$$\bar{u}(\mathbf{p}, s)u(\mathbf{p}, s') = -\bar{v}(\mathbf{p}, s)v(\mathbf{p}, s') = \delta_{ss'}, \tag{3.128}$$

$$u^{\dagger}(\mathbf{p}, s)u(\mathbf{p}, s') = v^{\dagger}(\mathbf{p}, s)v(\mathbf{p}, s') = \frac{E(\mathbf{p})}{m}\delta_{ss'}, \tag{3.129}$$

$$\bar{u}(\mathbf{p}, s)v(\mathbf{p}, s') = \bar{v}(\mathbf{p}, s)u(\mathbf{p}, s') = 0, \tag{3.130}$$

as well as the completeness relation

$$\sum_{s}\left[u(\mathbf{p}, s)\bar{u}(\mathbf{p}, s) - v(\mathbf{p}, s)\bar{v}(\mathbf{p}, s)\right] = \mathbb{1}_{4}, \tag{3.131}$$

since the spinors form a complete set of solutions to the Dirac equation, which can also be written in component form as

$$\sum_s \left[u_\alpha(\mathbf{p}, s) \bar{u}_\beta(\mathbf{p}, s) - v_\alpha(\mathbf{p}, s) \bar{v}_\beta(\mathbf{p}, s) \right] = \delta_{\alpha\beta} \mathbb{1}_4, \tag{3.132}$$

where α and β are Dirac spinor indices.

In analogy to the spin projection operators (3.109), we can define projection operators for positive and negative energies. Acting with the operator for positive-energy spinors $\not{p} - m\mathbb{1}_4$ on any spinor that has been multiplied by $\not{p} + m\mathbb{1}_4$, yields zero, since $p^2 = m^2$. Similarly, acting with the operator for negative-energy spinors $\not{p} + m\mathbb{1}_4$ on any spinor that has been multiplied by $\not{p} - m\mathbb{1}_4$, yields also zero, since again $p^2 = m^2$. Therefore, let us introduce the normalized *energy projection operators* as

$$\Lambda_\pm = \frac{\pm\not{p} + m\mathbb{1}_4}{2m}. \tag{3.133}$$

Thus, acting with the two operators in Eq. (3.133) on the Dirac spinors, we find the properties

$$\Lambda_+ u(\mathbf{p}, s) = u(\mathbf{p}, s), \tag{3.134}$$

$$\Lambda_+ v(\mathbf{p}, s) = 0, \tag{3.135}$$

$$\Lambda_- u(\mathbf{p}, s) = 0, \tag{3.136}$$

$$\Lambda_- v(\mathbf{p}, s) = v(\mathbf{p}, s), \tag{3.137}$$

where we have used that $\not{p} u(\mathbf{p}, s) = m u(\mathbf{p}, s)$ and $\not{p} v(\mathbf{p}, s) = -m v(\mathbf{p}, s)$, cf. Eq. (3.39). Note that Λ_+ is the projection operator that eliminates the antiparticle part of a wave function by projecting out the particle part, whereas Λ_- is the projection operator that eliminates the particle part of a wave function by projecting out the antiparticle part. Of course, these operators fulfil the relations for projection operators, i.e. $\Lambda_\pm^2 = \Lambda_\pm$, $\Lambda_\pm \Lambda_\mp = 0$, and $\Lambda_\pm + \Lambda_\mp = \mathbb{1}_4$, since we have

$$\Lambda_\pm^2 = \left(\frac{\pm\not{p} + m\mathbb{1}_4}{2m} \right)^2 = \frac{p^2 \pm 2m\not{p} + m^2\mathbb{1}_4}{4m^2} = \{p^2 = m^2\mathbb{1}_4\}$$

$$= \frac{\pm\not{p} + m\mathbb{1}_4}{2m} = \Lambda_\pm, \tag{3.138}$$

$$\Lambda_\pm \Lambda_\mp = \frac{\pm\not{p} + m\mathbb{1}_4}{2m} \frac{\mp\not{p} + m\mathbb{1}_4}{2m} = \frac{-p^2 + m^2\mathbb{1}_4}{4m^2} = \{p^2 = m^2\mathbb{1}_4\} = 0, \tag{3.139}$$

$$\Lambda_\pm + \Lambda_\mp = \frac{\pm\not{p} + m\mathbb{1}_4}{2m} + \frac{\mp\not{p} + m\mathbb{1}_4}{2m} = \mathbb{1}_4. \tag{3.140}$$

In addition, the operators $\Lambda_\pm$ are Hermitian, i.e. $\Lambda_\pm^\dagger = \Lambda_\pm$. Furthermore, note that, in calculating unpolarized cross-sections, one normally encounters the following polarization sums:

$$\sum_s u(\mathbf{p}, s)\bar{u}(\mathbf{p}, s) = \frac{\not{p} + m\mathbb{1}_4}{2m}, \tag{3.141}$$

$$-\sum_s v(\mathbf{p}, s)\bar{v}(\mathbf{p}, s) = \frac{-\not{p} + m\mathbb{1}_4}{2m}, \tag{3.142}$$

which constitute the individual terms in the left-hand side of the completeness relation (3.131) due to Eq. (3.140). Finally, for the 'helicity' operator $\boldsymbol{\Sigma} \cdot \hat{\mathbf{n}}$, where $\hat{\mathbf{n}} = \mathbf{p}/|\mathbf{p}|$, one can show that

$$\left(\boldsymbol{\Sigma} \cdot \hat{\mathbf{n}}\right) u(\mathbf{p}, \pm s) = \pm u(\mathbf{p}, \pm s), \tag{3.143}$$

$$\left(\boldsymbol{\Sigma} \cdot \hat{\mathbf{n}}\right) v(\mathbf{p}, \pm s) = \mp v(\mathbf{p}, \pm s). \tag{3.144}$$

3.6 Relativistic invariance of the Dirac equation

In this section, we investigate the transformation law for a Dirac spinor such that

$$\psi'(x') = S(\Lambda)\psi(x), \tag{3.145}$$

where Λ is a Lorentz transformation, which transforms x to x', i.e. $x' = \Lambda x$. Thus, we want to find $S \equiv S(\Lambda)$. From covariance of the Dirac equation

$$(i\gamma^\mu \partial_\mu - m\mathbb{1}_4)\psi(x) = 0, \tag{3.146}$$

it follows that

$$\left(i\gamma^\mu \partial'_\mu - m\mathbb{1}_4\right)\psi'(x') = 0, \tag{3.147}$$

where

$$\partial'_\mu \equiv \frac{\partial}{\partial x'^\mu}. \tag{3.148}$$

Note that $m' = m$ follows from the fact that the rest mass is itself a Lorentz invariant quantity and it can also be proven that the gamma matrices are Lorentz invariants, i.e. $\gamma'^\mu = \gamma^\mu$. Thus, we can use the same set of gamma matrices in Eqs. (3.146) and (3.147). In addition, the derivative transforms as a 4-vector under Lorentz transformations

$$\frac{\partial}{\partial x^\mu} = \frac{\partial x'^\nu}{\partial x^\mu}\frac{\partial}{\partial x'^\nu} = \Lambda^\nu{}_\mu\frac{\partial}{\partial x'^\nu}. \tag{3.149}$$

Using the inverse transformation law $\psi(x) = S^{-1}\psi'(x')$ as well as the fact that derivatives transform as vectors under Lorentz transformations, we can write the Dirac equation as

The Dirac equation

$$\left(i\gamma^{\mu}\Lambda^{\nu}{}_{\mu}\frac{\partial}{\partial x'^{\nu}} - m\mathbb{1}_4\right)S^{-1}\psi'(x') = 0. \tag{3.150}$$

Multiplying with S from the left, we have

$$iS\gamma^{\mu}\Lambda^{\nu}{}_{\mu}S^{-1}\frac{\partial}{\partial x'^{\nu}}\psi'(x') - m\psi'(x') = 0. \tag{3.151}$$

Thus, comparing Eqs. (3.147) and (3.151), we obtain

$$\gamma^{\nu} = S\gamma^{\mu}\Lambda^{\nu}{}_{\mu}S^{-1} \quad\Leftrightarrow\quad \Lambda^{\nu}{}_{\mu}\gamma^{\mu} = S^{-1}\gamma^{\nu}S. \tag{3.152}$$

In order to find an explicit expression for S, we assume that $\Lambda^{\mu}{}_{\nu}$ is an infinitesimal Lorentz transformation, which can be written as

$$\Lambda^{\mu}{}_{\nu} = \delta^{\mu}_{\nu} + \epsilon\omega^{\mu}{}_{\nu}, \tag{3.153}$$

where $\omega^{\mu\nu} = -\omega^{\nu\mu}$. For S, we make the Ansatz

$$S = \mathbb{1}_4 - \frac{i}{4}\epsilon\sigma_{\mu\nu}\omega^{\mu\nu} + \mathcal{O}(\epsilon^2), \tag{3.154}$$

which has the inverse (to order ϵ^2)

$$S^{-1} = \mathbb{1}_4 + \frac{i}{4}\epsilon\sigma_{\mu\nu}\omega^{\mu\nu}. \tag{3.155}$$

Thus, we want to search for $\sigma_{\mu\nu} = -\sigma_{\nu\mu}$ such that $S^{-1}\gamma^{\nu}S = \Lambda^{\nu}{}_{\mu}\gamma^{\mu}$. Inserting the infinitesimal versions of S, S^{-1}, and $\Lambda^{\mu}{}_{\nu}$ into Eq. (3.152), we obtain

$$\gamma^{\mu} - \frac{i}{4}\epsilon\omega^{\lambda\rho}\left(\gamma^{\mu}\sigma_{\lambda\rho} - \sigma_{\lambda\rho}\gamma^{\mu}\right) = \gamma^{\mu} + \epsilon\omega^{\mu}{}_{\nu}\gamma^{\nu}, \tag{3.156}$$

which leads to

$$[\gamma^{\mu}, \sigma_{\lambda\nu}] = 2i\left(g^{\mu}{}_{\lambda}\gamma_{\nu} - g^{\mu}{}_{\nu}\gamma_{\lambda}\right). \tag{3.157}$$

This has the solution

$$\sigma_{\mu\nu} = \frac{i}{2}[\gamma_{\mu}, \gamma_{\nu}]. \tag{3.158}$$

Thus, we obtain the following infinitesimal expression for S

$$S = \mathbb{1}_4 - \frac{i}{4}\epsilon\sigma_{\mu\nu}\omega^{\mu\nu} = \mathbb{1}_4 + \frac{\epsilon}{8}[\gamma_{\mu}, \gamma_{\nu}]\omega^{\mu\nu}. \tag{3.159}$$

Let us look at a specific example of an infinitesimal expression for S. Consider a boost in the x-direction with $v = \tanh\xi$. Then, we have

$$\Lambda^{(01)} = \left(\Lambda^{\nu}{}_{\mu}\right) = \begin{pmatrix} \cosh\xi & -\sinh\xi & 0 & 0 \\ -\sinh\xi & \cosh\xi & 0 & 0 \\ 0 & 0 & 1 & 0 \\ 0 & 0 & 0 & 1 \end{pmatrix}, \tag{3.160}$$

which, using Eq. (3.153), implies that

$$\epsilon\left(\omega^{\nu}{}_{\mu}\right) = \epsilon\xi \begin{pmatrix} 0 & -1 & 0 & 0 \\ -1 & 0 & 0 & 0 \\ 0 & 0 & 0 & 0 \\ 0 & 0 & 0 & 0 \end{pmatrix} = \epsilon\xi\left(M^{\nu}{}_{\mu}\right). \tag{3.161}$$

Indeed, the matrix $M = \left(M^{\nu}{}_{\mu}\right)$ satisfies $M^2 = \operatorname{diag}(1, 1, 0, 0)$, $M^3 = M$, $M^4 = M^2$, ... Note that, in this case, $M = M^{01}$, where M^{01} is one of the generators of the Poincaré group. Thus, we have

$$x'^{\nu} = \Lambda^{\nu}{}_{\mu} x^{\mu} = \lim_{N \to \infty} \left(\delta + \frac{\xi}{N}M\right)^{\nu}_{\alpha_1} \cdots \left(\delta + \frac{\xi}{N}M\right)^{\alpha_N}_{\mu} x^{\mu} = \left(e^{\xi M}\right)^{\nu}{}_{\mu} x^{\mu}, \tag{3.162}$$

which means that

$$e^{\xi M} = \mathbb{1}_4 + \sum_{n=1}^{\infty} \frac{\xi^n M^n}{n!} = \mathbb{1}_4 - M^2 + M^2 \cosh\xi + M \sinh\xi = \Lambda^{(01)}. \tag{3.163}$$

Note that the series contains only terms proportional to $\mathbb{1}_4$, M, and M^2 and that it cuts thereafter, which is due to the special form of the matrix M. Now, we have for a boost in the x-direction

$$S(\Lambda^{(01)}(\xi)) = e^{\frac{i}{2}\xi\sigma_{01}} = \left\{\sigma_{01} = \frac{i}{2}[\gamma_0, \gamma_1] = i\gamma_0\gamma_1\right\} = e^{-\frac{\xi}{2}\gamma_0\gamma_1}, \tag{3.164}$$

where $(-\gamma_0\gamma_1)^2 = \gamma_0\gamma_1\gamma_0\gamma_1 = -\gamma_0^2\gamma_1^2 = \mathbb{1}_4$, since $\gamma_0\gamma_1 = -\gamma_1\gamma_0$, $\gamma_0^2 = \mathbb{1}_4$, and $\gamma_1^2 = -\mathbb{1}_4$. Therefore, using Eq. (3.163), we can write the expression for S as

$$S(\Lambda^{(01)}(\xi)) = \mathbb{1}_4 \cosh\frac{\xi}{2} - \gamma_0\gamma_1 \sinh\frac{\xi}{2}. \tag{3.165}$$

In general, for a boost in the x^i-direction, we have

$$S(\Lambda^{(0i)}(\xi)) = \mathbb{1}_4 \cosh\frac{\xi}{2} - \gamma_0\gamma_i \sinh\frac{\xi}{2}. \tag{3.166}$$

Instead of boosts, we could have rotations. One example is a rotation about the z-axis, which has

$$S(R_z(\theta)) = e^{\frac{i}{2}\theta\sigma_{12}} = \left\{\sigma_{12} = \frac{i}{2}[\gamma_1, \gamma_2] = i\gamma_1\gamma_2\right\} = e^{i\frac{\theta}{2}\gamma_1\gamma_2}, \tag{3.167}$$

where $(i\gamma_1\gamma_2)^2 = -\gamma_1\gamma_2\gamma_1\gamma_2 = \gamma_1^2\gamma_2^2 = \mathbb{1}_4$. Thus, in this case, we obtain

$$S(R_z(\theta)) = \mathbb{1}_4 \cosh\frac{i\theta}{2} + i\gamma_1\gamma_2 \sinh\frac{i\theta}{2} = \mathbb{1}_4 \cos\frac{\theta}{2} - \gamma_1\gamma_2 \sin\frac{\theta}{2}, \tag{3.168}$$

where we in the last step have used the identities $\cosh(ix) = \cos(x)$ and $\sinh(ix) = i\sin(x)$ for x real. Similarly, for a rotation about the x^k-axis, we find (for cyclic permutations of $i, j, k = 1, 2, 3$)

$$S(R_k(\theta)) = \mathbb{1}_4 \cos\frac{\theta}{2} - \gamma_i\gamma_j \sin\frac{\theta}{2}. \tag{3.169}$$

In addition to the Dirac equation, one can show that the 'conjugate' Dirac equation for the Dirac adjoint $\bar{\psi}$, i.e.

$$i\partial_\mu\bar{\psi}(x)\gamma^\mu - m\bar{\psi}(x) = 0 \tag{3.170}$$

is relativistically invariant if it holds that

$$\bar{\psi}'(x') = \bar{\psi}(x)S^{-1}. \tag{3.171}$$

3.7 Bilinear covariants

To this end, we have established the transformation matrix S that determines how the four components of a Dirac spinor change under a Lorentz transformation Λ. It is important to note that S is **not** the same as Λ. Therefore, the Dirac spinors are **not** 4-vectors. However, using Eqs. (3.145), (3.152), and (3.171), we observe that the quantity

$$V^\mu(x) = \bar{\psi}(x)\gamma^\mu\psi(x) \tag{3.172}$$

transforms as a 4-vector under Lorentz transformations, i.e.

$$V'^\mu(x') = \Lambda^\mu{}_\nu V^\nu(x), \tag{3.173}$$

since we have

$$V'^\mu(x') = \bar{\psi}'(x')\gamma'^\mu\psi'(x') = \bar{\psi}(x)S^{-1}\gamma^\mu S\psi(x)$$
$$= \Lambda^\mu{}_\nu\bar{\psi}(x)\gamma^\nu\psi(x) = \Lambda^\mu{}_\nu V^\nu(x). \tag{3.174}$$

Similarly, for the scalar quantity

$$s(x) = \bar{\psi}(x)\psi(x), \tag{3.175}$$

we find that

$$s'(x') = \bar{\psi}'(x')\psi'(x') = \bar{\psi}(x)S^{-1}S\psi(x) = \bar{\psi}(x)\psi(x) = s(x). \tag{3.176}$$

Next, we investigate if it is possible to find an operator P acting on Dirac spinors that corresponds to the parity transformation of spacetime points, i.e. $x' = \Lambda_P x = (x^0, -\mathbf{x})$, where $\Lambda_P = \mathrm{diag}(1, -1, -1, -1)$, cf. the discussion on parity transformation in Sections 1.2 and 2.1. The Lorentz transformation Λ_P has $\det\Lambda_P = -1$, which means that it is not a pure Lorentz transformation. The requirement that the Dirac equation is covariant [cf. Eq. (3.152)] can be fulfilled by choosing $P = \eta_P \gamma^0$, where η_P is a phase factor. Thus, we obtain

$$P^{-1}\gamma^5 P = -\gamma^5 = (\det \Lambda_P)\gamma^5, \tag{3.177}$$

since $\{\gamma^5, \gamma^0\} = 0$. Therefore, for a general Lorentz transformation Λ, using the definition of γ^5 (3.44) and Eq. (3.152), γ^5 transforms as

$$\begin{aligned}
S^{-1}\gamma^5 S &= -\frac{i}{4!}\epsilon_{\mu\nu\rho\sigma} S^{-1}\gamma^\mu S S^{-1}\gamma^\nu S S^{-1}\gamma^\rho S S^{-1}\gamma^\sigma S \\
&= -\frac{i}{4!}\epsilon_{\mu\nu\rho\sigma}\Lambda^\mu{}_\alpha\Lambda^\nu{}_\beta\Lambda^\rho{}_\gamma\Lambda^\sigma{}_\delta\gamma^\alpha\gamma^\beta\gamma^\gamma\gamma^\delta \\
&= -\frac{i}{4!}\epsilon_{\alpha\beta\gamma\delta}(\det \Lambda)\gamma^\alpha\gamma^\beta\gamma^\gamma\gamma^\delta = (\det \Lambda)\gamma^5, \tag{3.178}
\end{aligned}$$

since $\epsilon_{\alpha\beta\gamma\delta}\det \Lambda = \epsilon_{\mu\nu\rho\sigma}\Lambda^\mu{}_\alpha\Lambda^\nu{}_\beta\Lambda^\rho{}_\gamma\Lambda^\sigma{}_\delta$. In conclusion, if Λ is a Lorentz transformation corresponding to a parity transformation, then $\det \Lambda = -1$ in Eq. (3.178), but if Λ is a pure Lorentz transformation, then $\det \Lambda = 1$.

Now, we have the quantities

$$s^5(x) = \bar{\psi}(x)\gamma^5\psi(x) \tag{3.179}$$

and

$$A^\mu(x) = \bar{\psi}(x)\gamma^\mu\gamma^5\psi(x), \tag{3.180}$$

respectively, which transform under Lorentz transformations in the following way:

$$s^{5'} = \bar{\psi}'(x')\gamma^{5'}\psi'(x') = \bar{\psi}(x)S^{-1}\gamma^5 S\psi(x) = (\det \Lambda)\bar{\psi}(x)\gamma^5\psi(x), \tag{3.181}$$

$$\begin{aligned}
A'^\mu(x') &= \bar{\psi}'(x')\gamma'^\mu\gamma^{5'}\psi'(x') = \bar{\psi}(x)S^{-1}\gamma^\mu\gamma^5 S\psi(x) \\
&= \Lambda^\mu{}_\nu\bar{\psi}(x)\gamma^\nu S^{-1}\gamma^5 S\psi(x) = (\det \Lambda)\Lambda^\mu{}_\nu\bar{\psi}(x)\gamma^\nu\gamma^5\psi(x) \\
&= (\det \Lambda)\Lambda^\mu{}_\nu A^\nu(x). \tag{3.182}
\end{aligned}$$

Thus, the quantity s^5 transforms as a pseudoscalar, which means that it transforms as a scalar under pure Lorentz transformations, but as a pseudoscalar under parity transformations, whereas the quantity A^μ (which is not the electromagnetic 4-vector current) transforms as a 4-vector under pure Lorentz transformations, but as a so-called *pseudovector* (or *dual vector*) under parity transformations. Finally, the quantity

$$T^{\mu\nu}(x) = \bar{\psi}(x)\sigma^{\mu\nu}\psi(x) \tag{3.183}$$

transforms as a second-rank tensor, i.e.

$$T'^{\mu\nu}(x') = \Lambda^\mu{}_\lambda\Lambda^\nu{}_\omega T^{\lambda\omega}(x), \tag{3.184}$$

which is antisymmetric. Note that the five quantities defined in Eqs. (3.172), (3.175), (3.179), (3.180), and (3.183) are usually called *bilinear covariants* (or *bilinear forms*). In fact, there are only five basic bilinear covariants of the Dirac theory.

Exercise 3.6 Show that the bilinear covariant $T^{\mu\nu}$ transforms as a second-rank antisymmetric tensor with respect to Lorentz transformations.

3.8 Electromagnetic structure of Dirac particles and charge conjugation

In order to investigate the structure of Dirac particles, we will study electromagnetic interaction coupled to these particles. We perform the minimal coupling replacement

$$p_\mu \mapsto p_\mu - eA_\mu \quad \Leftrightarrow \quad i\partial_\mu \mapsto i\partial_\mu - eA_\mu, \tag{3.185}$$

where e is the charge of the electron, in the Dirac equation (cf. the Klein–Gordon equation), which leads to the following equation

$$\gamma^\mu \left[i\partial_\mu - eA_\mu(x) \right] \psi(x) = m\psi(x). \tag{3.186}$$

Note that the minimal coupling replacement is probably the easiest way to introduce electromagnetism. Equation (3.186) is equivalent to

$$\left(\not{p} - e\not{A} - m\mathbb{1}_4 \right) \psi = 0. \tag{3.187}$$

Setting $\psi = \frac{1}{m} \left(\not{p} - e\not{A} + m\mathbb{1}_4 \right) \chi$, we obtain

$$\left(\not{p} - e\not{A} \right)^2 \chi = m^2 \chi. \tag{3.188}$$

The left-hand side of Eq. (3.188) should be interpreted as

$$\left(\not{p} - e\not{A} \right)^2 \chi = \gamma^\mu \gamma^\nu \left(p_\mu - eA_\mu \right) \left(p_\nu - eA_\nu \right) \chi. \tag{3.189}$$

Now, renaming the summation indices according to $\mu \leftrightarrow \nu$ in half of this expression, we have

$$\left(\not{p} - e\not{A} \right)^2 \chi = \frac{1}{2} \left[\gamma^\mu \gamma^\nu \left(p_\mu - eA_\mu \right) \left(p_\nu - eA_\nu \right) + \gamma^\nu \gamma^\mu \left(p_\nu - eA_\nu \right) \right. $$
$$\left. \times \left(p_\mu - eA_\mu \right) \right] \chi. \tag{3.190}$$

Next, expanding the parentheses in the right-hand side of the above expression, we find

$$\left(\not{p} - e\not{A} \right)^2 \chi = \frac{1}{2} \gamma^\mu \gamma^\nu \left[p_\mu p_\nu + e^2 A_\mu A_\nu - e \left(p_\mu A_\nu + A_\mu p_\nu \right) \right] \chi$$
$$+ \frac{1}{2} \gamma^\nu \gamma^\mu \left[p_\nu p_\mu + e^2 A_\nu A_\mu - e \left(p_\nu A_\mu + A_\nu p_\mu \right) \right] \chi. \tag{3.191}$$

Obviously, $p_\mu p_\nu = p_\nu p_\mu$ and $A_\mu A_\nu = A_\nu A_\mu$. However, we have to be careful with the cross-terms such as $p_\mu A_\nu$. Therefore, we add and subtract the same quantity in the last parenthesis in the last bracket in order to be able to factorize out the gamma matrices, i.e. we write $p_\nu A_\mu + A_\nu p_\mu = p_\nu A_\mu + A_\nu p_\mu + p_\mu A_\nu + A_\mu p_\nu - p_\mu A_\nu - A_\mu p_\nu$. Then, this yields

$$
(\not{p} - eA)^2 \chi = \frac{1}{2} (\gamma^\mu \gamma^\nu + \gamma^\nu \gamma^\mu) \left[p_\mu p_\nu + e^2 A_\mu A_\nu - e \left(p_\mu A_\nu + A_\mu p_\nu \right) \right] \chi
$$
$$
- \frac{e}{2} \gamma^\nu \gamma^\mu \left(p_\nu A_\mu + A_\nu p_\mu - p_\mu A_\nu - A_\mu p_\nu \right) \chi. \tag{3.192}
$$

Now, what about the last parenthesis? Collecting terms with the same indices for p and A, we have $-p_\mu A_\nu + A_\nu p_\mu$ (second and third terms) and $p_\nu A_\mu - A_\mu p_\nu$ (first and fourth terms), and using the correspondence principle $p \leftrightarrow i\partial$ for these expressions, we find that

$$
\left(-p_\mu A_\nu + A_\nu p_\mu \right) \chi = -i\partial_\mu (A_\nu \chi) + A_\nu i \partial_\mu \chi
$$
$$
= -i \left(\partial_\mu A_\nu \right) \chi - i A_\nu \partial_\mu \chi + i A_\nu \partial_\mu \chi = -i \left(\partial_\mu A_\nu \right) \chi, \tag{3.193}
$$

and similarly,

$$
\left(p_\nu A_\mu - A_\mu p_\nu \right) \chi = i \left(\partial_\nu A_\mu \right) \chi. \tag{3.194}
$$

Thus, the commutator of p and A acts as a derivative, i.e.

$$
[p_\mu, A_\nu] = i\partial_\mu A_\nu. \tag{3.195}
$$

Using $\{\gamma^\mu, \gamma^\nu\} = 2g^{\mu\nu}\mathbb{1}_4$ and inserting Eqs. (3.193) and (3.194) into Eq. (3.192), we obtain

$$
(\not{p} - eA)^2 \chi = \left[p_\mu p^\mu + e^2 A_\mu A^\mu - e \left(p_\mu A^\mu + A_\mu p^\mu \right) \right] \chi
$$
$$
- \frac{e}{2} \gamma^\nu \gamma^\mu i \left(\partial_\nu A_\mu - \partial_\mu A_\nu \right) \chi
$$
$$
= (p - eA)^2 \chi + i\frac{e}{2} \gamma^\nu \gamma^\mu \left(\partial_\mu A_\nu - \partial_\nu A_\mu \right) \chi, \tag{3.196}
$$

where the expression in the parenthesis in the last term is the *electromagnetic field strength tensor*, i.e.[15]

$$
F_{\mu\nu} = \partial_\mu A_\nu - \partial_\nu A_\mu. \tag{3.197}
$$

However, note that $F_{\mu\nu}$ is not itself a derivative at all, but just a multiplicative factor that is invariant, which means that $F_{\mu\nu}$ should not be thought of acting on χ other

---

[15] For a detailed discussion on the electromagnetic field strength tensor in particular and classical electromagnetism in general, see Chapter 8.

than multiplicatively, see an excellent discussion in the book by M. E. Peskin and D. V. Schroeder, pp. 482–486. In addition, we can write the factor $\gamma^\mu \gamma^\nu$ as

$$\gamma^\mu \gamma^\nu = \frac{1}{2} \left[\underbrace{(\gamma^\mu \gamma^\nu + \gamma^\nu \gamma^\mu)}_{2g^{\mu\nu}\mathbb{1}_4} + \underbrace{(\gamma^\mu \gamma^\nu - \gamma^\nu \gamma^\mu)}_{-2i\sigma^{\mu\nu}} \right], \tag{3.198}$$

from which it follows that

$$\gamma^\mu \gamma^\nu = g^{\mu\nu}\mathbb{1}_4 - i\sigma^{\mu\nu}. \tag{3.199}$$

Thus, inserting the above into Eq. (3.196), we find that

$$(\not{p} - eA)^2 \chi = (p - eA)^2 \chi + i\frac{e}{2} (g^{\nu\mu}\mathbb{1}_4 - i\sigma^{\nu\mu}) F_{\mu\nu}\chi. \tag{3.200}$$

Finally, using the facts that $g^{\mu\nu}$ is symmetric as well as $\sigma^{\mu\nu}$ and $F_{\mu\nu}$ are antisymmetric, which means that $g^{\nu\mu}F_{\mu\nu} = 0$ and $\sigma^{\nu\mu} = -\sigma^{\mu\nu}$, we arrive at

$$(\not{p} - eA)^2 \chi = (p - eA)^2 \chi - \frac{e}{2}\sigma^{\mu\nu} F_{\mu\nu}\chi. \tag{3.201}$$

Thus, it follows that we can write the Dirac equation with electromagnetism coupled to the Dirac particles as

$$\left[(p_\mu - eA_\mu)(p^\mu - eA^\mu)\mathbb{1}_4 - \frac{e}{2}\sigma^{\mu\nu} F_{\mu\nu} \right] \chi = m^2\chi. \tag{3.202}$$

The new term $\frac{e}{2}\sigma^{\mu\nu} F_{\mu\nu}$ can now be written [by using Eqs. (3.98) and (8.8) as well as $\Sigma = \gamma^5\alpha$] as

$$\frac{e}{2}\sigma^{\mu\nu} F_{\mu\nu} = -e \left(\Sigma \cdot \mathbf{B} - i\alpha \cdot \mathbf{E} \right), \tag{3.203}$$

where $\Sigma \cdot \mathbf{B}$ is the magnetic dipole interaction between the spin and the external field and $\alpha \cdot \mathbf{E}$ is the electric monopole term.

The natural question to ask is if there exists another possibility of a gauge-invariant coupling, similar to the minimal coupling leading to Eq. (3.202). In fact, there is a gauge-invariant **direct** (and non-minimal) coupling to the electromagnetic field strength tensor $F_{\mu\nu}$ by so-called *Pauli terms*

$$\frac{\mu}{2}\sigma^{\mu\nu} F_{\mu\nu},$$

which give rise to an anomalous magnetic moment

$$\mu\frac{e}{2m}$$

in addition to the normal magnetic moment $\frac{e}{2m}$. Thus, the anomalous magnetic moment immediately gives the equation

$$\left[\gamma^\mu (i\partial_\mu - eA_\mu) + \frac{\mu}{2}\sigma^{\mu\nu} F_{\mu\nu} \right] \psi = m\psi. \tag{3.204}$$

Exercise 3.7   Show that $\frac{1}{2}\sigma^{\mu\nu}F_{\mu\nu} = i\boldsymbol{\alpha}\cdot\mathbf{E} - \boldsymbol{\Sigma}\cdot\mathbf{B}$.

Finally, we answer the following question: how should positrons (i.e. holes in the Dirac sea), which are the antiparticles of electrons, be described by the Dirac equation?[16] A negatively charged electron is described by the Dirac equation [cf. Eq. (3.186)]

$$\left(i\slashed{\partial} - e\slashed{A} - m\mathbb{1}_4\right)\psi = 0. \tag{3.205}$$

Since a positron is positively charged and should have the same mass as the electron, the *charge conjugation* of the Dirac equation, i.e.

$$\left(i\slashed{\partial} + e\slashed{A} - m\mathbb{1}_4\right)\psi^c = 0, \tag{3.206}$$

where ψ^c is the *charge-conjugated* Dirac wave function, should describe the positron. Using the fact that A_μ is real and taking the complex conjugate of Eq. (3.205), we obtain the equation

$$\left[\left(i\partial_\mu + eA_\mu\right)(\gamma^\mu)^* + m\mathbb{1}_4\right]\psi^* = 0. \tag{3.207}$$

Now, introducing the matrix S_c such that

$$S_c(\gamma^\mu)^* S_c^{-1} = -\gamma^\mu, \tag{3.208}$$

Eq. (3.207) can be written as Eq. (3.206). Note that $S_c = i\gamma^2$ fulfils Eq. (3.208). Thus, the charge-conjugated Dirac wave function can be written as

$$\psi^c = S_c\psi^* = i\gamma^2\psi^*. \tag{3.209}$$

In addition, introducing the matrix C such that $S_c = C\gamma^0$, which means that $C = i\gamma^2\gamma^0 = -i\gamma^0\gamma^2$. Hence, we arrive at the relation

$$\psi^c = C\gamma^0\psi^* = C\bar{\psi}^T, \tag{3.210}$$

where again $\bar{\psi} = \gamma^0\psi^\dagger$ is the Dirac adjoint and $\psi^\dagger = (\psi^*)^T$ is the conjugate to the Dirac wave function ψ. In conclusion, under charge conjugation, a Dirac wave function transforms as

$$\psi(x) \mapsto \psi^c(x) = C\bar{\psi}(x)^T, \tag{3.211}$$

where C satisfies the relations

$$C^{-1} = C^T = C^\dagger = -C, \tag{3.212}$$

$$C\gamma^\mu C^{-1} = -(\gamma^\mu)^T. \tag{3.213}$$

Thus, charge conjugation (or more generally, particle–antiparticle conjugation) transforms a particle (here: a negatively charged electron) into its antiparticle

---

[16] Note that we will be able to abandon the holes in the Dirac sea, but the existence of positrons is indisputable, cf. the discussion in Section 3.2.

(here: a positively charged positron), and vice versa. Note that, in Section 7.5.3, we will discuss the symmetry of charge conjugation for Dirac fields.

In analogy to charge conjugation of the Dirac wave function, we can define charge conjugation of the free Dirac spinors. For the plane-wave solutions to the Dirac equation, we have [cf. Eq. (3.39)]

$$(\not{p} - m\mathbb{1}_4)\, u(p) = 0, \tag{3.214}$$

$$(\not{p} + m\mathbb{1}_4)\, v(p) = 0. \tag{3.215}$$

In addition, we have [cf. Eq. (3.56)]

$$\bar{u}(p)\, (\not{p} - m\mathbb{1}_4) = 0, \tag{3.216}$$

$$\bar{v}(p)\, (\not{p} + m\mathbb{1}_4) = 0. \tag{3.217}$$

Now, taking the transpose of Eq. (3.217), which is the Dirac equation for the conjugated Dirac spinor $\bar{v}$ describing a negative-energy solution, we find that

$$\left[\bar{v}(p)\, (\not{p} + m\mathbb{1}_4)\right]^T = \left(\gamma^T \cdot p + m\mathbb{1}_4\right) \bar{v}(p)^T = 0. \tag{3.218}$$

Next, multiplying this equation with the matrix C and inserting the identity $C^{-1}C = \mathbb{1}_4$, yields

$$\left(C\gamma^T C^{-1} \cdot p + m\mathbb{1}_4\right) C\bar{v}(p)^T = 0, \tag{3.219}$$

which is equal to

$$(\not{p} - m\mathbb{1}_4)\, C\bar{v}(p)^T = 0, \tag{3.220}$$

since $C\gamma^T C^{-1} = -\gamma$ [cf. Eq. (3.213)]. Then, introducing the quantity

$$u^c(p) = C\bar{v}(p)^T, \tag{3.221}$$

which is the *charge-conjugated Dirac spinor*, we obtain the equation

$$(\not{p} - m\mathbb{1}_4)\, u^c(p) = 0, \tag{3.222}$$

which means that $u^c(p)$ is a positive-energy solution, since it fulfils Eq. (3.214).

Exercise 3.8   Use the explicit representation $C = -i\gamma^0\gamma^2$ to show that $C^{-1} = C^T = C^\dagger = -C$. In addition, show that $v(p) = C\left[u^c(p)\right]^T$.

3.9 Constants of motion

The following four statements are equivalent for an operator $\mathcal{O}$

(1) The operator $\mathcal{O}$ commutes with the Hamiltonian H, i.e. $[\mathcal{O}, H] = 0$.
(2) The current of the operator $\mathcal{O}$ is conserved.
(3) The operators $\mathcal{O}$ and H can be simultaneously diagonalized.
(4) The operator $\mathcal{O}$ is a constant of motion.

Thus, constants of motion are dynamical variables that **commute** with the Hamiltonian.

For example, consider the Hamiltonian for a free Dirac particle

$$H_0 = \beta m + \boldsymbol{\alpha} \cdot \mathbf{p}. \tag{3.223}$$

The orbital angular momentum $\mathbf{L} = \mathbf{x} \times \mathbf{p} = (L^1, L^2, L^3)$ does not commute with the Hamiltonian H_0, i.e. $[L^k, H_0] \neq 0$, which means that $\mathbf{L}$ is not a constant of motion. In addition, the spin $\mathbf{S} = \frac{1}{2}\boldsymbol{\Sigma}$ is not a constant of motion. However, the total angular momentum $\mathbf{J} = \mathbf{L} + \mathbf{S}$ commutes with the Hamiltonian H_0, which means that $\mathbf{J}$ is a constant of motion. Therefore, $\mathbf{J}^2$, J^3, and H_0 can be simultaneously diagonalized. This is not possible for $\mathbf{L}^2$, $\mathbf{S}^2$, and the spin–orbit coupling $\mathbf{L} \cdot \mathbf{S}$.

Next, look at the operator $K := \beta(\boldsymbol{\Sigma} \cdot \mathbf{L} + \mathbb{1}_4)$, which is called the *Runge–Lenz* (or *eccentricity*) *vector* and measures the alignment of the angular momentum and the spin that are parallel or antiparallel for different signs of K. This operator commutes with the total angular momentum operator $\mathbf{J}$ as well as the Hamiltonian H_0. Thus, K is a constant of motion with conserved eigenvalues. Squaring K, we have

$$
\begin{aligned}
K^2 &= \beta(\boldsymbol{\Sigma} \cdot \mathbf{L} + \mathbb{1}_4)\beta(\boldsymbol{\Sigma} \cdot \mathbf{L} + \mathbb{1}_4) \\
&= \beta(\boldsymbol{\Sigma} \cdot \mathbf{L})\beta(\boldsymbol{\Sigma} \cdot \mathbf{L}) + \beta(\boldsymbol{\Sigma} \cdot \mathbf{L})\beta + \beta^2(\boldsymbol{\Sigma} \cdot \mathbf{L}) + \beta^2. \tag{3.224}
\end{aligned}
$$

Now, it holds that $[\sigma^{\mu\nu}, \gamma^5] = 0$, where $(\sigma^{23}, \sigma^{31}, \sigma^{12}) = (\sigma^1, \sigma^2, \sigma^3)$ is the vector of Pauli matrices, which means that $[\boldsymbol{\Sigma}, \gamma^5] = 0$. Actually, $\boldsymbol{\Sigma} = \gamma^5\boldsymbol{\alpha} = \boldsymbol{\alpha}\gamma^5$, and since $\boldsymbol{\alpha} = \beta\boldsymbol{\gamma} = \gamma^0\boldsymbol{\gamma}$, we find that $\boldsymbol{\Sigma} = \gamma^5\gamma^0\boldsymbol{\gamma}$. Therefore, using $\{\gamma^0, \boldsymbol{\gamma}\} = 0$ and $\{\gamma^0, \gamma^5\} = 0$ as well as $\gamma^{0^2} = \mathbb{1}_4$, we can write

$$\beta(\boldsymbol{\Sigma} \cdot \mathbf{L})\beta = \gamma^0\gamma^5\gamma^0\boldsymbol{\gamma} \cdot \mathbf{L}\gamma^0 = -\gamma^0\gamma^5\gamma^0\gamma^0\boldsymbol{\gamma} \cdot \mathbf{L} = \gamma^5\gamma^0\boldsymbol{\gamma} \cdot \mathbf{L} = \boldsymbol{\Sigma} \cdot \mathbf{L}. \tag{3.225}$$

Thus, inserting Eq. (3.225) into Eq. (3.224) and using $\beta^2 = \gamma^{0^2} = \mathbb{1}_4$, we obtain

$$K^2 = (\boldsymbol{\Sigma} \cdot \mathbf{L})^2 + 2(\boldsymbol{\Sigma} \cdot \mathbf{L}) + \mathbb{1}_4. \tag{3.226}$$

Next, we can rewrite

$$\boldsymbol{\Sigma} \cdot \mathbf{L} = \begin{pmatrix} \boldsymbol{\sigma} & 0 \\ 0 & \boldsymbol{\sigma} \end{pmatrix} \cdot \mathbf{L} = \begin{pmatrix} \boldsymbol{\sigma} \cdot \mathbf{L} & 0 \\ 0 & \boldsymbol{\sigma} \cdot \mathbf{L} \end{pmatrix}, \tag{3.227}$$

which means that

$$(\boldsymbol{\Sigma} \cdot \mathbf{L})^2 = \begin{pmatrix} (\boldsymbol{\sigma} \cdot \mathbf{L})^2 & 0 \\ 0 & (\boldsymbol{\sigma} \cdot \mathbf{L})^2 \end{pmatrix}. \tag{3.228}$$

Furthermore, using the identity $(\boldsymbol{\sigma} \cdot \mathbf{a})(\boldsymbol{\sigma} \cdot \mathbf{b}) = \mathbf{a} \cdot \mathbf{b}\mathbb{1}_2 + i\boldsymbol{\sigma} \cdot (\mathbf{a} \times \mathbf{b})$, we have $(\boldsymbol{\sigma} \cdot \mathbf{L})^2 = \mathbf{L} \cdot \mathbf{L}\mathbb{1}_2 + i\boldsymbol{\sigma} \cdot (\mathbf{L} \times \mathbf{L})$. In addition, note that $\mathbf{L} \times \mathbf{L} = i\mathbf{L}$, which gives $(\boldsymbol{\sigma} \cdot \mathbf{L})^2 = \mathbf{L}^2\mathbb{1}_2 + i\boldsymbol{\sigma} \cdot i\mathbf{L} = \mathbf{L}^2\mathbb{1}_2 - \boldsymbol{\sigma} \cdot \mathbf{L}$. Thus, we find that

$$(\mathbf{\Sigma} \cdot \mathbf{L})^2 = \begin{pmatrix} \mathbf{L}^2 \mathbb{1}_2 - \boldsymbol{\sigma} \cdot \mathbf{L} & 0 \\ 0 & \mathbf{L}^2 \mathbb{1}_2 - \boldsymbol{\sigma} \cdot \mathbf{L} \end{pmatrix} = \mathbf{L}^2 \mathbb{1}_4 - \mathbf{\Sigma} \cdot \mathbf{L}, \qquad (3.229)$$

$$K^2 = (\mathbf{L}^2 \mathbb{1}_4 - \mathbf{\Sigma} \cdot \mathbf{L}) + 2(\mathbf{\Sigma} \cdot \mathbf{L}) + \mathbb{1}_4 = \mathbf{L}^2 \mathbb{1}_4 + \mathbf{\Sigma} \cdot \mathbf{L} + \mathbb{1}_4. \qquad (3.230)$$

On the other hand, we assume that

$$K^2 = \left(\mathbf{L} \mathbb{1}_4 + \frac{1}{2} \mathbf{\Sigma} \right)^2 + \frac{1}{4} \mathbb{1}_4 = \mathbf{L}^2 \mathbb{1}_4 + \mathbf{\Sigma} \cdot \mathbf{L} + \frac{1}{4} \mathbf{\Sigma}^2 + \frac{1}{4} \mathbb{1}_4. \qquad (3.231)$$

Now, according to Eq. (3.48), we have

$$\mathbf{\Sigma}^2 = \begin{pmatrix} \boldsymbol{\sigma}^2 & 0 \\ 0 & \boldsymbol{\sigma}^2 \end{pmatrix} = 3 \begin{pmatrix} \mathbb{1}_2 & 0 \\ 0 & \mathbb{1}_2 \end{pmatrix} = 3 \mathbb{1}_4. \qquad (3.232)$$

This means that Eq. (3.231) is simplified to

$$K^2 = \mathbf{L}^2 \mathbb{1}_4 + \mathbf{\Sigma} \cdot \mathbf{L} + \frac{3}{4} \mathbb{1}_4 + \frac{1}{4} \mathbb{1}_4 = \mathbf{L}^2 \mathbb{1}_4 + \mathbf{\Sigma} \cdot \mathbf{L} + \mathbb{1}_4. \qquad (3.233)$$

Thus, Eqs. (3.230) and (3.233) are equivalent, and therefore, it follows that the quantity K^2 can be written as

$$K^2 = \mathbf{L}^2 \mathbb{1}_4 + \mathbf{\Sigma} \cdot \mathbf{L} + \mathbb{1}_4 = \left(\mathbf{L} \mathbb{1}_4 + \frac{1}{2} \mathbf{\Sigma} \right)^2 + \frac{1}{4} \mathbb{1}_4 = \left(\mathbf{J}^2 + \frac{1}{4} \right) \mathbb{1}_4, \qquad (3.234)$$

which has the eigenvalues (each eigenvalue with multiplicity 2)

$$j(j+1) + \frac{1}{4} = \left(j + \frac{1}{2} \right)^2 \quad \Leftrightarrow \quad \pm \left(j + \frac{1}{2} \right). \qquad (3.235)$$

3.10 Central potentials

The Dirac equation with electromagnetic field interactions can be solved **exactly** for several cases. For example, we have the following cases.

- The Coulomb potential

$$V(r) = -\frac{Ze^2}{4\pi r}. \qquad (3.236)$$

- A homogeneous, static magnetic field, which leads to *Landau levels* (cf. the non-relativistic case).
- A plane electromagnetic wave, which leads to *Volkov solutions*.

In the case of *central potentials*, i.e. spherically symmetric potentials, which are potentials of the form $V = V(r)$ (e.g. the Coulomb potential), the Hamiltonian is given by

$$H = H_0 + V \mathbb{1}_4 = \beta m + \boldsymbol{\alpha} \cdot \mathbf{p} + V(r) \mathbb{1}_4. \qquad (3.237)$$

In fact, if the Hamiltonian is of the form of Eq. (3.237), one can show that

$$[\mathbf{L}, H] = i\boldsymbol{\alpha} \times \mathbf{p} \tag{3.238}$$

and

$$[\mathbf{S}, H] = \left[\tfrac{1}{2}\boldsymbol{\Sigma}, H\right] = -i\boldsymbol{\alpha} \times \mathbf{p}, \tag{3.239}$$

which means that the angular momentum $\mathbf{L}$ and the spin $\mathbf{S}$ do not commute with the Hamiltonian and they are therefore not constants of motion. For this case, the constants of motion are the total angular momentum $\mathbf{J} = \mathbf{L} + \mathbf{S} = \mathbf{x} \times \mathbf{p} + \tfrac{1}{2}\boldsymbol{\Sigma}$ (since $[\mathbf{J}, H] = [\mathbf{L}, H] + [\mathbf{S}, H] = i\boldsymbol{\alpha} \times \mathbf{p} - i\boldsymbol{\alpha} \times \mathbf{p} = 0$) and the parity P. Hence, we can find simultaneous eigenstates to the Hamiltonian as well as the operators $\mathbf{J}^2 = J^2$, J_z, and P, which are given by the following equations

$$H \begin{pmatrix} \phi \\ \chi \end{pmatrix} = E \begin{pmatrix} \phi \\ \chi \end{pmatrix}, \tag{3.240}$$

$$J^2 \begin{pmatrix} \phi \\ \chi \end{pmatrix} = j(j+1) \begin{pmatrix} \phi \\ \chi \end{pmatrix}, \tag{3.241}$$

$$J_z \begin{pmatrix} \phi \\ \chi \end{pmatrix} = M \begin{pmatrix} \phi \\ \chi \end{pmatrix}, \tag{3.242}$$

$$P \begin{pmatrix} \phi \\ \chi \end{pmatrix} = (-1)^{j+\varpi/2} \begin{pmatrix} \phi \\ \chi \end{pmatrix}, \quad \text{where } \varpi = \begin{cases} 1 & \text{if } P = (-1)^{j+1/2} \\ -1 & \text{if } P = (-1)^{j-1/2} \end{cases}, \tag{3.243}$$

and give rise to the eigenstates $|j, M\rangle$. The solutions to the angular parts of these equations are $\mathcal{Y}_{\ell j}^M(\theta, \varphi)$ with parity equal to $(-1)^L$. The possible L values are

$$\mathcal{Y}_{\ell j}^M(\theta, \varphi): \quad L = \ell = j + \varpi/2, \quad P = (-1)^{j+\varpi/2} = (-1)^\ell, \tag{3.244}$$

$$\mathcal{Y}_{\ell' j}^M(\theta, \varphi): \quad L = \ell' = j - \varpi/2, \quad P = (-1)^{j-\varpi/2} = (-1)^{\ell'}, \tag{3.245}$$

which have opposite parity. Thus, the explicit solutions are given by

$$\mathcal{Y}_{\ell j}^M = \begin{pmatrix} \sqrt{\frac{j+M}{2j}} Y_{j-1/2, M-1/2} \\ \sqrt{\frac{j-M}{2j}} Y_{j-1/2, M+1/2} \end{pmatrix}, \quad j = \ell + 1/2, \tag{3.246}$$

$$\mathcal{Y}_{\ell j}^M = \begin{pmatrix} \sqrt{\frac{j-M+1}{2(j+1)}} Y_{j+1/2, M-1/2} \\ -\sqrt{\frac{j+M+1}{2(j+1)}} Y_{j+1/2, M+1/2} \end{pmatrix}, \quad j = \ell - 1/2, \quad \ell > 0, \tag{3.247}$$

which fulfil the eigenvalue equation

$$J^2 \mathcal{Y}_{\ell j}^M = j(j+1) \mathcal{Y}_{\ell j}^M. \tag{3.248}$$

In order to find the full solutions to the equation

$$H\psi^M_{\varpi j} = E\psi^M_{\varpi j}, \tag{3.249}$$

we make the Ansätze

$$\phi = \frac{1}{r}F(r)\mathcal{Y}^M_{\ell j}, \tag{3.250}$$

$$\chi = \frac{1}{r}G(r)\mathcal{Y}^M_{\ell' j}, \tag{3.251}$$

which mean that the solutions should have the form

$$\psi^M_{\varpi j} = \frac{1}{r}\begin{pmatrix} F(r)\mathcal{Y}^M_{\ell j} \\ G(r)\mathcal{Y}^M_{\ell' j} \end{pmatrix}. \tag{3.252}$$

Furthermore, using spherical coordinates, we find that for any f

$$\mathbf{x}\cdot\mathbf{p}f = \frac{1}{i}r\frac{\partial f}{\partial r}, \tag{3.253}$$

and introducing the operators

$$p_r f = \frac{1}{ir}\frac{\partial}{\partial r}(rf) = \frac{1}{r}(\mathbf{x}\cdot\mathbf{p} - i)f, \tag{3.254}$$

$$\alpha_r = \frac{1}{r}\boldsymbol{\alpha}\cdot\mathbf{x}, \tag{3.255}$$

one can show that

$$\boldsymbol{\alpha}\cdot\mathbf{p} = \alpha_r\left(p_r\mathbb{1}_4 + \frac{i}{r}\beta K\right), \tag{3.256}$$

where $K = \beta(\boldsymbol{\Sigma}\cdot\mathbf{L} + \mathbb{1}_4) = -\varpi(j + 1/2)\mathbb{1}_4$. Finally, we can write the Hamiltonian as

$$H = \alpha_r\left(p_r\mathbb{1}_4 - \frac{i\varpi(j + 1/2)}{r}\beta\right) + \beta m + V(r)\mathbb{1}_4, \tag{3.257}$$

which means that we have the following eigenvalue equation

$$\left[\begin{pmatrix} 0 & -d/dr \\ d/dr & 0 \end{pmatrix} + \begin{pmatrix} 0 & \varpi(j + 1/2)/r \\ \varpi(j + 1/2)/r & 0 \end{pmatrix}\right]\begin{pmatrix} F \\ G \end{pmatrix}$$
$$= \begin{pmatrix} E - m - V & 0 \\ 0 & E + m - V \end{pmatrix}\begin{pmatrix} F \\ G \end{pmatrix} \tag{3.258}$$

that can be written as a system of equations

$$\left[-\frac{d}{dr} + \frac{\varpi(j + 1/2)}{r}\right]G = (E - m - V)F, \tag{3.259}$$

$$\left[\frac{d}{dr} + \frac{\varpi(j + 1/2)}{r}\right]F = (E + m - V)G. \tag{3.260}$$

Thus, Eqs. (3.259) and (3.260) are the coupled differential equations for the radial functions F and G of the Dirac equation in the case of central potentials $V = V(r)$. Note that the correct normalization of the functions F and G is given by

$$\int_0^\infty \left(|F|^2 + |G|^2\right) dr = 1. \tag{3.261}$$

Exercise 3.9 Show the relation in Eq. (3.256) and that the Hamiltonian can be written as in Eq. (3.257).

3.11 The hydrogenic atom

Next, we want to solve the Dirac equation for a hydrogenic atom. In order to solve this equation, we assume that the potential is a Coulomb potential of the form

$$V(r) = -\frac{Ze^2}{4\pi r}, \tag{3.262}$$

which is a central potential. Therefore, we can use the system of equations (3.259) and (3.260). Introducing the new parameters $Z' = Ze^2/(4\pi)$, $\kappa = \sqrt{m^2 - E^2}$, $\tau = \varpi(j+1/2)$, and $\nu = \sqrt{(m-E)/(m+E)}$ as well as the new variable $\rho = \kappa r$, we have

$$\left(-\frac{d}{d\rho} + \frac{\tau}{\rho}\right) G = \left(-\nu + \frac{Z'}{\rho}\right) F, \tag{3.263}$$

$$\left(\frac{d}{d\rho} + \frac{\tau}{\rho}\right) F = \left(\frac{1}{\nu} + \frac{Z'}{\rho}\right) G. \tag{3.264}$$

Behaviour at $\rho \to 0$. For small values of the variable ρ, i.e. in the limit $\rho \to 0$, the system of equations (3.263) and (3.264) can approximately be written as

$$\left(-\frac{d}{d\rho} + \frac{\tau}{\rho}\right) G \simeq \frac{Z'}{\rho} F, \tag{3.265}$$

$$\left(\frac{d}{d\rho} + \frac{\tau}{\rho}\right) F \simeq \frac{Z'}{\rho} G. \tag{3.266}$$

Making the Ansätze

$$F(\rho) \sim a\rho^\mu \quad \text{and} \quad G(\rho) \sim b\rho^\mu, \tag{3.267}$$

where a and b are arbitrary constants, we obtain

$$\begin{cases} -\mu b\rho^{\mu-1} + \tau b\rho^{\mu-1} = Z'a\rho^{\mu-1} \\ \mu a\rho^{\mu-1} + \tau a\rho^{\mu-1} = Z'b\rho^{\mu-1} \end{cases} \tag{3.268}$$

or

$$\begin{pmatrix} \mu + \tau & -Z' \\ Z' & \mu - \tau \end{pmatrix} \begin{pmatrix} a \\ b \end{pmatrix} = 0, \tag{3.269}$$

which yields $(\mu + \tau)(\mu - \tau) + (Z')^2 = 0$, since the determinant of the coefficients must vanish in order to yield a and b that are non-zero. Thus, we find that

$$\mu = \pm\sqrt{\tau^2 - (Z')^2}, \tag{3.270}$$

where we have to choose the positive solution for μ, i.e. $\mu = \sqrt{\tau^2 - (Z')^2}$, since the wave function has to be normalizable according to Eq. (3.261). The negative solution for μ, i.e. $\mu = -\sqrt{\tau^2 - (Z')^2}$, yields $|F|^2 + |G|^2 \sim \rho^{-2\sqrt{\tau^2 - (Z')^2}}$ close to $\rho = 0$, which means that the integral in the definition of the normalization (3.261) will be divergent if $\sqrt{\tau^2 - (Z')^2} < 1/2$. In addition, in the case that $\tau^2 < (Z')^2$, the exponents in the Ansätze (3.267) become imaginary. Thus, the assumed ρ-dependence in the Ansätze makes only the positive solution possible.

Behaviour at $\rho \to \infty$. Asymptotically, for large values of the variable ρ, the system of equations reduces to

$$-\frac{dG}{d\rho} \simeq -\nu F, \tag{3.271}$$

$$\frac{dF}{d\rho} \simeq \frac{1}{\nu} G. \tag{3.272}$$

For example, inserting Eq. (3.271) into Eq. (3.272), implies that

$$\frac{d^2 F}{d\rho^2} \simeq F, \tag{3.273}$$

which has the characteristic equation $\lambda^2 = 1$ with the solutions $\lambda_\pm = \pm 1$. This means that the general solution to the ordinary differential equation (3.273) is given by

$$F(\rho) = C_- e^{\lambda_- \rho} + C_+ e^{\lambda_+ \rho} = C_- e^{-\rho} + C_+ e^{\rho}, \tag{3.274}$$

where C_- and C_+ are arbitrary constants. However, in particular, the solution to Eq. (3.273) has to be bounded when $\rho \to \infty$, which means that $C_+ = 0$. Thus, we obtain $F(\rho) \sim e^{-\rho}$, and similarly for $G(\rho)$. In conclusion, in the case when $\rho \to \infty$, i.e. in the asymptotic case, the solutions to Eqs. (3.263) and (3.264) behave as

$$F(\rho) \sim e^{-\rho} \quad \text{and} \quad G(\rho) \sim e^{-\rho}. \tag{3.275}$$

Therefore, we make the Ansätze

$$F(\rho) = e^{-\rho} f(\rho), \tag{3.276}$$

$$G(\rho) = e^{-\rho} g(\rho), \tag{3.277}$$

which separate the exponentially decreasing behaviour from the rest of the solutions, and obtain

$$\rho \frac{df}{d\rho} = (\rho - \tau)f + \left(\frac{\rho}{\nu} + Z'\right)g, \tag{3.278}$$

$$\rho \frac{dg}{d\rho} = (\rho + \tau)g + \left(\rho\nu - Z'\right)f. \tag{3.279}$$

Introducing the notations

$$w = \begin{pmatrix} f \\ g \end{pmatrix}, \quad A = \begin{pmatrix} -\tau & Z' \\ -Z' & \tau \end{pmatrix}, \quad \text{and} \quad B = \begin{pmatrix} 1 & 1/\nu \\ \nu & 1 \end{pmatrix}, \tag{3.280}$$

we can write the above system of equations as a matrix equation

$$\rho \frac{dw}{d\rho} = (A + \rho B)w, \tag{3.281}$$

where A has the eigenvalues $\pm\lambda = \pm\sqrt{\tau^2 - (Z')^2}$ and B has the eigenvalues 0 and 2. Now, using the *method of Frobenius*, i.e. we formulate the solution to the matrix equation as a series solution

$$w = \rho^\mu \sum_{s=0}^{N} w_s \rho^s, \tag{3.282}$$

where the factor ρ^μ describes the behaviour of the solution at $\rho \to 0$ (that we have discussed in detail above) and N is an integer (that we will determine below), we find the following expressions for the different terms in the matrix equation

$$\rho \frac{dw}{d\rho} = \sum_{s=0}^{N} (s + \mu)w_s \rho^{s+\mu}, \tag{3.283}$$

$$Aw = \sum_{s=0}^{N} (Aw)_s \rho^{s+\mu} = \sum_{s=0}^{N} Aw_s \rho^{s+\mu}, \tag{3.284}$$

$$\rho Bw = \sum_{s=1}^{N+1} (Bw)_{s-1} \rho^{s+\mu} = \sum_{s=1}^{N+1} Bw_{s-1} \rho^{s+\mu}, \tag{3.285}$$

where

$$w_s = \begin{pmatrix} f_s \\ g_s \end{pmatrix}. \tag{3.286}$$

Inserting Eqs. (3.283)–(3.285) into Eq. (3.281), we obtain

$$\sum_{s=0}^{N} \left[(s + \mu)w_s - Aw_s\right] \rho^{s+\mu} - \sum_{s=1}^{N+1} Bw_{s-1} \rho^{s+\mu} = 0, \tag{3.287}$$

which means that each coefficient for every power $\rho^{s+\mu}$ ($s \geq 0$) in the left-hand side of Eq. (3.287) must be equal to zero, since in general $\rho^{s+\mu} \neq 0$. Thus, for $s = 0$, we have the eigenvalue equation

$$\mu w_0 = A w_0, \tag{3.288}$$

where $\mu = \pm \lambda = \pm \sqrt{\tau^2 - (Z')^2}$, which has the solution (for $\mu = \lambda$)

$$w_0 = \begin{pmatrix} f_0 \\ g_0 \end{pmatrix} = \frac{1}{\Delta} \begin{pmatrix} Z' \\ \Delta^2/(2\tau) \end{pmatrix}, \tag{3.289}$$

where $\Delta = 2\tau \left(\tau + \sqrt{\tau^2 - (Z')^2} \right)$. Note that in order for the solution to be normalizable we must have $\mu = \lambda$. However, for a general value of $s \geq 1$, using $\mu = \lambda$, we have the recursive equation

$$[(s + \lambda)\mathbb{1}_2 - A]\, w_s = B w_{s-1}, \quad s \geq 1, \tag{3.290}$$

which has the formal solution

$$w_s = [(s + \lambda)\mathbb{1}_2 - A]^{-1} B w_{s-1} = \frac{1}{s} \left(\mathbb{1}_2 + \frac{\lambda \mathbb{1}_2 - A}{s} \right)^{-1} B w_{s-1}, \tag{3.291}$$

since $\det\left[(s + \lambda)\mathbb{1}_2 - A \right] \neq 0$ for $s \geq 1$. Nevertheless, it holds that

$$\| w_s \| \leq \frac{c}{s} \| w_{s-1} \|, \tag{3.292}$$

where c is a constant, which means that the series converges, and therefore, it has an upper bound and is cut. Let us find the explicit solution for w_s. In order to do so, we bring the matrix B to diagonal form using a similarity transformation in the following way

$$B' = U B U^{-1} = \text{diag}(2, 0), \tag{3.293}$$

where the change-of-basis matrix U, relating the unprimed basis and the primed basis, is given by

$$U = \begin{pmatrix} 1 & 1/\nu \\ 1 & -1/\nu \end{pmatrix} \quad \text{and} \quad U^{-1} = \frac{1}{2} \begin{pmatrix} 1 & 1 \\ \nu & -\nu \end{pmatrix}. \tag{3.294}$$

Note that the matrices B and B' are similar matrices. Thus, we can also write the quantities A and w in the primed basis (in which the matrix B is diagonal) as

$$A' = UAU^{-1} = \frac{1}{2} \begin{pmatrix} 1 & 1/v \\ 1 & -1/v \end{pmatrix} \begin{pmatrix} -\tau & Z' \\ -Z' & \tau \end{pmatrix} \begin{pmatrix} 1 & 1 \\ v & -v \end{pmatrix}$$

$$= \frac{1}{2} \begin{pmatrix} Z'\left(v - \frac{1}{v}\right) & -2\tau - Z'\left(v + \frac{1}{v}\right) \\ -2\tau + Z'\left(v + \frac{1}{v}\right) & Z'\left(\frac{1}{v} - v\right) \end{pmatrix}, \tag{3.295}$$

$$w' = Uw = \begin{pmatrix} f + \frac{1}{v}g \\ f - \frac{1}{v}g \end{pmatrix} = \begin{pmatrix} f' \\ g' \end{pmatrix} \quad \text{or}$$

$$w = U^{-1}w' = \frac{1}{2} \begin{pmatrix} f' + g' \\ v\left(f' - g'\right) \end{pmatrix} = \begin{pmatrix} f \\ g \end{pmatrix}. \tag{3.296}$$

Using Eq. (3.295), we have

$$(s + \lambda)\mathbb{1}_2 - A' = \begin{pmatrix} s + \lambda - \frac{Z'}{2}\left(v - \frac{1}{v}\right) & \tau + \frac{Z'}{2}\left(v + \frac{1}{v}\right) \\ \tau - \frac{Z'}{2}\left(v + \frac{1}{v}\right) & s + \lambda - \frac{Z'}{2}\left(\frac{1}{v} - v\right) \end{pmatrix}, \tag{3.297}$$

which means that

$$[(s + \lambda)\mathbb{1}_2 - A']^{-1} = \frac{1}{s(s + 2\lambda)} \begin{pmatrix} s + \lambda - \frac{Z'}{2}\left(\frac{1}{v} - v\right) & -\tau - \frac{Z'}{2}\left(v + \frac{1}{v}\right) \\ -\tau + \frac{Z'}{2}\left(v + \frac{1}{v}\right) & s + \lambda - \frac{Z'}{2}\left(v - \frac{1}{v}\right) \end{pmatrix}. \tag{3.298}$$

Therefore, in the primed basis, the solution is given by

$$w'_s = [(s + \lambda)\mathbb{1}_2 - A']^{-1} B'w'_{s-1} = \frac{2}{s(s + 2\lambda)} \begin{pmatrix} s + \lambda - \frac{Z'}{2}\left(\frac{1}{v} - v\right) \\ -\tau + \frac{Z'}{2}\left(v + \frac{1}{v}\right) \end{pmatrix} f'_{s-1} = \begin{pmatrix} f'_s \\ g'_s \end{pmatrix}. \tag{3.299}$$

In addition, in the primed basis, we have the initial condition

$$\lambda w'_0 = A'w'_0, \tag{3.300}$$

where

$$w'_0 = Uw_0 = \frac{1}{\Delta} \begin{pmatrix} Z' + \frac{\Delta^2}{2v\tau} \\ Z' - \frac{\Delta^2}{2v\tau} \end{pmatrix} = \begin{pmatrix} f'_0 \\ g'_0 \end{pmatrix}. \tag{3.301}$$

Note that N is an integer such that $w'_N \neq 0$ and $w'_{N+1} = 0$, which means that such N exists if and only if $f'_{N-1} \neq 0$ and $f'_N = 0$. Therefore, N is the order of the polynomial in the series solution (3.282). Thus, using $f'_{N-1} \neq 0$ and $f'_N = 0$ together with Eq. (3.299), we obtain the order N as

$$N + \lambda - \frac{Z'}{2}\left(\frac{1}{v} - v\right) = 0, \quad N \geq 1, \tag{3.302}$$

which can be written as

$$N + \lambda = \frac{1 - v^2}{2v}\frac{Ze^2}{4\pi} = \frac{E}{\sqrt{m^2 - E^2}}\frac{Ze^2}{4\pi}. \tag{3.303}$$

Now, we can find the radial solution to the Dirac equation for a hydrogenic atom. Again, using Eq. (3.299), we obtain two formulas for the coefficients in the series solution

$$f'_s = \frac{2}{s(s+2\lambda)}\left[s+\lambda - \frac{Z'}{2}\left(\frac{1}{\nu}-\nu\right)\right]f'_{s-1}, \tag{3.304}$$

$$g'_s = \frac{-\tau + \frac{Z'}{2}\left(\nu+\frac{1}{\nu}\right)}{s+\lambda - \frac{Z'}{2}\left(\frac{1}{\nu}-\nu\right)}f'_s, \tag{3.305}$$

where the first formula is recursive for the coefficients of the function f' and the second formula relates the coefficients of the functions f' and g'. Introducing the quantity

$$s_0 = \frac{Z'}{2}\left(\frac{1}{\nu}-\nu\right)-\lambda \tag{3.306}$$

and using Eq. (3.304), we find that

$$\begin{aligned}
f'_s &= \frac{2}{s(s+2\lambda)}(s-s_0)f'_{s-1}\\
&= \frac{2\cdot 2(s-1-s_0)(s-s_0)}{(s-1)s(s-1+2\lambda)(s+2\lambda)}f'_{s-2} = \cdots\\
&= \frac{2^s(1-s_0)(2-s_0)\cdots(s-s_0)}{s!(2\lambda+1)(2\lambda+2)\cdots(2\lambda+s)}f'_0,
\end{aligned} \tag{3.307}$$

where

$$f'_0 = \frac{1}{\Delta}\left(Z'+\frac{\Delta^2}{2\nu\tau}\right) \tag{3.308}$$

is directly obtained from the intial condition given in Eq. (3.301). Furthermore, inserting Eq. (3.307) into Eq. (3.305), it follows that

$$\begin{aligned}
g'_s &= \frac{-\tau + \frac{Z'}{2}\left(\nu+\frac{1}{\nu}\right)}{s-s_0}f'_s\\
&= \frac{\tau - \frac{Z'}{2}\left(\nu+\frac{1}{\nu}\right)}{s_0}\frac{2^s(-s_0)(1-s_0)(2-s_0)\cdots(s-1-s_0)}{s!(2\lambda+1)(2\lambda+2)\cdots(2\lambda+s)}f'_0.
\end{aligned} \tag{3.309}$$

One can relate the power series corresponding to the coefficients f'_s and g'_s with the *confluent hypergeometric function* (or *Kummer's function of the first kind*) given by

$$_1F_1(a,b;z) = 1 + \frac{a}{b}z + \frac{a(a+1)}{b(b+1)}\frac{z^2}{2!} + \cdots = \sum_{k=0}^{\infty}\frac{(a)_k}{(b)_k}\frac{z^k}{k!}, \tag{3.310}$$

where $(a)_k$ and $(b)_k$ are so-called *Pochhammer symbols*, or equivalently, with the *generalized* (or *associated*) *Laguerre polynomials*

$$L_n^{(\alpha)}(x) = \frac{(\alpha+1)_n}{n!} {}_1F_1(-n, \alpha+1; x). \tag{3.311}$$

Thus, using the definition (3.310) as well as Eqs. (3.307) and (3.309), it holds that

$$f'(\rho) = \rho^\lambda \sum_{s=0}^{N} f'_s \rho^s = \rho^\lambda {}_1F_1(1-s_0, 2\lambda+1; 2\rho) f'_0, \tag{3.312}$$

$$g'(\rho) = \rho^\lambda \sum_{s=0}^{N} g'_s \rho^s = \rho^\lambda \frac{\tau - \frac{Z'}{2}\left(\nu+\frac{1}{\nu}\right)}{s_0} {}_1F_1(-s_0, 2\lambda+1; 2\rho) f'_0. \tag{3.313}$$

Finally, inserting the expressions for f' and g' into Eq. (3.296) and the result of that into Eqs. (3.276) and (3.277), we obtain the radial solution to the Dirac equation for a hydrogenic atom as

$$F(\rho) = e^{-\rho}\rho^\lambda \Bigg[{}_1F_1(1-s_0, 2\lambda+1; 2\rho)$$

$$+ \frac{\tau - \frac{Z'}{2}\left(\nu+\frac{1}{\nu}\right)}{s_0} {}_1F_1(-s_0, 2\lambda+1; 2\rho) \Bigg] \frac{f'_0}{2}, \tag{3.314}$$

$$G(\rho) = e^{-\rho}\rho^\lambda \Bigg[{}_1F_1(1-s_0, 2\lambda+1; 2\rho)$$

$$- \frac{\tau - \frac{Z'}{2}\left(\nu+\frac{1}{\nu}\right)}{s_0} {}_1F_1(-s_0, 2\lambda+1; 2\rho) \Bigg] \frac{\nu f'_0}{2}. \tag{3.315}$$

Next, introducing the normalized energy

$$\varepsilon = \frac{E}{m} \tag{3.316}$$

in Eq. (3.303), we obtain

$$N + \lambda = \frac{\varepsilon}{\sqrt{1-\varepsilon^2}} \frac{Ze^2}{4\pi}. \tag{3.317}$$

In addition, introducing *Sommerfeld's fine-structure constant*[17]

$$\alpha = \frac{e^2}{4\pi} \simeq 1/137.035\,999\,679(94) \tag{3.318}$$

and solving for ε, we obtain

$$\varepsilon = \frac{1}{\sqrt{1 + \dfrac{Z^2\alpha^2}{\left[N+\sqrt{(j+1/2)^2 - Z^2\alpha^2}\right]^2}}}, \quad N+(j+1/2) = 1, 2, 3, \ldots, n. \tag{3.319}$$

[17] Note that we are using rationalized units or Heaviside–Lorentz units, where $\alpha = e^2/(4\pi)$, instead of Gaussian units, where $\alpha = \hat{e}^2$. Thus, the two system of units are related to each other as $e = \hat{e}\sqrt{4\pi}$. In Gaussian units, the Coulomb potential has the form $V(r) = -Z\hat{e}^2/r$, but instead the factor of 4π appears in Maxwell's equations.

Since $\alpha \lesssim 1/137$, we have in general that $Z\alpha \ll 1$, which means that we can use $Z\alpha$ as an expansion parameter. Performing an expansion of Eq. (3.319), we obtain

$$\varepsilon = 1 - \frac{1}{2}\frac{Z^2\alpha^2}{(N+j+1/2)^2} + \cdots = 1 - \frac{1}{2}\frac{Z^2\alpha^2}{n^2}\left[1 + \frac{Z^2\alpha^2}{n^2}\left(\frac{n}{j+1/2} - \frac{3}{4}\right) + \cdots\right],$$

(3.320)

where the term

$$\frac{Z^2\alpha^2}{n^2}\left(\frac{n}{j+1/2} - \frac{3}{4}\right)$$

(3.321)

gives rise to the *fine structure*. Note that the form of the series expansion in Eq. (3.320) is the same as the series expansion in Eq. (2.77), except that ℓ is replaced by j. In addition, the *Bohr term* is given by

$$E' = E - m = m(\varepsilon - 1) \simeq -\frac{1}{2}m\frac{Z^2\alpha^2}{n^2},$$

(3.322)

which corresponds to the energy levels of an electron bound to a nucleus of charge Z in a hydrogen-like atom, or equivalently, the electron binding energies. Thus, the expansion in Eq. (3.320) leads to the correct non-relativistic result for the energy levels, since they are given by the formula for orbital state energy levels

$$E_n = -2\pi R_\infty \frac{Z^2}{n^2},$$

where R_∞ is the Rydberg constant.

In the formulas for ε, the parameters (or rather quantum numbers) can assume the following values: $n = 1, 2, \ldots$, $\ell = 0, 1, \ldots, n-1$, and

$$j = \begin{cases} \frac{1}{2} & \ell = 0 \\ \ell \pm \frac{1}{2} & \text{otherwise} \end{cases}.$$

Here n is the so-called *main* (or *principal*) *quantum number*, ℓ is the *orbital angular momentum quantum number*, and j is the *spin–orbital angular momentum quantum number*. Note that the non-relativistic Bohr term depends on the main quantum number n only, whereas the relativistic Bohr term (using the Dirac equation) depends on both n and j. However, both the non-relativistic and relativistic Bohr terms are independent of ℓ. In order to demonstrate relativistic effects in the spectrum, we will consider the hydrogen atom. Table 3.3 compares the non-relativistic $\left(E' \simeq -\frac{1}{2}m\alpha^2/n^2\right)$ and relativistic $[E' = m(\varepsilon - 1)]$ spectra of the hydrogen atom.[18] We observe that the non-relativistic and relativistic energies are hardly distinguishable. The reason is that the mean kinetic energy of the electron

---

[18] Note that we have used the reduced mass $m_{\text{red}} = mm_N/(m + m_N)$, where m_N is the mass of the nucleus (here: proton), instead of the electron mass m. In this case, we find that $m_{\text{red}} \simeq 0.999\,4557m$.

Table 3.3 *Binding energies of the electron in the hydrogen ($Z = 1$) atom.*

spectral notation	n	ℓ	j	non-relativistic binding energy [eV]	relativistic binding energy [eV]	database values of binding energy [eV]
$1s_{1/2}$	1	0	1/2	$-13.598\,29$	$-13.598\,47$	$-13.598\,43$
$2s_{1/2}$	2	0	1/2	$-3.399\,57$	$-3.399\,63$	$-3.399\,62$
$2p_{1/2}$	2	1	1/2	$\downarrow$	$\downarrow$	$-3.399\,63$
$2p_{3/2}$	2	1	3/2	$\downarrow$	$-3.399\,58$	$-3.399\,58$
$3s_{1/2}$	3	0	1/2	$-1.510\,921$	$-1.510\,941$	$-1.510\,940$
$3p_{1/2}$	3	1	1/2	$\downarrow$	$\downarrow$	$-1.510\,941$
$3p_{3/2}$	3	1	3/2	$\downarrow$	$-1.510\,927$	$-1.510\,927$
$3d_{3/2}$	3	2	3/2	$\downarrow$	$\downarrow$	$-1.510\,928$
$3d_{5/2}$	3	2	5/2	$\downarrow$	$-1.510\,923$	$-1.510\,923$

We have used $m = 0.510\,9989$ MeV, $m_N = m_p = 938.2720$ MeV, and $\alpha = 1/137.0360$. Degeneracies are denoted by $\downarrow$. The database values of the binding energies have been adopted from Y. Ralchenko, A. E. Kramida, J. Reader, and NIST ASD Team (2008). NIST Atomic Spectra Database (version 3.1.5), [Online]. Available: http://physics.nist.gov/asd3. Note that there exists an experimental value of the electron binding energy for the $1s_{1/2}$ state, $-13.598\,11$ eV, that has been adopted from J. E. Mack (1949) as given in C. E. Moore, *Atomic Energy Levels* (U.S. National Bureau of Standards, Washington D.C., 1949), vol. 1, p. 1.

in the hydrogen atom is of the order of 10 eV, which is much smaller than the rest mass of the electron, i.e. approximately 0.511 MeV. In conclusion, for the hydrogen atom, the relativistic corrections are small, but they have the property that states with larger values of j are at higher energies, which is due to the factor $1/(j+1/2)$ in Eq. (3.320). These so-called *fine-structure splittings* are in very good agreement with observations. In order to obtain an even more accurate spectrum, other small effects have to be added. First, the *hyperfine shift* (a triplet–singlet splitting of the ground state), which is due to the interaction between proton and electron spins, has to be added. Then, the *Lamb shift*,[19] which is an interaction between electrons and the fluctuating electromagnetic field, should be included. Finally, there is an effect coming from the zero-point energy of the infinitely many quantum harmonic oscillators describing the electromagnetic field.

However, in the case of heavier nuclei, i.e. larger values of Z, where Z is the nuclear charge, the kinetic energies of bound electrons in the ground state are proportional to Z^2. Thus, in the case of $Z = 100$, the kinetic energies are of the order of the rest mass of the electron, and therefore, relativistic effects are important. This can be clearly observed by comparing the non-relativistic and relativistic spectra of a hydrogen-like atom with $Z = 100$, which are presented in Table 3.4. In addition, we observe that the effect of the spin–orbit coupling removes the non-relativistic

[19] The Lamb shift is named after W. Lamb. In 1955, Lamb was awarded the Nobel Prize in physics 'for his discoveries concerning the fine structure of the hydrogen spectrum'.

Table 3.4 *Binding energies of the electron in the hydrogen-like*
(Z = 100) atom.

spectral notation	n	ℓ	j	non-relativistic binding energy [keV]	relativistic binding energy [keV]
$1s_{1/2}$	1	0	1/2	-136.1	-161.6
$2s_{1/2}$	2	0	1/2	-34.0	-42.1
$2p_{1/2}$	2	1	1/2	$\downarrow$	$\downarrow$
$2p_{3/2}$	2	1	3/2	$\downarrow$	-35.2
$3s_{1/2}$	3	0	1/2	-15.1	-17.9
$3p_{1/2}$	3	1	1/2	$\downarrow$	$\downarrow$
$3p_{3/2}$	3	1	3/2	$\downarrow$	-15.8
$3d_{3/2}$	3	2	3/2	$\downarrow$	$\downarrow$
$3d_{5/2}$	3	2	5/2	$\downarrow$	-15.3

We have used $m = 0.511\,00$ MeV and $\alpha = 1/137.04$. Degeneracies are denoted by $\downarrow$.

degeneracy for the six $2p$ states of both the hydrogen atom and the hydrogen-like atom, since in the relativistic case, these six states split into energetically different $2p_{1/2}$ (two) and $2p_{3/2}$ (four) states. Similarly, this is also true for the $3p_{1/2}$, $3p_{3/2}$, $3d_{3/2}$, and $3d_{5/2}$ states. For the interested reader, I would like to recommend the book G. W. Series (Ed.), *The Spectrum of Atomic Hydrogen – Advances*, World Scientific (1988).

3.12 The Weyl equation

In the limit $m \to 0$, the Dirac equation reduces to

$$i\gamma^\mu \partial_\mu \psi(x) = 0, \tag{3.323}$$

which is called the *Weyl equation*.[20] Using the following representation for the gamma matrices

$$\gamma^0 = \begin{pmatrix} 0 & \mathbb{1}_2 \\ \mathbb{1}_2 & 0 \end{pmatrix}, \quad \gamma = \begin{pmatrix} 0 & \sigma \\ -\sigma & 0 \end{pmatrix}, \quad \gamma^5 = \begin{pmatrix} -\mathbb{1}_2 & 0 \\ 0 & \mathbb{1}_2 \end{pmatrix}, \tag{3.324}$$

which is known as the *Weyl* (or *chiral*) *representation*, and in analogy with Eq. (3.50), writing the 4-spinor in terms of two 2-spinors

$$\psi = \begin{pmatrix} \psi_1 \\ \psi_2 \end{pmatrix}, \tag{3.325}$$

---

[20] Note that the Weyl equation has sometimes been called the *neutrino equation*. However, since neutrinos are massive particles, although with small masses, this name is not justified. Anyway, the Weyl equation gives still a fairly good description of the evolution of neutrinos.

we find the uncoupled system of equations

$$i\partial_0\psi_2 + i\boldsymbol{\sigma} \cdot \nabla\psi_2 = 0, \tag{3.326}$$

$$i\partial_0\psi_1 - i\boldsymbol{\sigma} \cdot \nabla\psi_1 = 0, \tag{3.327}$$

where the so-called *chiral states* fulfil

$$\gamma^5 \begin{pmatrix} \psi_1 \\ 0 \end{pmatrix} = - \begin{pmatrix} \psi_1 \\ 0 \end{pmatrix}, \quad \gamma^5 \begin{pmatrix} 0 \\ \psi_2 \end{pmatrix} = \begin{pmatrix} 0 \\ \psi_2 \end{pmatrix} \tag{3.328}$$

with chiralities (i.e. eigenvalues) -1 and 1, respectively. Note that if we had used another representation of the gamma matrices, then the system of equations would not have decoupled, which is essentially the reason for using the chiral representation. Thus, using the correspondence principle (for the spatial components), it holds that Eqs. (3.326) and (3.327) for the two 2-spinors can be written as

$$i\partial_0\psi_i = (-1)^i \boldsymbol{\sigma} \cdot \mathbf{p}\psi_i, \quad i = 1, 2. \tag{3.329}$$

Introducing $\sigma^0 = \sigma_0 = \mathbb{1}_2$ and using again the correspondence principle (but now for the time component), we can write

$$\sigma^\mu p_\mu \psi_2 = 0 \quad \text{and} \quad \bar{\sigma}^\mu p_\mu \psi_1 = 0, \tag{3.330}$$

where $(\sigma^\mu) = (\mathbb{1}_2, \boldsymbol{\sigma})$ and $(\bar{\sigma}^\mu) = (\mathbb{1}_2, -\boldsymbol{\sigma})$. Inserting plane-wave solutions $\psi(x) = e^{-ip\cdot x}u(p)$ [cf. Eq. (3.38)] into the Weyl equation, we obtain for the upper component (negative chirality) ψ_1

$$p^0 u(p) = -\boldsymbol{\sigma} \cdot \mathbf{p}u(p), \tag{3.331}$$

which leads to

$$(p^{0^2} - \mathbf{p}^2)u(p) = 0, \quad p^0 = \pm|\mathbf{p}|, \tag{3.332}$$

whereas we obtain for the lower component (positive chirality) ψ_2

$$p^0 u(p) = \boldsymbol{\sigma} \cdot \mathbf{p}u(p), \tag{3.333}$$

which also leads to

$$(p^{0^2} - \mathbf{p}^2)u(p) = 0, \quad p^0 = \pm|\mathbf{p}|. \tag{3.334}$$

If $p^0 = |\mathbf{p}|$, then the solution is a Weyl state, whereas if $p^0 = -|\mathbf{p}|$, then the solution is an anti-Weyl state.

Introducing the 4-spinor representation of the chiral states

$$\psi_- = \begin{pmatrix} \psi_1 \\ 0 \end{pmatrix} \quad \text{and} \quad \psi_+ = \begin{pmatrix} 0 \\ \psi_2 \end{pmatrix}, \tag{3.335}$$

these states can be written in terms of the *chirality operator* γ^5 as follows

$$\psi_- = \frac{1}{2}\left(\mathbb{1}_4 - \gamma^5\right)\psi \quad \text{and} \quad \psi_+ = \frac{1}{2}\left(\mathbb{1}_4 + \gamma^5\right)\psi, \qquad (3.336)$$

where the operators

$$P_\pm = \frac{1}{2}\left(\mathbb{1}_4 \pm \gamma^5\right) \qquad (3.337)$$

are called the *chirality projection operators* (or *chirality projectors*), cf. the spin projection operators (3.109). Note that for a Weyl state, the spin is antiparallel to the 3-momentum **p**, whereas for an anti-Weyl state, the spin is parallel to the 3-momentum **p**.

3.12.1 Graphene

In order to study the Weyl equation for a physical system, we consider the material *graphene* as the physical system. Graphene is a two-dimensional layer (in fact, a one-atom-thick layer) of carbon atoms, where the atoms are organized in a so-called *honeycomb lattice*.[21] Actually, graphene is around 200 times stronger than normal steel and is also transparent, flexible, and has a very good electrical conductivity, which increases the possible uses of the material. In graphene, electrons and holes behave as relativistic massless particles described by two two-dimensional Weyl equations, which are given as functions of the 2-position vector $\mathbf{r} = (x, y)$ by

$$-iv_F\boldsymbol{\sigma} \cdot \nabla\psi(\mathbf{r}) = E\psi(\mathbf{r}), \qquad (3.338)$$

$$-iv_F\boldsymbol{\sigma}^* \cdot \nabla\psi'(\mathbf{r}) = E'\psi'(\mathbf{r}), \qquad (3.339)$$

where $v_F \sim 10^6$ m/s is the *Fermi velocity* and replaces the velocity of light $c = 1$,[22]

$$\boldsymbol{\sigma} = (\sigma^1, \sigma^2) = \left(\begin{pmatrix} 0 & 1 \\ 1 & 0 \end{pmatrix}, \begin{pmatrix} 0 & -i \\ i & 0 \end{pmatrix}\right) \qquad (3.340)$$

is the two-dimensional vector of Pauli matrices [cf. Eq. (1.53)], $\boldsymbol{\sigma}^* = (\sigma^1, -\sigma^2)$, $\psi(\mathbf{r})$ and $\psi'(\mathbf{r})$ are the two-component wave functions of the system, and finally, E and E' are their corresponding energy eigenvalues. For example, it can be shown that the energy–momentum dispersion relation of Eq. (3.338) is

$$E = \pm v_F|\mathbf{k}| = \pm v_F\sqrt{k_x^2 + k_y^2}, \qquad (3.341)$$

---

[21] In 2010, A. Geim and K. Novoselov were awarded the Nobel Prize in physics 'for groundbreaking experiments regarding the two-dimensional material graphene'.

[22] In graphene, electrons and holes have masses equal to zero and velocities that are around 300 times slower than the velocity of light.

which means that the energy spectrum has both positive- and negative-energy solutions as well as it is linear, since it depends linearly on the wave vector (or 2-momentum) $\mathbf{k} = (k_x, k_y)$. As usual, the positive-energy solutions describe electrons, whereas the negative-energy solutions describe holes. Note that a similar energy-dispersion relation holds for Eq. (3.339).

Indeed, since electrons and holes in graphene are governed by the Weyl equations (3.338) and (3.339), it means that they have an intrinsic degree of freedom that resembles the spin degree of freedom in the ordinary Weyl equation. This degree of freedom, which is described by the Pauli matrices, is called *pseudospin* in order to distinguish it from normal spin. In connection to the pseudospin, there is a quantity, which is used to characterize the wave functions and known as the *helicity* that is defined as the projection of the 2-momentum operator along the pseudospin direction, i.e. $\hat{h} = \frac{1}{2}\boldsymbol{\sigma} \cdot \mathbf{k}/|\mathbf{k}|$. Thus, we obtain $\hat{h}\psi(\mathbf{r}) = \pm\frac{1}{2}\psi(\mathbf{r})$ and $\hat{h}\psi'(\mathbf{r}) = \mp\frac{1}{2}\psi'(\mathbf{r})$. Therefore, electrons have positive helicity, whereas holes have negative helicity. The helicity is only a good quantum number as long as Eqs. (3.338) and (3.339) are valid. Note that electrons and holes in graphene are sometimes called 'chiral', since their direction of motion is connected to the direction of the pseudospin.

3.13 Helicity and chirality

In this section, we would like to clarify the difference between helicity and chirality, since it can be difficult to distinguish them.

Definition *Helicity* is defined as the relation (scalar product) between a particle's spin $\mathbf{s}$ and direction of motion $\mathbf{p}$, i.e. $\mathbf{s} \cdot \mathbf{p}$ is the component of angular momentum along the momentum.

The helicity operator projects out **two** physical states:

(1) a physical state with the spin along the direction of motion and
(2) a physical state with the spin opposite the direction of motion.

This is true independently of the particle being massive or massless. The physical state that has spin parallel to its direction of motion has been chosen to be called *right helicity*, whereas the physical state that has spin antiparallel to its direction of motion has been chosen to be called *left helicity*. In the case of massless particles, the helicity is Lorentz invariant, since the helicity scalar product is invariant under space rotations.

Definition *Chirality* is defined as left-chiral and right-chiral objects (cf. the discussion in Section 3.12). Note that chirality can only be defined for massless particles.

For a massless fermion, we have the Dirac equation (or more specifically the Weyl equation)

$$i\gamma^\mu \partial_\mu \psi = 0, \tag{3.342}$$

which is also satisfied by $\gamma^5 \psi$, i.e.

$$i\gamma^\mu \partial_\mu (\gamma^5 \psi) = 0. \tag{3.343}$$

Using the chirality projection operators

$$P_L \equiv P_- = \frac{1}{2}(\mathbb{1}_4 - \gamma^5) \quad \text{and} \quad P_R \equiv P_+ = \frac{1}{2}(\mathbb{1}_4 + \gamma^5), \tag{3.344}$$

we have

$$\gamma^5 \psi_L = -\psi_L \quad \text{and} \quad \gamma^5 \psi_R = \psi_R, \tag{3.345}$$

where $\psi_L = P_L \psi$, $\psi_R = P_R \psi$, and $\psi = \psi_L + \psi_R$, which means that a left-handed (right-handed) chiral state ψ_L (ψ_R) is an eigenstate of γ^5 with eigenvalue -1 (1). Thus, chiral states (or fields) transform among themselves under Lorentz transformations. Chirality refers to the property that negative helicity means 'left-hand turning' (of the spin with respect to the direction of motion), whereas positive helicity means 'right-hand turning'. Actually, the helicity of the chiral states ψ_L and ψ_R are given by the following equation [cf. Eqs. (3.331) and (3.333)]

$$\sigma \cdot \mathbf{p} \left[\frac{1}{2}(\mathbb{1}_4 \pm \gamma^5)\psi \right] = \pm p^0 \frac{1}{2}(\mathbb{1}_4 \pm \gamma^5)\psi, \tag{3.346}$$

which is again the Weyl equation. In the Standard Model of particle physics, describing weak interactions of fermions (i.e. quarks and leptons), chiral states are the fundamental states of the theory. Finally, in the case of massless particles (fermions), it can be shown that chirality is equivalent to helicity.

Problems

(1) (a) Show that

$$\text{tr}(\gamma^\mu \gamma^\nu \gamma^\rho \gamma^\sigma) = 4(g^{\mu\nu}g^{\rho\sigma} + g^{\mu\sigma}g^{\nu\rho} - g^{\mu\rho}g^{\nu\sigma}).$$

(b) Compute $\gamma^\mu \gamma_\mu$.
(c) Compute $\gamma^\mu \gamma^\nu \gamma_\mu$.
(d) Compute $\not{p}\gamma^\mu \not{p}$.
(e) Compute $\gamma^\mu \gamma^5 \gamma_\mu \gamma_5$.
(f) Compute $\sigma^{\mu\nu}\sigma_{\mu\nu}$.
(g) Compute $\text{tr}(\gamma^\mu \gamma^\nu \gamma^5)$.
(h) Compute $\text{tr}(\gamma^\mu \gamma^\nu \gamma^\rho \gamma^\sigma \gamma^5)$.
(i) Compute $\text{tr}(\gamma^\mu \gamma^\nu \gamma^\rho \gamma^\sigma \gamma^\alpha \gamma^\beta \gamma^5)$.

(2) (a) Determine the Hamiltonian $H = \beta m + \boldsymbol{\alpha} \cdot \mathbf{p}$ using the Dirac equation for a free particle $(i\gamma^\mu \partial_\mu - m\mathbb{1}_4)\psi(x) = 0$.

(b) Does this Hamiltonian commute with any of the three operators $\mathbf{L} = \mathbf{x} \times \mathbf{p}$ (angular momentum), $\mathbf{S} = \boldsymbol{\Sigma}/2$ (spin), where $\boldsymbol{\Sigma} = i\boldsymbol{\gamma} \times \boldsymbol{\gamma}/2$, and $\mathbf{J} = \mathbf{L} + \mathbf{S}$ (total angular momentum)?

(c) If any of the operators in (b) commuted with the Hamiltonian, then what is the physical meaning of this result?

(3) Prove that $\bar{\psi}\sigma^{\mu\nu}\psi$ transforms as an antisymmetric tensor under Lorentz transformations.

(4) Construct explicitly the four linearly independent solutions to the Dirac equation [for $A = (A_\mu) = 0$] corresponding to an arbitrary momentum vector $\mathbf{p} = (p_1, p_2, p_3)$.

(5) *The Klein paradox.* Solve the Dirac equation for a step-function potential on the form

$$V(z) = \begin{cases} 0 & z < 0 \\ V_0 & z > 0 \end{cases},$$

where V_0 is a positive constant. Evaluate the current reflected and transmitted.

(6) Assume a Dirac electron of mass m in an attractive electrostatic potential

$$V(z) = \begin{cases} 0 & z < 0,\ z > a \\ -V_0 & 0 < z < a \end{cases},$$

where V_0 is a positive constant.

(a) Find the energy levels, i.e. compute the eigenvalue spectrum of the Dirac electron.

(b) Solve the problem of scattering of such an electron with momentum $\mathbf{p}$ off this potential.

(7) Consider an electron in a uniform and constant magnetic field $\mathbf{B}$ along the z-axis.

(a) Obtain the most general four-component positive-energy eigenfunctions.

(b) Show that the energy eigenvalues are given by

$$E_n = \sqrt{m^2 + p_3^2 + 2n|e\mathbf{B}|},$$

where $n = 0, 1, 2, \ldots$ List all constants of motion.

(8) Show that the solutions to the free Dirac equation for positive energies, which are normalized to $\bar{u}u = 1$, may be written in the form

$$u = \sqrt{\frac{E + M}{2M}} \left[\mathbb{1}_4 - \frac{1}{E + M}\gamma^0 \boldsymbol{\alpha} \cdot \mathbf{p} \right] u_0,$$

where

$$u_0 = \begin{pmatrix} u_+ \\ 0 \end{pmatrix} \quad \text{or} \quad u_0 = \begin{pmatrix} u_- \\ 0 \end{pmatrix}$$

with u_+ and u_- the usual two-component Pauli spinors corresponding to spin up or spin down.

(9) Find the solutions to the Dirac equation for a spherical 'square-well' potential of the form

$$V(r) = \begin{cases} -V_0 & r \leq R \\ 0 & r > R \end{cases},$$

where $V_0 > 0$.

(10) An electron scatters from a repulsive spherical Coulomb potential of the form

$$A^0(r) = \begin{cases} 0 & r > R \\ U & r < R \end{cases},$$

where U is a constant.

    (a) Calculate the unpolarized cross-section in the first Born approximation (lowest order in A^0). Use the Dirac formalism.

    (b) Compare your relativistic result [from (a) above] with the result you would obtain from the Schrödinger equation in the first Born approximation.

(11) Find the solutions to the Dirac equation for a mixed potential consisting of a Coulomb potential $V_1 = -\alpha/r$ and a scalar potential $V_2 = -\alpha'/r$, where α and α' are the respective coupling constants. Note that V_1 is an electrostatic potential and couples into the Dirac equation by using minimal coupling, whereas V_2 is a Newtonian potential and couples into the Dirac equation by adding it to the mass term.

(12) Find the stationary continuum states of a Dirac particle in a Coulomb potential.

(13) Obtain the representations of the Lorentz group $S(\Lambda)$ that are required to maintain the form invariance of the Weyl equations.

(14) Show that $P_L \equiv \frac{1}{2}(\mathbb{1}_4 - \gamma^5)$ and $P_R \equiv \frac{1}{2}(\mathbb{1}_4 + \gamma^5)$ are projectors.

Guide to additional recommended reading

The following books (see the indicated pages) and their authors have similar treatments of the content in the present chapter:

- H. A. Bethe and R. Jackiw, *Intermediate Quantum Mechanics*, 3rd edn., Westview Press (1997), pp. 349–389.
- A. Z. Capri, *Relativistic Quantum Mechanics and Introduction to Quantum Field Theory*, World Scientific (2002), pp. 24–52, 54–68, 68–77.
- W. Greiner, *Relativistic Quantum Mechanics – Wave Equations*, 3rd edn., Springer (2000), pp. 333–345, 351–352, 354–355, 383–387.
- F. Gross, *Relativistic Quantum Mechanics and Field Theory*, Wiley (1993), pp. 119–157, 157–159, 159–162, 163–177, 177–184.
- F. Halzen and A. D. Martin, *Quarks & Leptons: An Introductory Course in Modern Particle Physics*, Wiley (1984), pp. 100–119.
- R. H. Landau, *Quantum Mechanics II – A Second Course in Quantum Theory*, 2nd edn., Wiley-VCH (2004), pp. 219–263.
- F. Schwabl, *Advanced Quantum Mechanics*, Springer (1999), pp. 120–130, 135–151, 168–179, 239–243.
- S. S. Schweber, *An Introduction to Relativistic Quantum Field Theory*, Dover (2005), pp. 65–117.
- G. W. Series (Ed.), *The Spectrum of Atomic Hydrogen – Advances*, World Scientific (1988), pp. 1–502.
- H. Snellman, *Elementary Particle Physics*, KTH (2004), pp. 26–29.
- F. J. Ynduráin, *Relativistic Quantum Mechanics and Introduction to Field Theory*, Springer (1996), pp. 35–62, 62–64, 64–66, 67–68, 69–88, 104–106.
- For the interested reader: V. Bargmann and E. P. Wigner, Group theoretical discussion of relativistic wave equations, *Proc. Nat. Acad. Sci. (USA)* **34**, 211–223 (1948); P. A. M. Dirac, The quantum theory of the electron, *Proc. R. Soc. (London)* A **117**,

610–624 (1928); The quantum theory of the electron. *Part II, ibid.* **118**, 351–361 (1928); A theory of electrons and protons, *ibid.* **126**, 360–365 (1930); R. H. Good, Jr., Properties of the Dirac matrices, *Rev. Mod. Phys.* **27**, 187–211 (1955); S. Kusaka, β-Decay with neutrino of spin $\frac{3}{2}$, *Phys. Rev.* **60**, 61–62 (1941); and W. Rarita and J. Schwinger, On a theory of particles with half-integral spin, *ibid.* **60**, 61 (1941).

4

Quantization of the non-relativistic string

In this chapter, we will consider the first example of a quantum field theory. The theory that will be investigated is a non-relativistic example. For this purpose, we have chosen the non-relativistic string. However, in order to extend ordinary non-relativistic quantum mechanics to describe relativistic systems as well as creation and annihilation of particles, we need field theory, which we will study later. Nevertheless, quantum mechanics is the particle mechanics, which describes the dynamics of microscopic particles such as molecules, atoms, or subatomic particles. In the transformation from quantum mechanics to quantum field theory, we perform the limit $N \to \infty$, where N is the number of degrees of freedom of the system.

4.1 Equation of motion for the non-relativistic string

In the example of a one-dimensional non-relativistic string, we discretize the string into N oscillators and let $\bar{\phi}_i = \bar{\phi}_i(t)$ be the displacement of the ith oscillator from equilibrium, see Fig. 4.1. In addition, we impose periodic boundary conditions such that $\bar{\phi}_0 = \bar{\phi}_N$ (and, therefore, it follows automatically that $d\bar{\phi}_0/dt = d\bar{\phi}_N/dt$), which naturally follow from the fact that the string is closed. Thus, we find the kinetic and potential energies of the discretized version of the string

$$T = \frac{1}{2} m \sum_{i=0}^{N-1} \left(\frac{d\bar{\phi}_i}{dt} \right)^2, \tag{4.1}$$

$$V = \frac{1}{2} k \sum_{i=0}^{N-1} \left(\bar{\phi}_{i+1} - \bar{\phi}_i \right)^2, \tag{4.2}$$

where m is the mass of one of the N points (or oscillators) on the (massless) string and k is the spring constant between two consecutive points. In addition, the equilibrium distance between two points is given by a. Hence, we have N identical

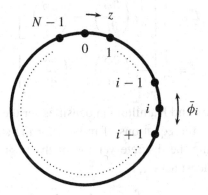

Figure 4.1 The one-dimensional non-relativistic string.

point particles, each of mass m, equispaced with the ith particle at a distance $z_i \equiv ia$ along the string of length ℓ, i.e. a system of N degrees of freedom.

In the continuum limit $N \to \infty$ (or $a \to 0$) such that $\ell = Na$, $\mu \equiv m/a$, and $\tau \equiv ka$ are fixed, we have that the spatial coordinate z_i is replaced by a continuous space variable z and the displacement is replaced as follows

$$\bar{\phi}_i(t) = \bar{\phi}(t, z_i) \to \bar{\phi}(t, z). \tag{4.3}$$

Thus, the kinetic and potential energies are replaced in the following ways

$$T = \frac{1}{2}m \sum_{i=0}^{N-1} \left(\frac{d\bar{\phi}_i}{dt}\right)^2 \to \frac{1}{2}\mu \int_0^\ell \left[\frac{\partial \bar{\phi}(t, z)}{\partial t}\right]^2 dz, \tag{4.4}$$

$$V = \frac{1}{2}k \sum_{i=0}^{N-1} \left(\bar{\phi}_{i+1} - \bar{\phi}_i\right)^2 \to \frac{1}{2}\tau \int_0^\ell \left[\frac{\partial \bar{\phi}(t, z)}{\partial z}\right]^2 dz, \tag{4.5}$$

which imply that the Hamiltonian and Lagrangian functions can be written as

$$L = T - V = \int_0^\ell \left[\frac{1}{2}\mu \left(\frac{\partial \bar{\phi}}{\partial t}\right)^2 - \frac{1}{2}\tau \left(\frac{\partial \bar{\phi}}{\partial z}\right)^2\right] dz = \int_0^\ell \mathcal{L}\, dz, \tag{4.6}$$

$$H = T + V = \int_0^\ell \left[\frac{1}{2}\mu \left(\frac{\partial \bar{\phi}}{\partial t}\right)^2 + \frac{1}{2}\tau \left(\frac{\partial \bar{\phi}}{\partial z}\right)^2\right] dz = \int_0^\ell \mathcal{H}\, dz. \tag{4.7}$$

Redefining the displacement according to $\bar{\phi} \to \sqrt{\tau}\bar{\phi} \equiv \phi$ and introducing the quantity $v^2 \equiv \tau/\mu$, we find that

$$\mathcal{L} = \frac{1}{2} \left[\frac{1}{v^2} \left(\frac{\partial \phi}{\partial t} \right)^2 - \left(\frac{\partial \phi}{\partial z} \right)^2 \right],$$ (4.8)

$$\mathcal{H} = \frac{1}{2} \left[\frac{1}{v^2} \left(\frac{\partial \phi}{\partial t} \right)^2 + \left(\frac{\partial \phi}{\partial z} \right)^2 \right],$$ (4.9)

which are the Lagrangian and Hamiltonian densities for the string, respectively.

Now, we want to find the equations of motion for the string. This will be performed by again utilizing the discrete version of the Lagrangian function. Thus, we use the principle of least action, i.e.

$$\delta A = \delta \int L(\bar{\phi}_i, \dot{\bar{\phi}}_i, t) \, \mathrm{d}t = 0,$$ (4.10)

which implies the *Euler–Lagrange equations*

$$\frac{\mathrm{d}}{\mathrm{d}t} \frac{\partial L}{\partial \dot{\bar{\phi}}_i} - \frac{\partial L}{\partial \bar{\phi}_i} = 0, \quad i = 0, 1, \ldots, N - 1.$$ (4.11)

In the continuum limit ($N \to \infty$), we obtain

$$\frac{\partial L}{\partial \bar{\phi}_i} = -k \left[-(\bar{\phi}_{i+1} - \bar{\phi}_i) + (\bar{\phi}_i - \bar{\phi}_{i-1}) \right] \to -a\sqrt{\tau} \frac{\partial}{\partial z} \frac{\partial \mathcal{L}}{\partial \left(\frac{\partial \phi}{\partial z} \right)},$$ (4.12)

$$\frac{\mathrm{d}}{\mathrm{d}t} \frac{\partial L}{\partial \dot{\bar{\phi}}_i} = m\ddot{\bar{\phi}}_i = \mu a \frac{\partial^2 \bar{\phi}_i}{\partial t^2} \to a\sqrt{\tau} \frac{\partial}{\partial t} \frac{\partial \mathcal{L}}{\partial \left(\frac{\partial \phi}{\partial t} \right)},$$ (4.13)

which means that

$$\frac{\partial}{\partial t} \frac{\partial \mathcal{L}}{\partial \left(\frac{\partial \phi}{\partial t} \right)} + \frac{\partial}{\partial z} \frac{\partial \mathcal{L}}{\partial \left(\frac{\partial \phi}{\partial z} \right)} = 0,$$ (4.14)

using Euler–Lagrange equations. Finally, inserting the Lagrangian density (4.8) into Eq. (4.14), we find that

$$\frac{1}{v^2} \frac{\partial^2 \phi}{\partial t^2} - \frac{\partial^2 \phi}{\partial z^2} = 0,$$ (4.15)

which is the wave equation that describes the displacement (or the amplitude of vibration) of the string. Here $v = \sqrt{\tau/\mu}$ is to be interpreted as the wave speed, and τ is the tension in the string. Note that this is a canonical example of an N-particle system transforming into a continuous field in the continuum limit ($N \to \infty$). In conclusion, Eq. (4.11) gives the discrete equations of motion for the string, whereas Eq. (4.15) is the continuous equation of motion for the string.

Comparing Eq. (4.15) and the Klein–Gordon equation (2.10) for a massless particle, we observe that both equations are different wave equations. In the first case, Eq. (4.15) is a 1+1-dimensional non-relativistic wave equation with speed v, whereas in the second case, the Klein–Gordon equation for a massless particle is a 1+3-dimensional relativistic wave equation with speed $c = 1$. However, replacing v by c in the first equation or vice versa in the second equation and assuming the Klein–Gordon equation to be 1+1-dimensional instead of 1+3-dimensional, the two physical systems, i.e. the Klein–Gordon wave function and the non-relativistic string, have the same equation of motion, although the two systems have completely different origins. Nonetheless, this similarity is something that we are going to use to our advantage throughout the rest of this book, especially for the quantization of the Klein–Gordon field and the electromagnetic field.

Exercise 4.1 Show that Eqs. (4.12) and (4.13) are obtained in the continuum limit.

4.2 Solutions to the wave equation: normal modes

Solutions to the wave equation with periodic boundary conditions are given by

$$\phi \sim e^{\pm i(k_n z - \omega_n t)}, \tag{4.16}$$

where the periodic boundary conditions imply that

$$k_n = \frac{2\pi n}{\ell}, \quad n = 0, \pm 1, \pm 2, \ldots, \tag{4.17}$$

which means that

$$\omega_n^2 = v^2 k_n^2. \tag{4.18}$$

Thus, assuming that $\omega_n > 0$, we have positive-frequency states

$$\phi_n(t, z) = \frac{1}{\sqrt{\ell}} e^{i(k_n z - \omega_n t)} \tag{4.19}$$

and negative-frequency states

$$\phi_n^*(t, z) = \frac{1}{\sqrt{\ell}} e^{i(-k_n z + \omega_n t)}, \tag{4.20}$$

since k_n can be **both** positive and negative. In addition, the normalization condition is

$$\int_0^\ell \phi_n^*(t, z) \phi_m(t, z) \, dz = \delta_{nm}. \tag{4.21}$$

Thus, the general solution to the wave equation, i.e. the field, is given by an expansion in the normal modes

$$\phi(t, z) = \sum_{n=-\infty}^{\infty} c_n \left[a_n(0)\phi_n(t, z) + a_n^*(0)\phi_n^*(t, z) \right]$$

$$= \sum_{n=-\infty}^{\infty} \frac{c_n}{\sqrt{\ell}} \left[a_n(t)e^{ik_n z} + a_n^*(t)e^{-ik_n z} \right], \tag{4.22}$$

where $a_n(t) = a_n(0)e^{-i\omega_n t}$, which fulfils the equation for a simple harmonic oscillator

$$\ddot{a}_n(t) + \omega_n^2 a_n(t) = 0. \tag{4.23}$$

In order to quantize the field $\phi(t, z)$, it is **only** necessary to quantize the simple harmonic oscillators $a_n(t)$.

Using $\dot{a}_n(t) = -i\omega_n a_n(t)$, we obtain the Hamiltonian

$$H = \sum_{n=-\infty}^{\infty} c_n^2 \frac{2\omega_n^2}{v^2} a_n^*(t)a_n(t), \tag{4.24}$$

which should have the dimension of energy. We want to choose the coefficients c_n to make the simple harmonic oscillators a_n dimensionless. If we choose

$$c_n = \sqrt{\frac{v^2}{2\omega_n}}, \tag{4.25}$$

then we obtain

$$H = \sum_{n=-\infty}^{\infty} \omega_n a_n^*(t)a_n(t), \tag{4.26}$$

which has the dimension of energy, since the ω_ns have the dimension of energy (in natural units) and the a_ns are dimensionless. In addition, note that the Hamiltonian H is time-independent.

Exercise 4.2   Show that $a_n(t) = a_n(0)e^{-i\omega_n t}$ fulfils Eq. (4.23) and that the Hamiltonian can be written as in Eq. (4.24).

4.3 Generalized positions and momenta

In classical mechanics, the energy is always given by the Hamiltonian function

$$H = \sum_n p_n \dot{q}_n - L, \tag{4.27}$$

where q_n and p_n are the generalized coordinates and their corresponding generalized momenta, respectively, and L is the Lagrangian function. Note that Eq. (4.27) is called a *Legendre transformation* and is used to transfer between the Hamiltonian and Lagrangian formulations of classical mechanics. In field theory, the energy will also be given by the Hamiltonian and a similar Legendre transformation will also exist for the Hamiltonian and Lagrangian densities.

If we choose generalized positions and momenta as

$$q_n(t) = \frac{1}{\sqrt{2\omega_n}} \left[a_n(t) + a_n^*(t) \right], \tag{4.28}$$

$$p_n(t) = \frac{dq_n}{dt} = -\frac{i\omega_n}{\sqrt{2\omega_n}} \left[a_n(t) - a_n^*(t) \right] = -i\sqrt{\frac{\omega_n}{2}} \left[a_n(t) - a_n^*(t) \right], \tag{4.29}$$

then we can, of course, express the a_ns and a_n^*s in terms of the q_ns and p_ns as

$$a_n = \frac{ip_n + \omega_n q_n}{\sqrt{2\omega_n}}, \tag{4.30}$$

$$a_n^* = \frac{-ip_n + \omega_n q_n}{\sqrt{2\omega_n}}, \tag{4.31}$$

which mean that the Hamiltonian is given by

$$H = \sum_{n=-\infty}^{\infty} \frac{1}{2} \left(p_n^2 + \omega_n^2 q_n^2 \right). \tag{4.32}$$

In addition, *Hamilton's equations of motion* give

$$\dot{q}_n = \frac{\partial H}{\partial p_n} = p_n, \tag{4.33}$$

$$\dot{p}_n = -\frac{\partial H}{\partial q_n} = -\omega_n^2 q_n. \tag{4.34}$$

4.4 Quantization

In this book, we will use *canonical quantization*[1] in order to quantize classical fields into operator fields acting on quantum states of the field theory. Historically, canonical quantization was the first method developed to quantize field theories and it is convenient to use this method for rather simple theories such as the non-relativistic string, the Klein–Gordon field (see Chapter 6), and the Dirac field (see Chapter 7), since the method is relatively transparent. However, in other situations, other methods of quantization are much more powerful than canonical quantization, such as *path integral quantization* [see the book by M. E. Peskin

[1] In the concept of canonical quantization, the word 'canonical' refers to the strong connection between classical field theory and classical mechanics that is preserved in this case of quantization.

and D. V. Schroeder for the method of path integral (or functional) quantization]. Other methods of quantization include mathematical quantization such as *deformation quantization* (Weyl quantization, Moyal bracket, star product), *geometric quantization*, and *loop quantization*. Note that we will also use a covariant version of the canonical quantization method to quantize the electromagnetic field (see Chapter 8), but in this case, the procedure might not be as simple as one could naively expect.

Thus, we use the canonical quantization procedure to quantize the generalized positions and momenta. In this procedure, we have the canonical commutation relations

$$[q_n, p_m] = i\delta_{nm}, \tag{4.35}$$

$$[q_n, q_m] = [p_n, p_m] = 0, \tag{4.36}$$

which lead to the commutation relations for a_n and $a_n^\dagger$

$$[a_n, a_m^\dagger] = \delta_{nm}, \tag{4.37}$$

$$[a_n, a_m] = [a_n^\dagger, a_m^\dagger] = 0, \tag{4.38}$$

which are harmonic oscillator commutation relations. Hence, in the quantization procedure, the classical quantities a_n and a_n^* are replaced by the quantum operators a_n and $a_n^\dagger$. However, note that there is some abuse of notation, since we use the same symbol for the classical quantity a_n and the quantum operator a_n. We hope that this usage will not cause any confusion. Importantly, a_n and a_n^* do not fulfil the harmonic oscillator commutation relations, since they are just complex numbers, but a_n and $a_n^\dagger$ do. Now, imposing the commutation relation for the operators a_n and $a_n^\dagger$, we have the quantum field[2]

$$\phi(t, z) = v \sum_{n=-\infty}^{\infty} \frac{1}{\sqrt{2\omega_n \ell}} \left[a_n e^{i(k_n z - \omega_n t)} + a_n^\dagger e^{-i(k_n z - \omega_n t)} \right] = \phi^{(+)}(t, z) + \phi^{(-)}(t, z), \tag{4.39}$$

where $\phi^{(+)}(t, z)$ contains the so-called *annihilation operators* a_n and $\phi^{(-)}(t, z)$ contains the so-called *creation operators* $a_n^\dagger$. The interpretation of the annihilation and creation operators will be discussed below. The Hamiltonian is given by

$$H = \sum_{n=-\infty}^{\infty} \frac{1}{2}\omega_n \left(a_n^\dagger a_n + a_n a_n^\dagger \right) = \sum_{n=-\infty}^{\infty} \omega_n \left(a_n^\dagger a_n + \frac{1}{2} \right), \tag{4.40}$$

where the term $1/2$ in the last expression leads to an infinite energy that actually can be removed. Redefining the zero level of the energy, we have the Hamiltonian

---

[2] Note that the transformation from many single-particles to a continuous field by replacing a wave function for N particles with a continuous quantum field is often called *second quantization*.

$$H = \sum_{n=-\infty}^{\infty} \omega_n a_n^\dagger a_n. \tag{4.41}$$

Now, introducing the *conjugate* (or *canonical*) *momentum* conjugate to the field ϕ

$$\pi(t, z) \equiv \frac{\partial \mathcal{L}}{\partial \left(\frac{\partial \phi}{\partial t}\right)} = \frac{1}{v^2} \frac{\partial \phi}{\partial t}, \tag{4.42}$$

we can generalize the commutation relations (4.35) and (4.36) [or Eqs. (4.37) and (4.38)] to the *canonical commutation relations* for the field ϕ and its conjugate momentum π as

$$[\phi(t, z), \pi(t, z')] = i\delta(z - z'), \tag{4.43}$$

$$[\phi(t, z), \phi(t, z')] = [\pi(t, z), \pi(t, z')] = 0, \tag{4.44}$$

where one should note the equal time t. Thus, in principle, the generalized positions and momenta are replaced by the field and its conjugate momentum, respectively, or in other words, the commutation relations between the a_ns and the $a_n^\dagger$s imply commutation relations between the fields.

Next, we discuss the *number operator* for the nth state

$$N_n = a_n^\dagger a_n, \tag{4.45}$$

where N_n is Hermitian, which means that it has a complete set of orthonormal eigenstates $|m_n\rangle$, where m_n is a non-negative integer. The following relations hold

$$N_n|m_n\rangle = m_n|m_n\rangle, \quad \langle m_{n'}|m_n\rangle = \delta_{m_{n'}m_n}, \quad \mathbb{1} = \sum_{m_n} |m_n\rangle\langle m_n| \quad \text{(completeness)}. \tag{4.46}$$

In addition, we have

$$a_n|m_n\rangle = \sqrt{m_n}|m_n - 1\rangle, \tag{4.47}$$

$$a_n^\dagger|m_n\rangle = \sqrt{m_n + 1}|m_n + 1\rangle. \tag{4.48}$$

The interpretation of the operators a_n and $a_n^\dagger$ is the following:

- a_n annihilates (destroys) a phonon with frequency ω_n,
- $a_n^\dagger$ creates a phonon with frequency ω_n.

4.5 Quanta as particles

The quanta associated with the quantum fields of the non-relativistic string (i.e. phonons) are physical particles with both energy and momentum. The kinetic energy (per unit length) of the non-relativistic string is given by

$$T = \frac{1}{2}\rho\,(z - vt)\,v^2, \tag{4.49}$$

where $\rho = \rho(z - vt)$ is the local mass density of the string. The virial theorem says that an equal amount of energy comes from the potential energy, which means that the total energy (per unit length) of the non-relativistic string is given by

$$E = \rho\,(z - vt)\,v^2. \tag{4.50}$$

However, the momentum density of the non-relativistic string is $\mathcal{P}^z = \rho(z - vt)v$, which is related to the total energy by

$$\frac{1}{v^2}\frac{\partial E}{\partial t} = -v\rho' = -\frac{\partial}{\partial z}\mathcal{P}^z. \tag{4.51}$$

Now, using the Hamiltonian density (4.9) for the total energy E as well as the wave equation (4.15), we have

$$\begin{aligned}
\frac{1}{v^2}\frac{\partial E}{\partial t} &= \frac{1}{v^2}\left[\frac{1}{v^2}\frac{\partial\phi}{\partial t}\frac{\partial^2\phi}{\partial t^2} + \frac{\partial\phi}{\partial z}\frac{\partial^2\phi}{\partial t\partial z}\right] \\
&= \frac{1}{v^2}\left(\frac{\partial\phi}{\partial t}\frac{\partial^2\phi}{\partial z^2} + \frac{\partial\phi}{\partial z}\frac{\partial^2\phi}{\partial t\partial z}\right) = \frac{1}{v^2}\frac{\partial}{\partial z}\left(\frac{\partial\phi}{\partial t}\frac{\partial\phi}{\partial z}\right),
\end{aligned} \tag{4.52}$$

which implies that

$$\mathcal{P}_z = -\frac{1}{v^2}\frac{\partial\phi}{\partial t}\frac{\partial\phi}{\partial z}. \tag{4.53}$$

However, the order between the two derivatives of the field is important, and therefore, we have to introduce the concept of a *normal-ordered product* for bosons, which is given by

$$\begin{aligned}
:\phi_1\phi_2: &= \phi_1\phi_2 - \langle 0|\phi_1\phi_2|0\rangle \\
&= \phi_1^{(+)}\phi_2^{(+)} + \phi_1^{(-)}\phi_2^{(+)} + \phi_2^{(-)}\phi_1^{(+)} + \phi_1^{(-)}\phi_2^{(-)} + \left[\phi_1^{(+)}, \phi_2^{(-)}\right] - \langle 0|\phi_1\phi_2|0\rangle \\
&= \phi_1^{(+)}\phi_2^{(+)} + \phi_1^{(-)}\phi_2^{(+)} + \phi_2^{(-)}\phi_1^{(+)} + \phi_1^{(-)}\phi_2^{(-)},
\end{aligned} \tag{4.54}$$

where the commutator and vacuum expectation value cancel each other exactly. In principle, Eq. (4.40) was our first example of *normal-ordering*. Note that taking the vacuum expectation value of the definition (4.54), it follows immediately that

$$\langle 0| :\phi_1\phi_2: |0\rangle = \langle 0|\phi_1\phi_2|0\rangle - \langle 0|\phi_1\phi_2|0\rangle = 0, \tag{4.55}$$

since the vacuum expectation value of a vacuum expectation value is just the vacuum expectation value itself. In general, this implies that a normal-ordered product (or even more generally a normal-ordered operator) has a vanishing vacuum expectation value. Thus, we define the momentum of the non-relativistic string as

$$P_z = \int_0^\ell \, : P_z : \, dz = -\int_0^\ell \frac{1}{v^2} : \frac{\partial \phi}{\partial t} \frac{\partial \phi}{\partial z} : \, dz. \qquad (4.56)$$

Inserting the expansion of the field (4.39) into Eq. (4.56), we obtain after some computations

$$P_z = -\frac{1}{2} \sum_{n=-\infty}^{\infty} \left(k_{-n} a_n a_{-n} e^{-2i\omega_n t} + k_{-n} a_n^\dagger a_{-n}^\dagger e^{2i\omega_n t} - 2k_n a_n^\dagger a_n \right), \qquad (4.57)$$

but $k_n = -k_{-n}$ and $\omega_n = \omega_{-n}$, which mean that

$$P_z = \sum_{n=-\infty}^{\infty} k_n a_n^\dagger a_n, \qquad (4.58)$$

since the first two terms in the parenthesis in Eq. (4.57) are zero when summed over n. Thus, phonons are particles with energies ω_n and momenta k_n.

Exercise 4.3 Show that the momentum of the non-relativistic string can be written as in Eq. (4.57).

Problem

(1) Consider the Lagrangian density

$$\mathcal{L} = \frac{1}{2} \left[\left(\frac{\partial \phi}{\partial t} \right)^2 - \left(\frac{\partial \phi}{\partial z} \right)^2 - m^2 \phi^2 \right],$$

where $\phi = \phi(t, z)$ is a generalized coordinate.
(a) Find the conjugate momentum π to the field ϕ.
(b) Find the equations of motion for the fields and their solutions. Use periodic boundary conditions.
(c) Suppose the field is expanded in normal modes

$$\phi(t, z) = \sum_{n=-\infty}^{\infty} c_n \left[a_n \phi_n(t, z) + a_n^\dagger \phi_n^*(t, z) \right],$$

where the operators a_n satisfy the commutation relations

$$[a_n, a_{n'}] = \left[a_n^\dagger, a_{n'}^\dagger \right] = 0, \quad \left[a_n, a_{n'}^\dagger \right] = \delta_{nn'}.$$

Find the coefficients c_n, which will ensure that the canonical commutation relations assume the standard form

$$[\phi(t, z), \pi(t, z')] = i\delta(z - z').$$

(d) What is the physical interpretation of the field ϕ?

Guide to additional recommended reading

The following books (see the indicated pages) and their authors have similar treatments of the content in the present chapter:

- F. Gross, *Relativistic Quantum Mechanics and Field Theory*, Wiley (1993), pp. 3–25.
- W. D. McComb, *Renormalization Methods – A Guide for Beginners*, Oxford (2007), pp. 33–35.

5

Introduction to relativistic quantum field theory: propagators, interactions, and all that

We start this chapter with a brief history of quantum field theory. In the 1920s, Dirac attempted to quantize the electromagnetic field (see Chapter 8). This attempt was the beginning of the history of quantum field theory. Then, in 1926, M. Born,[1] W. Heisenberg,[2] and P. Jordan invented canonical quantization, which we will use in Chapters 6 and 7 in order to quantize the Klein–Gordon field and the Dirac field, respectively, that will bring us from relativistic quantum mechanics to relativistic quantum field theory. Next, in 1927, Dirac created and presented the first reasonably complete theory of quantum electrodynamics (QED), and in the following year, 1928, he formulated the Dirac equation (cf. Chapter 3). In addition, in 1928, E. Wigner[3] found that the quantum field describing electrons, or other (spin-1/2) fermions, had to be expanded using anticommuting creation and annihilation operators due to the *Pauli exclusion principle*. After World War II, in the 1940s, H. Bethe,[4] F. Dyson, R. Feynman, J. Schwinger, and S.-I. Tomonaga solved the so-called 'divergence problem' through renormalization (see Chapter 13).[5] This was the start of the modern theory of QED. In the 1950s, C.-N. Yang and R. Mills generalized QED to gauge theories – known as Yang–Mills theories – which we will discuss classically (i.e. without quantization) in Chapter 9.

In particular, in this chapter, we will present the propagators and Lagrangians for the Klein–Gordon field, the Dirac field, and the electromagnetic field. We will

---

[1] In 1954, Born was awarded half of the Nobel Prize in physics 'for his fundamental research in quantum mechanics, especially for his statistical interpretation of the wavefunction'.

[2] In 1932, Heisenberg was awarded the Nobel Prize in physics 'for the creation of quantum mechanics, the application of which has, inter alia, led to the discovery of the allotropic forms of hydrogen'.

[3] In 1963, Wigner was awarded the Nobel Prize in physics 'for his contributions to the theory of the atomic nucleus and the elementary particles, particularly through the discovery and application of fundamental symmetry principles'.

[4] In 1967, Bethe was awarded the Nobel Prize in physics 'for his contributions to the theory of nuclear reactions, especially his discoveries concerning the energy production in stars'.

[5] In 1965, Feynman, Schwinger, and Tomonaga were awarded the Nobel Prize in physics 'for their fundamental work in quantum electrodynamics, with deep-ploughing consequences for the physics of elementary particles'.

also present the Lagrangian of QED. In addition, we will give a formal discussion on the treatment of the S operator including symmetries and conservation laws.

5.1 Propagators

Quantum field theory (QFT) is the quantum theory of fields. In general, QFT has applications in

- particle physics (relativistic QFT)
- condensed matter physics (non-relativistic QFT; relativistic QFT normally leads to effective field theories)
- statistical physics (many-particle QFT)
- mathematical physics (axiomatic development of QFT).

In QFT, operators are generated by operator-valued fields, i.e. operators that are parametrized by spacetime points. This means that operators can be localized. The interpretation of field operators is creation/annihilation of a particle at a given spacetime point.

We will consider amplitudes as $\langle 0|\hat{F}(x)|\psi\rangle$, where $\hat{F}$ is a generic field operator and $|\psi\rangle$ is a generic state. For example, we have

$$\langle 0|\phi(x)|\Phi\rangle = \Phi(x), \tag{5.1}$$

$$\langle 0|\psi(x)|\Psi\rangle = \Psi(x), \tag{5.2}$$

$$\langle 0|A^{\mu}(x)|\Psi\rangle = \Psi^{\mu}(x), \tag{5.3}$$

where $\Phi(x)$ is a Klein–Gordon wave function, $\Psi(x)$ is a Dirac wave function, and finally, $\Psi^{\mu}(x)$ is a wave function for the electromagnetic field (or more generally a Yang–Mills field).

A *propagator* represents the amplitude (of probability) for a particle to propagate from one spacetime point to another. Furthermore, the time-ordered product T for field operators is given by

$$T\phi_1(x_1)\phi_2(x_2) = \theta(t_1 - t_2)\phi_1(x_1)\phi_2(x_2) + \theta(t_2 - t_1)\phi_2(x_2)\phi_1(x_1) \tag{5.4}$$

for bosonic fields ϕ_1 and ϕ_2, whereas it is given by

$$T\psi_1(x_1)\psi_2(x_2) = \theta(t_1 - t_2)\psi_1(x_1)\psi_2(x_2) - \theta(t_2 - t_1)\psi_2(x_2)\psi_1(x_1) \tag{5.5}$$

for fermionic fields ψ_1 and ψ_2, where $\theta(t)$ is the Heaviside step-function. Thus, for a (charged) Klein–Gordon (spin-0) particle, we have the *Feynman propagator*

$$y \bullet\!-\!-\!-\!\blacktriangleright\!-\!-\!-\!\bullet\, x \quad i\Delta_F(x - y) \equiv \langle 0|T\phi(x)\phi^\dagger(y)|0\rangle, \qquad (5.6)$$

where a particle is created at $\mathbf{y}$ at time $t = y^0$, propagated from $\mathbf{y}$ to $\mathbf{x}$ in the time $t = x^0 - y^0$, and annihilated at $\mathbf{x}$ at time $t = x^0$, or equivalently, an antiparticle is created at $\mathbf{x}$ at time $t = x^0$, propagated from $\mathbf{x}$ to $\mathbf{y}$, and annihilated at $\mathbf{y}$ at time $t = y^0$. Therefore, $y = (y^0, \mathbf{y})$ and $x = (x^0, \mathbf{x})$ are creation and annihilation spacetime points of the particle, respectively, and vice versa for the antiparticle. The difference $x - y$ is the propagation spacetime interval for both the particle and the antiparticle. Note that the imaginary unit i in the definition of the propagator (5.6) is a convention. In addition, in the case of a neutral Klein–Gordon field, the dagger in Eq. (5.6) can be removed, since the field operator is self-adjoint (or Hermitian) in this case. In Section 6.3, we will present in great detail the derivation of the Klein–Gordon Feynman propagator.

The propagator acts as a *Green's function* for the equation of motion. In the case of a free Klein–Gordon particle the equation of motion is

$$\left(\Box + m^2\right)\Delta_F(x - y) = -\delta(x - y), \qquad (5.7)$$

which is an inhomogeneous Klein–Gordon equation. In Fourier representation, one can express the propagator as

$$-\!-\!-\!-\!\blacktriangleright\!-\!-\!-\!-\!-\!_{k} \quad i\Delta_F(x - y) = \int \frac{\mathrm{d}^4k}{(2\pi)^4} e^{-ik\cdot(x-y)} \underbrace{\frac{i}{k^2 - m^2 + i0}}_{i\Delta_F(k)}, \qquad (5.8)$$

where i0 means $\lim_{\epsilon\to 0} i\epsilon$ and $\Delta_F(k)$ is the momentum space propagator for $\Delta_F(x - y)$, i.e. the Fourier transform of the Feynman propagator. Here, the quantity k is the 4-momentum that corresponds to the propagation spacetime interval $x - y$, which connects the spacetime points x and y. Without loss of generality, we can set $y = 0$, and thus, x is the propagation spacetime interval. Note that since $\Delta_F(x) = \Delta_F(-x)$ and $\Delta_F(k) = \Delta_F(-k)$, i.e. $\Delta_F(x)$ and $\Delta_F(k)$ are even functions with respect to their corresponding arguments, the direction of k in Eq. (5.8) is arbitrary, and therefore, the direction of the 4-momentum is not important.

Similarly, for a Dirac (spin-1/2) particle, we have the Feynman propagator

$$y, \beta \bullet\!\longrightarrow\!\bullet\, x, \alpha \quad iS_{F\alpha\beta}(x - y) \equiv \langle 0|T\psi_\alpha(x)\bar\psi_\beta(y)|0\rangle, \qquad (5.9)$$

where α and β are Dirac spinor indices. Again, in Fourier representation, one can express the propagator as

$$\beta \xrightarrow[\quad k \quad]{} \alpha \quad iS_{F\alpha\beta}(x-y) = \int \frac{d^4k}{(2\pi)^4} e^{-ik\cdot(x-y)} \left(\frac{i}{\not{k} - m\mathbb{1}_4 + i0} \right)_{\alpha\beta}$$

$$= \int \frac{d^4k}{(2\pi)^4} e^{-ik\cdot x} \underbrace{\frac{i(\not{k} + m\mathbb{1}_4)_{\alpha\beta}}{k^2 - m^2 + i0}}_{iS_{F\alpha\beta}(k)} . \tag{5.10}$$

Note that in this case, since e.g. $S_{F\alpha\beta}(k) \neq S_{F\alpha\beta}(-k)$, the direction of k in Eq. (5.10) is not arbitrary, and therefore, the direction of the 4-momentum is important. In Section 7.7, we will investigate the Green's functions of the Dirac field and derive the Dirac Feynman propagator. In addition, we will find that the Green's functions and propagators for the Dirac and Klein–Gordon fields are highly related to each other. However, it is important to note that the Klein–Gordon propagator is proportional to $1/k^2$ for $k^2 \gg m^2$, whereas the Dirac propagator is proportional to $1/k$ for $k^2 \gg m^2$.

Finally, for a photon, we have the Feynman propagator[6]

$$y, \nu \sim\!\!\sim\!\!\sim\!\!\bullet \, x, \mu \quad iD_{F\mu\nu}(x-y) \equiv \langle 0 | T A_\mu(x) A_\nu(y) | 0 \rangle, \tag{5.11}$$

where μ and ν are Lorentz indices that should not be confused with the Dirac spinor indices in the Dirac Feynman propagator (5.9).[7] In Fourier representation, one has the propagator in the so-called *Lorenz gauge* (see the discussion on different gauges in Section 8.2)

$$\nu \sim\!\!\sim\!\!\sim \mu \quad iD_{F\mu\nu}(x-y) = \int \frac{d^4k}{(2\pi)^4} e^{-ik\cdot(x-y)} \underbrace{\frac{-ig_{\mu\nu}}{k^2 + i0}}_{iD_{F\mu\nu}(k)}, \tag{5.12}$$

where $D_{F\mu\nu}(k)$ is proportional to $1/k^2$ (as the Klein–Gordon propagator) and an even function with respect to k, i.e. $D_{F\mu\nu}(k) = D_{F\mu\nu}(-k)$. In other gauges, we have to make the replacement

$$-g_{\mu\nu} \mapsto -\left[g_{\mu\nu} + k_\mu f_\nu(k) + k_\nu f_\mu(k) \right], \tag{5.13}$$

where the function f depends on the chosen gauge. In addition, note that, comparing Eqs. (5.8) and (5.12), it holds that

---

[6] Note that, in the case of the Feynman propagator for the photon, the two field operators in this propagator appear without daggers or bars, which is different from the cases for a Klein–Gordon particle or a Dirac particle, cf. Eqs. (5.6) and (5.9). The reason is that the operator for the electromagnetic field is self-adjoint (or Hermitian).

[7] As will be discussed in Chapter 8, one has to fix a gauge when quantizing the electromagnetic field due to the redundancy in the 4-vector potential A_μ, although physical results have to be independent of the choice of gauge.

$$D_{F\mu\nu}(x - y) = -g_{\mu\nu} \lim_{m \to 0} \Delta_F(x - y), \qquad (5.14)$$

which means that the propagators for the Klein–Gordon and photon (or electro-magnetic) fields are also connected to each other.

In Section 11.7, we will continue to discuss the three propagators for Klein–Gordon and Dirac particles as well as photons in connection with four different models for the so-called *interaction theory*, in which propagators are used to represent contributions of virtual particles on the internal lines of so-called *Feynman diagrams*.

5.2 Lagrangians

In comparison to classical mechanics, where the equations of motion are given by the Euler–Lagrange equations through the principle of least action for the time integral of the Lagrangian function and the energy is always given by the Hamiltonian function, field theory is based on similar ideas, replacing the Lagrangian and Hamiltonian functions by the Lagrangian and Hamiltonian densities, respectively. Thus, let $\mathcal{L}$ denote the Lagrangian density (including interactions). Usually, the Lagrangian density is simply referred to as the Lagrangian. In addition, consider fields $\varphi_a(x)$, where a is an index that labels the different fields. Normally, we can use $a = 1, 2, \ldots, n$, where n is the number of fields. In covariant notation for relativistic fields, we then have the Lagrangian

$$\mathcal{L} = \mathcal{L}(\varphi_a(x), \partial_\mu \varphi_a(x)), \qquad (5.15)$$

which is obviously a function of the fields and their first-order derivatives. Note that higher-order derivatives would lead to so-called *non-local theories* and are therefore not considered. Thus, the Lagrangian function is given by

$$L = \int \mathcal{L} \, d^3x, \qquad (5.16)$$

and the action is

$$A = \int L \, dx^0 = \int \mathcal{L} \, d^4x. \qquad (5.17)$$

Note that relativistic invariance of A will be obtained if $\mathcal{L}$ is a Lorentz scalar.

Now, the principle of least action $\delta A = 0$ yields

$$0 = \delta A = \delta \int \mathcal{L}(\varphi_a(x), \partial_\mu \varphi_a(x)) \, d^4x$$

$$= \int \sum_{a=1}^{n} \left[\frac{\partial \mathcal{L}}{\partial \varphi_a} \delta \varphi_a + \frac{\partial \mathcal{L}}{\partial (\partial_\mu \varphi_a)} \delta(\partial_\mu \varphi_a) \right] d^4x = \{\text{integration by parts}\}$$

$$= \sum_{a=1}^{n} \int \left\{ \frac{\partial \mathcal{L}}{\partial \varphi_a} \delta \varphi_a - \partial_\mu \left[\frac{\partial \mathcal{L}}{\partial (\partial_\mu \varphi_a)} \right] \delta \varphi_a + \partial_\mu \left[\frac{\partial \mathcal{L}}{\partial (\partial_\mu \varphi_a)} \delta \varphi_a \right] \right\} d^4x, \quad (5.18)$$

where the last term in the integrand gives no contribution,[8] and since $\delta \varphi_a$ ($a = 1, 2, \ldots, n$) are arbitrary variations of the fields φ_a ($a = 1, 2, \ldots, n$), Eq. (5.18) leads to the equations of motion, i.e. the *Euler–Lagrange field equations*, which are generalizations of the 'particle' Euler–Lagrange equations (4.11), and they read

$$\partial_\mu \frac{\partial \mathcal{L}}{\partial (\partial_\mu \varphi_a(x))} - \frac{\partial \mathcal{L}}{\partial \varphi_a(x)} = 0, \quad a = 1, 2, \ldots, n. \quad (5.19)$$

Comparing Eqs. (4.11) and (5.19), we observe that, in Eq. (4.11), the fundamental quantity is the Lagrangian function L, whereas in Eq. (5.19), L is replaced by the Lagrangian density $\mathcal{L}$. In addition, in Eq. (4.11), the time variable t has a special meaning, whereas in Eq. (5.19), it has the same importance as all other coordinates of the spacetime point x. Since the integral of a 4-divergence vanishes, the Lagrangians $\mathcal{L}$ and $\mathcal{L} + \partial_\mu f^\mu$ are equivalent, where $f^\mu = f^\mu(\varphi_a, \partial_\nu \varphi_a)$. Thus, the free Lagrangians for charged Klein–Gordon and Dirac fields are given by

$$\mathcal{L}_{\text{KG}} = -\phi^\dagger(x) \left(\Box + m^2 \right) \phi(x) + \text{4-div.}, \quad (5.20)$$

$$\mathcal{L}_{\text{D}} = \bar{\psi}(x) \left(i\slashed{\partial} - m\mathbb{1}_4 \right) \psi(x) + \text{4-div.}, \quad (5.21)$$

respectively. Using Euler–Lagrange field equations, we find the equations of motion

$$\left(\Box + m^2 \right) \phi(x) = 0, \quad (5.22)$$

$$\left(i\slashed{\partial} - m\mathbb{1}_4 \right) \psi(x) = 0, \quad (5.23)$$

respectively, where Eq. (5.22) is the Klein–Gordon equation and Eq. (5.23) is the Dirac equation. For the electromagnetic case, the situation is more complicated because of gauge invariance. The free Lagrangian for the electromagnetic field is given by

$$\mathcal{L}_{\text{EM,0}} = \mathcal{L}_{\text{EM,0}}(A_\mu, \partial_\nu A_\mu) = -\frac{1}{4} F^{\mu\nu} F_{\mu\nu} + \text{g.t.}, \quad (5.24)$$

where it, of course, holds that $F^{\mu\nu} F_{\mu\nu} = F_{\mu\nu} F^{\mu\nu}$, and g.t. means gauge terms, which depend on the gauge in which we choose to quantize the 4-vector potential A_μ. In what follows of this book, we will normally not include the terms 4-div. and g.t. when we write down Lagrangians.

---

[8] Note that we assume that $\delta \varphi_a$ ($a = 1, 2, \ldots, n$) vanish on the boundary of the integration region.

5.3 Gauge interactions

The three fundamental interactions (i.e. strong, weak, and electromagnetic interactions) are so-called *gauge interactions*. These interactions are mediated by spin-1 particles. We denote the corresponding *coupling constant* and fields by g and $W_\mu(x)$, respectively. The interaction of matter with the W is obtained by the minimal coupling replacement

$$i\partial_\mu \mapsto i\partial_\mu - g W_\mu(x) \equiv i D_\mu^W(x), \tag{5.25}$$

where D_μ^W is the *(gauge) covariant derivative*. Note that g determines the strengh of the interaction. For example, in the electromagnetic case, we have $g W_\mu = e A_\mu$, where e is the charge of the electron[9] as well as the coupling constant for electromagnetic interactions, such that

$$i\partial_\mu \mapsto i\partial_\mu - e A_\mu(x) = i D_\mu. \tag{5.26}$$

Thus, using the free Lagrangian for n Dirac fields including the free Lagrangian for the electromagnetic field

$$\mathcal{L}_0 = \sum_{a=1}^n \bar\psi_a \left(i\partial\!\!\!/ - m_a \mathbb{1}_4 \right) \psi_a - \frac{1}{4} F^{\mu\nu} F_{\mu\nu} \tag{5.27}$$

and making the minimal coupling replacement, i.e. introducing the covariant derivative $D_\mu^a \equiv \partial_\mu + i e_a A_\mu(x)$, we obtain the Lagrangian

$$
\begin{aligned}
\mathcal{L} &= \sum_{a=1}^n \bar\psi_a \left(i D\!\!\!\!/^{\,a} - m_a \mathbb{1}_4 \right) \psi_a - \frac{1}{4} F^{\mu\nu} F_{\mu\nu} \\
&= \sum_{a=1}^n \bar\psi_a \left(i\partial\!\!\!/ - m_a \mathbb{1}_4 \right) \psi_a - \frac{1}{4} F^{\mu\nu} F_{\mu\nu} - \sum_{a=1}^n e_a \bar\psi_a A(x) \psi_a \equiv \mathcal{L}_0 + \mathcal{L}_{\text{int.}},
\end{aligned}
\tag{5.28}
$$

where the interaction Lagrangian is given by

$$\mathcal{L}_{\text{int.}} = -\sum_{a=1}^n e_a \bar\psi_a(x) A(x) \psi_a(x) = -A_\mu(x) \underbrace{\sum_{a=1}^n e_a \bar\psi_a(x) \gamma^\mu \psi_a(x)}_{j^\mu(x)} \equiv -j^\mu(x) A_\mu(x). \tag{5.29}$$

Here, j is the conserved 4-current, i.e. $\partial_\mu j^\mu(x) = 0$ [cf. Eq. (3.24)].

Now, we want to investigate how the Lagrangians transform with respect to the following transformation

$$A_\mu(x) \mapsto A_\mu'(x) = A_\mu(x) + \partial_\mu \chi(x), \tag{5.30}$$

---

[9] In fact, the electron charge is $-e$ (where $e > 0$), but we will anyway refer to e as the electron charge throughout this book.

which is a so-called *gauge transformation*, where $\chi = \chi(x)$ is an arbitrary function. Performing this gauge transformation in the Lagrangian (5.28), we find that

$$\mathcal{L} \mapsto \mathcal{L}' = \mathcal{L} - \sum_{a=1}^{n} e_a \bar{\psi}_a(x) \gamma^\mu \psi_a(x) \partial_\mu \chi(x). \tag{5.31}$$

Therefore, at first sight, we conclude that this Lagrangian is **not** gauge invariant. However, note that the free Lagrangian for the electromagnetic field, i.e. the term $-\frac{1}{4} F^{\mu\nu} F_{\mu\nu}$ is gauge invariant, which we will show later (see the discussion in Section 9.1). Thus, the following question naturally arises: how should we make the Lagrangian (5.28) gauge invariant? The answer to this question is as follows. Introducing the corresponding gauge transformations for the Dirac fields,[10] i.e.

$$\psi_a(x) \mapsto \psi_a'(x) = e^{-ie_a \chi(x)} \psi_a(x), \tag{5.32}$$

$$\bar{\psi}_a(x) \mapsto \bar{\psi}_a'(x) = \bar{\psi}_a(x) e^{ie_a \chi(x)}, \tag{5.33}$$

we have

$$\mathcal{L}_0 \mapsto \mathcal{L}_0' = \mathcal{L}_0 + \sum_{a=1}^{n} e_a \bar{\psi}_a(x) \gamma^\mu \psi_a(x) \partial_\mu \chi(x), \tag{5.34}$$

$$\mathcal{L}_{\text{int.}} \mapsto \mathcal{L}_{\text{int.}}' = \mathcal{L}_{\text{int.}} - \sum_{a=1}^{n} e_a \bar{\psi}_a(x) \gamma^\mu \psi_a(x) \partial_\mu \chi(x), \tag{5.35}$$

which are separately **not** gauge invariant. However, the sum of both $\mathcal{L}_0$ and $\mathcal{L}_{\text{int.}}$ is gauge invariant. In fact, adding to the sum also the free Lagrangian for the electromagnetic field, the result is, of course, still gauge invariant, since the free Lagrangian for the electromagnetic field is gauge invariant. This result is the extended version of the Lagrangian of QED with n Dirac fields. We stress that $\mathcal{L}_{\text{int.}}$ apparently leads to the correct interaction of QED, since it is in excellent agreement with experiment. Basically, changing $\mathcal{L}_{\text{int.}}$ to other interaction Lagrangians that are Lorentz and gauge invariant, one almost exclusively obtains interactions that are not renormalizable (see the discussions on renormalizability and renormalization in Chapters 11 and 13).

In conclusion, in QED, we have the gauge transformations

$$A_\mu(x) \mapsto A_\mu(x) + \partial_\mu \chi(x), \tag{5.36}$$

$$\psi(x) \mapsto e^{-ie\chi(x)} \psi(x), \tag{5.37}$$

$$\bar{\psi}(x) \mapsto \bar{\psi}(x) e^{ie\chi(x)} = e^{ie\chi(x)} \bar{\psi}(x). \tag{5.38}$$

---

[10] Note that the gauge transformations for the Dirac fields are known as *local* gauge transformations, since the function $\chi(x)$ depends on the spacetime point x. If $\chi(x) = $ const., then the gauge transformations for the Dirac fields are known as *global* gauge transformations.

$$D_\mu \psi(x) = [\partial_\mu + ieA_\mu(x)]\,\psi(x) \mapsto e^{-ie\chi(x)} D_\mu \psi(x), \qquad (5.39)$$

$$F_{\mu\nu}(x) \mapsto F_{\mu\nu}(x), \qquad (5.40)$$

which means that the full Lagrangian of QED (for one Dirac field), i.e.

$$\mathcal{L}_{\text{QED}} = \bar{\psi}\left(i\slashed{D} - m\mathbb{1}_4\right)\psi - \frac{1}{4}F^{\mu\nu}F_{\mu\nu} \qquad (5.41)$$

is gauge invariant (up to a 4-divergence and gauge terms). Thus, the form of the Lagrangian of QED is dependent on the gauge that is chosen. However, gauge invariance of QED means that physical observables are independent of the choice of gauge. Actually, this statement also holds for all other models and theories that are gauge invariant.

Exercise 5.1 Using the gauge transformations (5.36)–(5.40), show that the Lagrangian (5.41) is gauge invariant.

5.4 Scattering theory and Møller wave operators

In scattering theory, the prototype of a scattering problem is a single particle approaching an obstacle, which is mathematically described by a potential V, then interacting with V, and eventually, leaving the interaction region, becoming again asymptotically a free particle, described by a kinetic energy Hamiltonian H_0. (See Fig. 5.1.) Therefore, the total Hamiltonian of the system is given by $H = H_0 + V$. The information that one wants to extract is how the initial free time evolution of the incoming particles is transformed to a (different) free time evolution of the outgoing particle.

The time evolution of the one-particle system is given by $\psi(t) = e^{-iHt}\psi(0)$. The state being asymptotically free as time $t \to \pm\infty$ is written as the condition

$$\lim_{t \to \pm\infty} \|e^{-iHt}\psi(0) - e^{-iH_0 t}\psi_\pm\| = 0, \qquad (5.42)$$

where $\psi_\pm$ are some vectors in the Hilbert space $\mathcal{H}$ of the system, i.e. the asymptotically free states. In other words, we have

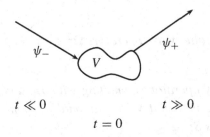

Figure 5.1 The prototype of a scattering problem.

$$\lim_{t\to\pm\infty} \|e^{iHt}e^{-iH_0t}\psi_\pm - \psi(0)\| = 0. \tag{5.43}$$

In general, the 'free' Hamiltonian H_0 might be more complicated than the kinetic energy Hamiltonian; in particular, it could have bound states. However, the bound states are never asymptotically free states, and for this reason, we will project out the *absolutely continuous spectrum*, i.e. the *scattering spectrum*, of H_0. Thus, we would like to prove the existence of the so-called *Møller wave operators*

$$\Omega^\pm = \text{s-lim}_{t\to\pm\infty} e^{iHt}e^{-iH_0t}P_{\text{ac}}(H_0), \tag{5.44}$$

where s-lim stands for strong limit and $P_{\text{ac}}(H)$ is the projection onto the absolutely continuous spectrum of H. In the case that the 'free' Hamiltonian H_0 is the kinetic energy Hamiltonian, we have $P_{\text{ac}}(H_0) = \mathbb{1}$. If the Møller wave operators exist, then define the spaces of incoming states as $\mathcal{H}_- = \mathcal{H}^{\text{in}} = \Omega^-\mathcal{H}$ and outgoing states as $\mathcal{H}_+ = \mathcal{H}^{\text{out}} = \Omega^+\mathcal{H}$. We say that $\Omega^\pm$ are *complete* if $\mathcal{H}_- = \mathcal{H}_+ = P_{\text{ac}}(H)\mathcal{H}$, whereas they are *asymptotically complete* if $\Omega^\pm$ are complete and H does not have any *singular spectrum*; in that case, $\mathcal{H}_\pm = \mathcal{H}_p^\perp$, where $\mathcal{H}_p$ is the subspace corresponding to the *point spectrum* (i.e. the bound states) of the total Hamiltonian.

The basic problem in scattering theory is: for time $t \ll 0$, a free state looks like $e^{-iH_0t}\alpha$. By the definition of the Møller wave operators, the true time evolution for all times is given by $e^{-iHt}\Omega^-\alpha$. What is the probability that this state is (asymptotically) the free state $e^{-iH_0t}\beta$? By the basic principles of quantum mechanics, the transition probability for $\alpha \to \beta$ is given by $|\langle\beta|\alpha\rangle|^2$, and therefore, the answer to the question is

$$P(\alpha \to \beta) = |\langle\Omega^+\beta|\Omega^-\alpha\rangle|^2 = |\langle\beta|(\Omega^+)^*\Omega^-|\alpha\rangle|^2. \tag{5.45}$$

Thus, we can define the S operator as $S = (\Omega^+)^*\Omega^-$, which means that

$$P(\alpha \to \beta) = |\langle\beta|S|\alpha\rangle|^2, \tag{5.46}$$

where the quantity $\langle\beta|S|\alpha\rangle$ is known as the S matrix. In the following discussion, we will make the standing assumption that $P_{\text{ac}}(H_0) = \mathbb{1}$. We now state the following theorem.

Theorem 5.1 *If the Møller wave operators $\Omega^\pm$ exist, then*

(1) $Se^{iH_0t} = e^{iH_0t}S \quad \forall t$;
(2) *if U is any unitary operator commuting with H_0 and H, then $[S, U] = 0$. In particular, if the potential V is rotationally symmetric, i.e. it is a central potential, then so is S;*
(3) *S is unitary if and only if $\mathcal{H}_- = \mathcal{H}_+$.*

The formulation of the scattering problem in terms of the S operator and the Møller wave operators is often called the *formal or time-dependent scattering theory*. Although this formulation is very elegant, the actual calculations are usually performed in the so-called *time-independent scattering theory*, i.e. in terms of static solutions to the Schrödinger equation.

5.5 The S operator

Definition The unitary scattering operator, the so-called S *operator*, is defined as the mapping $S : \mathcal{H}^{\text{out}} \to \mathcal{H}^{\text{in}}$, $\left|\psi_\alpha^{\text{out}}\right\rangle \mapsto \left|\psi_\alpha^{\text{in}}\right\rangle = S\left|\psi_\alpha^{\text{out}}\right\rangle$, where $\mathcal{H}^{\text{in}} = \{\left|\psi_\alpha^{\text{in}}\right\rangle\}$ denotes the set of all 'in'-states and $\mathcal{H}^{\text{out}} = \{\left|\psi_\alpha^{\text{out}}\right\rangle\}$ denotes the set of all 'out'-states. Note that since S is unitary, it holds that $S^\dagger = S^{-1}$.

The two Hilbert spaces $\mathcal{H}^{\text{in}}$ and $\mathcal{H}^{\text{out}}$ are two different Fock-space bases of the same state space $\mathcal{H}$, i.e. $\mathcal{H} = \mathcal{H}^{\text{in}} = \mathcal{H}^{\text{out}}$. Hence, every state in the Hilbert space $\mathcal{H}$ can be expressed as a superposition of either in- or out-states.

Let U^{in} and U^{out} denote unitary representations of a symmetry for the Hilbert spaces $\mathcal{H}^{\text{in}}$ and $\mathcal{H}^{\text{out}}$, respectively. These unitary representations are unitarily equivalent, and thus, we have the relation

$$S^{-1}U^{\text{in}}S = U^{\text{out}}. \tag{5.47}$$

Now, the *transition amplitude* from an in-state $\left|\psi_\alpha^{\text{in}}\right\rangle$ to an out-state $\left|\psi_\beta^{\text{out}}\right\rangle$ is given by

$$W_{\alpha\beta} = \left\langle\psi_\beta^{\text{out}}|\psi_\alpha^{\text{in}}\right\rangle. \tag{5.48}$$

Inserting the definition of the S operator into the transition amplitude, we obtain

$$W_{\alpha\beta} = \left\langle\psi_\beta^{\text{out}}|S|\psi_\alpha^{\text{out}}\right\rangle = \underbrace{\left\langle\psi_\beta^{\text{out}}|S^\dagger}_{\left\langle\psi_\beta^{\text{in}}\right|} S \underbrace{S|\psi_\alpha^{\text{out}}\right\rangle}_{|\psi_\alpha^{\text{in}}\rangle} = \left\langle\psi_\beta^{\text{in}}|S|\psi_\alpha^{\text{in}}\right\rangle. \tag{5.49}$$

Then, assume that the transition amplitude $W_{\alpha\beta}$ is invariant with respect to unitary symmetry transformations, i.e. $W_{\alpha'\beta'} = W_{\alpha\beta}$. Next, we have

$$W_{\alpha'\beta'} = \left\langle\psi_{\beta'}^{\text{out}}|\psi_{\alpha'}^{\text{in}}\right\rangle = \left\langle U^{\text{out}}\psi_\beta^{\text{out}}|U^{\text{in}}\psi_\alpha^{\text{in}}\right\rangle = \left\langle\psi_\beta^{\text{out}}|U^{\text{out}\dagger}U^{\text{in}}|\psi_\alpha^{\text{in}}\right\rangle, \tag{5.50}$$

which means that

$$U^{\text{out}\dagger}U^{\text{in}} = \mathbb{1}. \tag{5.51}$$

However, multiplying with U^{out} from the left, we find that

$$U^{\text{in}} = U^{\text{out}} \equiv U. \tag{5.52}$$

Thus, we obtain

$$S^{-1}US = U \quad \Leftrightarrow \quad [S, U] = 0, \tag{5.53}$$

which means that the S operator commutes with the unitary (symmetry) transformation. Note that if $W_{\alpha\beta}$ is invariant, then $U^{\text{in}} = U^{\text{out}}$, which implies that one can drop the labels 'in' and 'out'. From a physical point of view, it is actually enough to require that $|W_{\alpha'\beta'}|^2 = |W_{\alpha\beta}|^2$.

Finally, the S operator is a unitary operator, i.e. it holds that $S^{\dagger}S = SS^{\dagger} = 1$. Thus, the S operator can be written as

$$S = e^{i\Omega} = \sum_{n=0}^{\infty} \frac{(i\Omega)^n}{n!}, \tag{5.54}$$

where Ω is Hermitian, i.e. $\Omega^{\dagger} = \Omega$, since $e^{-i\Omega^{\dagger}}e^{i\Omega} = e^{i(-\Omega^{\dagger}+\Omega)} = 1$. Let the eigenvalues of the operator Ω be equal to 2δ, where δ is called the *phase shift*. Now, since the S operator commutes with invariance groups, it implies that S commutes with the Casimir operators of the invariance groups. Thus, the phase shift δ depends only on the eigenvalues of the Casimir operators.

For example, for the Poincaré group, we have that $[S, U(\Lambda, a)] = 0$, where $(\Lambda, a) \in \mathcal{P}_+^{\uparrow}$. This means that $[S, P^{\mu}] = 0$ and $[S, M^{\mu\nu}] = 0$. Thus, the phase shift δ is a function of the Casimir operators P^2 and w^2, i.e.

$$\delta_j(s) \equiv \delta(s, j) = f(m^2, -m^2 j(j+1)), \tag{5.55}$$

where s is a so-called *Mandelstam variable* [see Eqs. (11.142)–(11.144)].

5.5.1 Conservation laws

Let $\mathcal{G}$ be a Lie group (see Appendix A). Now, Eq. (5.53) implies that the whole representation commutes with the S operator, i.e.

$$[S, U(\mathcal{G})] = 0. \tag{5.56}$$

Thus, if a group element $g \in \mathcal{G}$ is parametrized by $\mathbf{t} = (t_1, t_2, \ldots, t_n)$ and the identity element is represented by $e = (0, 0, \ldots, 0)$, then we can write the representation

$$U(g) = \exp(i\mathbf{A} \cdot \mathbf{t}), \quad \text{where } \mathbf{A} \cdot \mathbf{t} = \sum_{i=1}^{n} A_i t_i \tag{5.57}$$

and the A_is are the infinitesimal generators of the representation (see Appendix A.4). Note that the A_is are the representing elements for the generators of the Lie algebra corresponding to the Lie group. For example, we have the canonical examples of unitary operators $U(\mathbb{1}_4, a) = \exp(\mathrm{i} P^\mu a_\mu)$ and $U(\Lambda, 0) = \exp\left(-\frac{\mathrm{i}}{2} M^{\mu\nu} \omega_{\mu\nu}\right)$. Note that

$$A_i = -\mathrm{i} \frac{\mathrm{d}}{\mathrm{d} t_i} U(g)\bigg|_{g=e} \quad \forall i = 1, 2, \ldots, n. \tag{5.58}$$

Differentiating Eq. (5.56) with respect to the parameter t_i, we obtain

$$-\mathrm{i} \frac{\mathrm{d}}{\mathrm{d} t_i} [S, U(\mathcal{G})]\bigg|_{g=e} = \left[S, -\mathrm{i} \frac{\mathrm{d}}{\mathrm{d} t_i} U(g)\bigg|_{g=e}\right] = [S, A_i] = 0. \tag{5.59}$$

Thus, in addition, the S operator commutes with all the generators of the representation. Furthermore, if $U(\mathcal{G})$ is a unitary representation, then $U(g)^\dagger U(g) = \mathbb{1}$. Differentiating with respect to the parameter t_i gives

$$-\mathrm{i} \frac{\mathrm{d}}{\mathrm{d} t_i} U(g)^\dagger U(g)\bigg|_{g=e} = -\mathrm{i}\left(-\mathrm{i} A_i^\dagger + \mathrm{i} A_i\right) = -A_i^\dagger + A_i = 0, \quad \text{i.e. } A_i^\dagger = A_i, \tag{5.60}$$

which means that the A_is are self-adjoint (or Hermitian) operators.

Let us assume that A is an infinitesimal generator such that $[S, A] = 0$ and the states $|\psi_\alpha^{\text{in}}\rangle$ and $|\psi_\beta^{\text{in}}\rangle$ are eigenstates of the operator A, i.e. $A |\psi_\alpha^{\text{in}}\rangle = a_\alpha |\psi_\alpha^{\text{in}}\rangle$ and $A |\psi_\beta^{\text{in}}\rangle = a_\beta |\psi_\beta^{\text{in}}\rangle$, where a_α and a_β are eigenvalues (i.e. quantum numbers) of the operator A. Taking the matrix element of the commutator $[S, A] = 0$, we find that (since the operator A is Hermitian)

$$\begin{aligned} 0 &= \langle \psi_\beta^{\text{in}} | [S, A] | \psi_\alpha^{\text{in}} \rangle = \langle \psi_\beta^{\text{in}} | SA - AS | \psi_\alpha^{\text{in}} \rangle \\ &= a_\alpha \langle \psi_\beta^{\text{in}} | S | \psi_\alpha^{\text{in}} \rangle - a_\beta \langle \psi_\beta^{\text{in}} | S | \psi_\alpha^{\text{in}} \rangle = (a_\alpha - a_\beta) \langle \psi_\beta^{\text{in}} | S | \psi_\alpha^{\text{in}} \rangle, \end{aligned} \tag{5.61}$$

which implies that $a_\alpha - a_\beta = 0$ or $\langle \psi_\beta^{\text{in}} | S | \psi_\alpha^{\text{in}} \rangle = 0$. However, if this should be non-trivial, i.e. $\langle \psi_\beta^{\text{in}} | S | \psi_\alpha^{\text{in}} \rangle \neq 0$, we must have $a_\alpha = a_\beta$, which means that the quantum numbers are *conserved*, i.e. the S operator does not change the quantum numbers. For example, we have for the Poincaré group that $[S, P^\mu] = 0$ and $[S, M^{\mu\nu}] = 0$. This implies conservation of 4-momentum (four conservation laws), angular momentum (three conservation laws), and three additional conservation laws. In total, we obtain ten conservation laws for the Poincaré group.

5.5.2 Discrete symmetries

Next, we investigate so-called *discrete symmetries*. Let us assume that the operator U has real eigenvalues, which means that U is Hermitian. Now, since U is unitary and Hermitian, it holds that $U^\dagger = U^{-1} = U$, which implies that $U^2 = \mathbb{1}$. Then, the operator U has eigenvalues ± 1, which are conserved multiplicatively. In order to show this, assume that the states $|\psi_\alpha^{\text{in}}\rangle$ and $|\psi_\beta^{\text{in}}\rangle$ are eigenstates to the operator U

with eigenvalues η_α and η_β, i.e. $U \left| \psi_\alpha^{in} \right\rangle = \eta_\alpha \left| \psi_\alpha^{in} \right\rangle$ and $U \left| \psi_\beta^{in} \right\rangle = \eta_\beta \left| \psi_\beta^{in} \right\rangle$. Thus, using Eq. (5.53), we obtain

$$\left\langle \psi_\beta^{in} |S| \psi_\alpha^{in} \right\rangle = \{ SU = US \Rightarrow S = U^\dagger SU \} = \left\langle \psi_\beta^{in} |U^\dagger SU| \psi_\alpha^{in} \right\rangle$$
$$= \bar{\eta}_\beta \eta_\alpha \left\langle \psi_\beta^{in} |S| \psi_\alpha^{in} \right\rangle = \{ \eta_\beta = \pm 1 \Rightarrow \bar{\eta}_\beta = \eta_\beta \} = \eta_\alpha \eta_\beta \left\langle \psi_\beta^{in} |S| \psi_\alpha^{in} \right\rangle,$$

$$(5.62)$$

which implies that $\eta_\alpha \eta_\beta = 1$, but since $|\eta_\beta| = 1$, it holds that $\eta_\alpha = \eta_\beta$.

Canonical examples of discrete symmetries are *space reflection* (or *parity*) and *time reflection* (or *time reversal*) as well as *charge conjugation*, which is a non-spacetime operation, that interchanges particles and antiparticles. These discrete symmetries are denoted by C (charge conjugation), P (parity), and T (time reversal), respectively. Section 7.5 is devoted to the symmetries C, P, and T and how they transform the Dirac field and the charged Klein–Gordon field. In addition, the CPT symmetry is discussed for the Dirac field. Experimentally, we have observed that three of the four fundamental forces of Nature – the gravitational, electromagnetic, and strong interactions – are symmetric with respect to C, P, and T. However, the weak interaction violates C and P separately, but normally preserves CP and T. Nevertheless, in some specific rare reactions, CP violation has been observed. Note that, assuming the CPT symmetry, the direct measurement of CP violation implies an indirect detection of T violation. In addition, all observations seem to indicate that CPT is a perfect symmetry of Nature. Thus, observations are consistent with the CPT *theorem*, which states that any Lorentz invariant local quantum field theory with a Hermitian Hamiltonian must have the CPT symmetry. It can also be stated as:

Theorem 5.2 (The CPT theorem)   *If the Lagrangian density of a theory is a Hermitian, normal-ordered Lorentz invariant operator constructed from quantized fields with the usual connection between spin and statistics, then the product of the C, P, and T transformations, i.e. CPT, is always a symmetry of the theory.*

The CPT theorem appeared for the first time, implicitly, in a work by Schwinger in 1951 to prove the connection between spin and statistics.[11] In 1954–1955, G. Lüders and Pauli derived more explicit proofs, which means that the theorem is sometimes known as the *Lüders–Pauli theorem*.[12] At about the same time and independently, the theorem was also proved by J. S. Bell. These proofs are based on the validity of Lorentz invariance and the principle of locality in the interaction

---

[11] For the interested reader, please see: J. Schwinger, The theory of quantized fields. I, *Phys. Rev.* **82**, 914–927 (1951).

[12] For the interested reader, please see: G. Lüders, Zum Bewegungsumkehr in quantisierten Feldtheorien, *Z. Phys.* **133**, 325–339 (1952); G. Lüders, *Kgl. Danske Videnskab. Selskab., Mat.-fys. Medd.* **28**, no. 5 (1954); and W. Pauli, *Exclusion Principle, Lorentz Group, and Reflection of Space-Time and Charge*, pp. 30–51 in *Niels Bohr and the Development of Physics*, eds. W. Pauli, L. Rosenfeld, and V. Weisskopf, Pergamon (1955).

of quantum fields. Note that violation of CPT automatically indicates violation of Lorentz invariance.

5.5.3 Noether's theorem

Finally, we state another important theorem, known as *Noether's theorem*,[13] that connects symmetries and conservation laws.

Theorem 5.3 (Noether's theorem) *To each continuous symmetry transformation of a local Lagrangian, which leaves the Lagrangian invariant (up to a 4-divergence), there exists a conservation law, and hence, a conserved quantity (or conserved current or constant of motion).*

In general, if the Lagrangian is given by $\mathcal{L} = \mathcal{L}(\varphi_a, \partial_\mu \varphi_a)$ [cf. Eq. (5.15)], where $a = 1, 2, \ldots, n$, then the variation of the Lagrangian (due to the variations $\delta \varphi_a$ of the fields φ_a) is

$$\delta \mathcal{L} = \sum_{a=1}^{n} \partial_\mu \left[\frac{\partial \mathcal{L}}{\partial(\partial_\mu \varphi_a)} \delta \varphi_a \right], \tag{5.63}$$

since the Euler–Lagrange field equations (5.19) for the fields φ_a are satisfied. Note that this variation of the Lagrangian can be used when we are studying specific variations $\delta \varphi_a$ of the fields φ_a.

Now, in particular, we utilize the abstract Noether's theorem for a concrete example. If the action A, which is derived from a Lagrangian $\mathcal{L}$ [which is **not** explicitly dependent on the spacetime point x, but only the fields $\varphi_a(x)$ and their first-order derivatives $\partial_\mu \varphi_a(x)$], is invariant with respect to continuous infinitesimal translation transformations on the following form

$$x \mapsto x' = x + \delta x, \tag{5.64}$$

where δx is an arbitrary infinitesimal translation, which induces the variations of the fields and the corresponding variation of the Lagrangian, namely

$$\delta \varphi_a(x) = \varphi_a'(x') - \varphi_a(x) = \delta x^\mu \partial_\mu \varphi_a(x), \tag{5.65}$$

$$\delta \mathcal{L} = \mathcal{L}'\left(\varphi_a'(x'), \partial_\mu' \varphi_a'(x')\right) - \mathcal{L}(\varphi_a(x), \partial_\mu \varphi_a(x)) = \delta x^\mu \partial_\mu \mathcal{L}(\varphi_a(x), \partial_\nu \varphi_a(x)), \tag{5.66}$$

then, using Noether's theorem, the 4-divergence of the current is equal to zero, i.e. $\partial_\mu j^\mu = 0$, where we need to find the expression for the current j. In this example, using Eqs. (5.63), (5.65), and (5.66), we find that

---

[13] Noether's theorem is named after the German-born mathematician E. Noether.

$$\delta x^\nu \left\{ \sum_{a=1}^n \partial_\mu \left[\frac{\partial \mathcal{L}}{\partial (\partial_\mu \varphi_a)} \partial_\nu \varphi_a \right] - \partial_\nu \mathcal{L} \right\} = 0, \tag{5.67}$$

which implies that

$$\partial_\mu \left[\sum_{a=1}^n \frac{\partial \mathcal{L}}{\partial (\partial_\mu \varphi_a)} \partial_\nu \varphi_a - \delta_\nu^\mu \mathcal{L} \right] = 0, \tag{5.68}$$

since δx is arbitrary and the Lagrangian does not explicitly depend on the spacetime point x. The expression in the brackets of Eq. (5.68) is the 'current', which we are searching for. Introducing the notation T^μ_ν for this expression, we obtain

$$T^{\mu\nu} = \sum_{a=1}^n \frac{\partial \mathcal{L}}{\partial (\partial_\mu \varphi_a)} \partial^\nu \varphi_a - g^{\mu\nu} \mathcal{L}, \tag{5.69}$$

which is the so-called *(canonical) energy–momentum tensor*. Thus, in this example, the four continuity equations $\partial_\mu T^{\mu\nu} = 0$ are the four conservation laws, and hence, the corresponding four conserved quantities are the Hamiltonian $H = \int T^{00} \, \mathrm{d}^3 x$ and the 3-momentum operator $P^i = \int T^{0i} \, \mathrm{d}^3 x$, where $i = 1, 2, 3$.

Exercise 5.2 Consider infinitesimal homogeneous Lorentz transformations of the following form

$$x^\mu \mapsto x'^\mu = (\Lambda x)^\mu = \left(\delta_\nu^\mu + \epsilon \lambda^\mu_{\ \nu} \right) x^\nu,$$

where ϵ is small and $\lambda^{\mu\nu} = -\lambda^{\nu\mu}$, which induce variations of the fields φ_a ($a = 1, 2, \ldots, n$) such that

$$\delta \varphi_a(x) = \sum_{b=1}^n \left(\delta_{ab} + \frac{1}{2} \epsilon S_{ab}^{\mu\nu} \lambda_{\mu\nu} \right) \varphi_b(x), \quad a = 1, 2, \ldots, n,$$

where $S_{ab}^{\mu\nu} = -S_{ab}^{\nu\mu}$, and show that they imply that the tensor

$$m^{\mu\nu\lambda} = \sum_{a=1}^n \sum_{b=1}^n \frac{\partial \mathcal{L}}{\partial (\partial_\mu \varphi_a(x))} S_{ab}^{\nu\lambda} \varphi_b(x) + x^\nu T^{\mu\lambda} - x^\lambda T^{\mu\nu} \tag{5.70}$$

is conserved, i.e. the six conservation laws $\partial_\mu m^{\mu\nu\lambda} = 0$ are satisfied. Note that $m^{\mu\nu\lambda} = -m^{\mu\lambda\nu}$.

In Exercise 5.2, the six conserved quantities are

$$M^{\mu\nu} = \int m^{0\mu\nu} \, \mathrm{d}^3 x = \int \left[\sum_{a=1}^n \sum_{b=1}^n \pi_a(x) S_{ab}^{\mu\nu} \varphi_b(x) + x^\mu T^{0\nu} - x^\nu T^{0\mu} \right] \mathrm{d}^3 x,$$

$$\tag{5.71}$$

which follow from the six conservation laws. Note that the first term in Eq. (5.70) corresponds to the (intrinsic) spin angular momentum density, whereas the second and third terms correspond to the orbital angular momentum density.

To this end, we observe that translation transformations give rise to four conserved quantities, i.e. $P^\mu = \int T^{0\mu}\,d^3x$, whereas homogeneous Lorentz transformations give rise to six conserved quantities, i.e. $M^{\mu\nu} = \int m^{0\mu\nu}\,d^3x$. Thus, in total, we have ten conserved quantities (or constants of motion) for inhomogeneous Lorentz transformations, i.e. Poincaré transformations (cf. the discussion in Chapter 1).

Note that in order to motivate the expressions for the 4-currents of the Klein–Gordon and Dirac equations, respectively, i.e. Eqs. (2.14), (2.15), and (2.17) as well as Eq. (3.24), we can use Noether's theorem. It follows that the 4-divergences of both 4-currents give rise to the continuity equation, which is therefore the conservation law in both cases. In Chapters 6, 7, and 9, we will use Noether's theorem, and especially Eq. (5.69), in order to find the energy–momentum tensor for the Klein–Gordon, Dirac, and electromagnetic fields, respectively, which in turn will yield the Hamiltonian and 3-momentum operator of the respective quantum fields.

Guide to additional recommended reading

The following books (see the indicated pages) and their authors have similar treatments of the content in the present chapter.

- W. Greiner and J. Reinhardt, *Quantum Electrodynamics*, 3rd edn., Springer (2003), pp. 12–13, 16–18.
- F. Gross, *Relativistic Quantum Mechanics and Field Theory*, Wiley (1993), pp. 187–194.
- F. Mandl and G. Shaw, *Quantum Field Theory*, rev. edn, Wiley (1994), pp. 33–40, 77–79.
- J. Mickelsson, edited by T. Ohlsson, *Advanced Quantum Mechanics*, KTH (2003), pp. 79–83.
- M. E. Peskin and D. V. Schroeder, *An Introduction to Quantum Field Theory*, Addison-Wesley (1995), pp. 15–19, 29–31, 62–63, 77–78.
- G. Scharf, *Finite Quantum Electrodynamics*, Springer (1989), pp. 69–80.
- F. Schwabl, *Advanced Quantum Mechanics*, Springer (1999), p. 266.
- S. S. Schweber, *An Introduction to Relativistic Quantum Field Theory*, Dover (2005), pp. 207–211.
- H. Snellman, *Elementary Particle Physics*, KTH (2004), pp. 49–54.
- F. J. Ynduráin, *Relativistic Quantum Mechanics and Introduction to Field Theory*, Springer (1996), pp. 210–223.
- For the interested reader: G. Lüders, Zum Bewegungsumkehr in quantisierten Feldtheorien, *Z. Phys.* **133**, 325–339 (1952); G. Lüders, *Kgl. Danske Videnskab. Selskab., Mat.-fys. Medd.* **28**, no. 5 (1954); W. Pauli, *Exclusion Principle, Lorentz Group, and Reflection of Space-Time and Charge*, pp. 30–51 in *Niels Bohr and the Development of Physics*, eds. W. Pauli, L. Rosenfeld, and V. Weisskopf, Pergamon (1955); and J. Schwinger, The theory of quantized fields. I, *Phys. Rev.* **82**, 914–927 (1951).

6

Quantization of the Klein–Gordon field

In ordinary non-relativistic quantum mechanics, quantizing a point particle, we interpret the generalized coordinates q_n (normally the three Cartesian spatial coordinates x, y, and z) and their corresponding 3-momenta p_n as operators, which act on the wave function Ψ that is a representation of the vector of states, and they fulfil the commutation relations $[q_n, p_m] = i\hbar\delta_{nm}$. However, in field theory, the position coordinates x^i have a different meaning and are used to label the infinitely many 'coordinates' $\phi(x)$ and their corresponding conjugate momenta $\pi(x)$. In order to quantize a field theory, we impose on the fields $\phi(x)$ and $\pi(x)$ the commutation relations $[\phi(t, \mathbf{x}), \pi(t, \mathbf{y})] = i\hbar\delta(\mathbf{x} - \mathbf{y})$, which are the natural generalizations of the canonical commutation relations for ordinary quantum mechanics. Note that, in what follows, we will again set $\hbar = 1$. In addition, the discussion in this chapter will be performed for a free neutral Klein–Gordon field. However, in the last two sections, we will present two natural extensions. First, in Section 6.5, we will extend the discussion with an example of interactions in the form of a classical external source, and second, in Section 6.6, we will describe how a free charged Klein–Gordon field can be treated.

6.1 Canonical quantization

Next, we want to develop the *canonical quantum field theory* for the neutral Klein–Gordon field.[1] This theory was invented in the 1930s and it is indeed a very successful theory. However, we will first treat the **classical** Klein–Gordon field, and then, we will quantize it using the method of *canonical quantization*, which usage was motivated in Section 4.4.

---

[1] Remember that we observed in Chapter 2 that the proper physical interpretation of the quantity ϕ is not that of a wave function, but a quantum field operator. Thus, in this chapter, we will solve the two problems (i.e. the indefiniteness of the density for the Klein–Gordon equation and the Klein paradox) presented in Chapter 2 by quantizing the quantity ϕ, which means that we will reject a single-particle theory described by the Klein–Gordon equation for the wave function and accept a many-particle theory described by the Klein–Gordon equation for the quantum field operator, i.e. a quantum field theory.

The Lagrangian density for a free real scalar field, i.e. a free neutral Klein–Gordon field, is given by

$$\mathcal{L}_{KG} = \frac{1}{2} \left(\partial_\mu \phi \partial^\mu \phi - m^2 \phi^2 \right).$$ (6.1)

Using Euler–Lagrange field equations for this Lagrangian

$$\partial_\mu \left(\frac{\partial \mathcal{L}_{KG}}{\partial \left(\partial_\mu \phi \right)} \right) - \frac{\partial \mathcal{L}_{KG}}{\partial \phi} = 0,$$ (6.2)

we find the Klein–Gordon equation

$$\left(\Box + m^2 \right) \phi = 0,$$ (6.3)

which of course should be the equation of motion for the Klein–Gordon field.

Exercise 6.1 Using Euler–Lagrange field equations for the Lagrangian (6.1), show that the Klein–Gordon equation (6.3) is obtained.

Expanding the Klein–Gordon field in a complete set of orthonormal functions $\{u_n\}$ yields

$$\phi(x) = \sum_n q_n(t) u_n(\mathbf{x}),$$ (6.4)

where the generalized coordinates are

$$q_n(t) = \int u_n(\mathbf{x}) \phi(x) \, d^3x.$$ (6.5)

Thus, for any complete set of orthonormal functions, we are able to obtain a particle interpretation. The drawback is that we have replaced the field ϕ with an infinite set of particles with generalized coordinates q_n. Note that ϕ satisfies the Euler–Lagrange field equations (6.2) if and only if the generalized coordinates satisfy the 'particle equations' [cf. Eq. (4.11)]

$$\frac{d}{dt} \frac{\partial L}{\partial \dot{q}_n} - \frac{\partial L}{\partial q_n} = 0.$$ (6.6)

The classical particles obeying the 'particle equations' can now be quantized by imposing the *canonical commutation relations* [cf. Eqs. (4.35) and (4.36)]

$$\left[q_n(t), p_m(t) \right] = i \delta_{nm},$$ (6.7)

$$\left[q_n(t), q_m(t) \right] = \left[p_n(t), p_m(t) \right] = 0,$$ (6.8)

where the momenta corresponding to the generalized coordinates are

$$p_m(t) = \int \underbrace{\frac{\partial \mathcal{L}_{KG}}{\partial \dot{\phi}}}_{\pi(x)} u_m(\mathbf{x}) \, d^3x.$$ (6.9)

The quantity

$$\pi(x) = \frac{\partial \mathcal{L}_{KG}}{\partial \dot{\phi}} \tag{6.10}$$

is again called the *conjugate momentum density operator* [cf. Eq. (4.42)]. For the Klein–Gordon field, the conjugate momentum is $\pi(x) = \dot{\phi}(x)$. Thus, inserting the expansions (6.5) and (6.9) into the commutation relation (6.7), we obtain

$$\left[q_n(t), p_m(t)\right] = \int u_n(\mathbf{x}) u_m(\mathbf{y}) \left[\phi(x), \pi(y)\right] \, \mathrm{d}^3 x \mathrm{d}^3 y = \mathrm{i}\delta_{nm}, \tag{6.11}$$

and similarly for the two other commutation relations given in Eq. (6.8), from which we find after some computations

$$\left[\phi(x), \pi(y)\right] = \mathrm{i}\delta(\mathbf{x} - \mathbf{y}), \tag{6.12}$$

$$\left[\phi(x), \phi(y)\right] = \left[\pi(x), \pi(y)\right] = 0, \tag{6.13}$$

i.e. the *canonical quantization* for the Klein–Gordon field. Note that Eqs. (6.11)–(6.13) are only valid for equal times t in the spacetime points x and y, i.e. $x = (t, \mathbf{x})$ and $y = (t, \mathbf{y})$. In addition, the quantized real Klein–Gordon field is represented by Hermitian operators, i.e. $\phi^\dagger(x) = \phi(x)$ and $\pi^\dagger(x) = \pi(x)$. Defining annihilation and creation operators (as simple harmonic oscillators), we have[2]

$$a_n(t) = \frac{1}{\sqrt{2}} \left[q_n(t) + \mathrm{i} p_n(t)\right], \tag{6.14}$$

$$a_n^\dagger(t) = \frac{1}{\sqrt{2}} \left[q_n(t) - \mathrm{i} p_n(t)\right]. \tag{6.15}$$

This can, of course, be expressed as

$$q_n(t) = \frac{1}{\sqrt{2}} \left[a_n^\dagger(t) + a_n(t)\right], \tag{6.16}$$

$$p_n(t) = \frac{\mathrm{i}}{\sqrt{2}} \left[a_n^\dagger(t) - a_n(t)\right]. \tag{6.17}$$

However, note that the operators $a_n(t)$ and $a_n^\dagger(t)$ fulfil the commutation relations [cf. Eqs. (4.37) and (4.38)]

$$\left[a_n(t), a_m^\dagger(t)\right] = \delta_{nm}, \tag{6.18}$$

$$\left[a_n(t), a_m(t)\right] = \left[a_n^\dagger(t), a_m^\dagger(t)\right] = 0, \tag{6.19}$$

and not the canonical commutation relations given in Eqs. (6.7) and (6.8). The state of zero quanta, i.e. the vacuum state $|0\rangle$, is defined as $a_n(t)|0\rangle = 0$, which is normalized such that $\langle 0|0\rangle = 1$. In general, the creation and annihilation operators

---

[2] It should be stressed that the quantum operators $a_n(t)$ and $a_n^\dagger(t)$ replace the classical coefficients $a_n(t)$ and $a_n^*(t)$, cf. the discussion on the non-relativistic string in Sections 4.2–4.4.

$a_n^\dagger(t)$ and $a_n(t)$ act on states and increase and decrease, respectively, the number of particles in a given state by one unit according to

$$a_n^\dagger(t)|\ldots, m_n, \ldots\rangle = \sqrt{m_n + 1}|\ldots, m_n + 1, \ldots\rangle, \tag{6.20}$$

$$a_n(t)|\ldots, m_n, \ldots\rangle = \sqrt{m_n}|\ldots, m_n - 1, \ldots\rangle, \tag{6.21}$$

where $|\ldots, m_n, \ldots\rangle$ is the given state. A normalized particle eigenstate with m_n particles in the state n is given by

$$|\ldots, m_n, \ldots\rangle = \frac{[a_n^\dagger(t)]^{m_n}}{\sqrt{m_n!}}|\ldots, 0, \ldots\rangle. \tag{6.22}$$

Since these states describe bosons, we can have several particles of the same kind in one state (see fermions in Chapter 7). In particular, the number operator $N_n(t) = a_n^\dagger(t)a_n(t)$ gives the number of particles in the state n, i.e.

$$N_n(t)|\ldots, m_n, \ldots\rangle = m_n|\ldots, m_n, \ldots\rangle, \tag{6.23}$$

where m_n is the number of particles [cf. Eq. (4.45)]. Since the simple harmonic oscillators describing the different particles are non-interacting, the many-particle eigenstates are direct products, i.e.

$$|m_1, m_2, \ldots\rangle = \bigotimes_{n=1}^{\infty} |0, \ldots, 0, m_n, 0, \ldots, 0, \ldots\rangle. \tag{6.24}$$

In addition, the operator for the total number of particles is given by

$$N(t) = \sum_{n=1}^{\infty} N_n(t) = \sum_{n=1}^{\infty} a_n^\dagger(t)a_n(t). \tag{6.25}$$

Therefore, we have

$$N(t)|m_1, m_2, \ldots\rangle = m|m_1, m_2, \ldots\rangle, \tag{6.26}$$

where

$$m = \sum_{n=1}^{\infty} m_n = m_1 + m_2 + \cdots \tag{6.27}$$

is the total number of particles. Thus, we obtain an infinite set of harmonic oscillators (cf. Chapter 4), whose basis eigenstates span an enormous Hilbert space known as a *Fock space*.

Exercise 6.2 Show Eqs. (6.12) and (6.13) using Eq. (6.11).

Now, we choose to quantize the fields $\phi(x)$ and $\pi(x)$ such that the commutation relation

$$\left[a(\mathbf{p}), a^\dagger(\mathbf{q})\right] = \omega(\mathbf{p})\delta(\mathbf{p} - \mathbf{q}), \tag{6.28}$$

where $\omega(\mathbf{p}) = \sqrt{m^2 + \mathbf{p}^2}$, is fulfilled. Here, the operator $a(\mathbf{p})$ annihilates a particle [having positive energy $\omega(\mathbf{p})$] with 3-momentum $\mathbf{p}$, whereas the operator $a^\dagger(\mathbf{p})$ creates a particle [having positive energy $\omega(\mathbf{p})$] with 3-momentum $\mathbf{p}$. Observe the difference of this commutation relation with the ones derived earlier in Eqs. (6.18) and (6.19).[3] In addition, we introduce the one-particle state with 3-momentum $\mathbf{p}$ and energy $\omega(\mathbf{p})$ as $|\mathbf{p}\rangle$, which is invariantly normalized such that

$$\langle \mathbf{q}|\mathbf{p}\rangle = \omega(\mathbf{p})\delta(\mathbf{p} - \mathbf{q}). \tag{6.29}$$

Many-particle states, consisting of n particles with 3-momenta $\mathbf{p}_1, \mathbf{p}_2, \ldots, \mathbf{p}_n$ and energies $\omega(\mathbf{p}_1), \omega(\mathbf{p}_2), \ldots, \omega(\mathbf{p}_n)$, are then denoted by $|\mathbf{p}_1, \mathbf{p}_2, \ldots, \mathbf{p}_n\rangle$. In particular for the one-particle states, the creation and annihilation operators $a^\dagger(\mathbf{p})$ and $a(\mathbf{p})$ fulfil the following relations

$$|\mathbf{p}\rangle = a^\dagger(\mathbf{p})|0\rangle, \tag{6.30}$$

$$a(\mathbf{p})|\mathbf{q}\rangle = \omega(\mathbf{p})\delta(\mathbf{p} - \mathbf{q})|0\rangle, \tag{6.31}$$

respectively, whereas in general for many-particle states, we have

$$|\mathbf{p}_1, \mathbf{p}_2, \ldots, \mathbf{p}_n\rangle = \frac{1}{\sqrt{n!}}a^\dagger(\mathbf{p}_1)a^\dagger(\mathbf{p}_2)\ldots a^\dagger(\mathbf{p}_n)|0\rangle. \tag{6.32}$$

Furthermore, the vacuum state $|0\rangle$ has the property

$$a(\mathbf{p})|0\rangle = 0 \quad \forall \mathbf{p}. \tag{6.33}$$

Finally, using the one-particle normalization condition (6.29), we must have

$$\langle \mathbf{q}|\mathbf{p}\rangle = \langle 0|a(\mathbf{q})a^\dagger(\mathbf{p})|0\rangle = \langle 0|[a(\mathbf{q}), a^\dagger(\mathbf{p})]|0\rangle = \omega(\mathbf{p})\delta(\mathbf{p} - \mathbf{q}), \tag{6.34}$$

which means that the commutation relation (6.28) is satisfied. Moreover, the commutation relations $[a(\mathbf{p}), a(\mathbf{q})] = 0$ and $[a^\dagger(\mathbf{p}), a^\dagger(\mathbf{q})] = 0$ are fulfilled.

6.2 Field operators and commutators

The Klein–Gordon field operator $\phi(x)$ can be decomposed into positive- and negative-frequency parts as

$$\phi(x) = \phi_+(x) + \phi_-(x), \tag{6.35}$$

---

[3] The reason to 'replace' the commutation relations in Eqs. (6.18) and (6.19) with the commutation relation in Eq. (6.28) is that we want to have commutation relations expressed in the continuous variable $\mathbf{p}$ instead of the discrete index n. In addition, the time dependence of Eqs. (6.18) and (6.19) is removed and instead combined with the orthonormal functions $u_n(\mathbf{x})$.

where

$$\phi_+(x) = \frac{1}{\sqrt{2(2\pi)^3}} \int_{k_0>0} \frac{d^3k}{k_0} a(\mathbf{k}) e^{-ik\cdot x}, \tag{6.36}$$

$$\phi_-(x) = \frac{1}{\sqrt{2(2\pi)^3}} \int_{k_0>0} \frac{d^3k}{k_0} a^\dagger(\mathbf{k}) e^{ik\cdot x} \tag{6.37}$$

such that it holds for the vacuum that $\phi_+(x)|0\rangle = 0$, which means that $\phi_+(x)$ acts as an annihilation operator for a particle, whereas $\phi_-(x)$ is the corresponding creation operator. Thus, in total for the Klein–Gordon field operator, we have the expansion

$$\phi(x) = \frac{1}{\sqrt{2(2\pi)^3}} \int_{k_0>0} \frac{d^3k}{k_0} \left[a(\mathbf{k}) e^{-ik\cdot x} + a^\dagger(\mathbf{k}) e^{ik\cdot x}\right], \tag{6.38}$$

which obeys the Klein–Gordon equation (cf. Chapter 2). In fact, since the operator $a(\mathbf{k})$ comes in pair with its adjoint operator $a^\dagger(\mathbf{k})$ in Eq. (6.38), the Klein–Gordon field operator is indeed Hermitian. Now, evaluate the commutator $[\phi(x), \phi(y)]$, which gives

$$[\phi(x), \phi(y)] = \frac{1}{2(2\pi)^3} \int_{k_0,q_0>0} \frac{d^3k\, d^3q}{k_0\ q_0} \left[a(\mathbf{k}) e^{-ik\cdot x} + a^\dagger(\mathbf{k}) e^{ik\cdot x}, a(\mathbf{q}) e^{-iq\cdot y}\right.$$

$$\left. + a^\dagger(\mathbf{q}) e^{iq\cdot y}\right]$$

$$= \{\text{using Eq. (6.28)}\}$$

$$= \frac{1}{2(2\pi)^3} \int_{k_0,q_0>0} \frac{d^3k\, d^3q}{k_0\ q_0} \omega(\mathbf{q}) \delta(\mathbf{k}-\mathbf{q}) \left[e^{-i(k\cdot x - q\cdot y)} - e^{i(k\cdot x - q\cdot y)}\right]$$

$$= \{q_0 = \omega(\mathbf{q}), \text{ since } q_0 > 0\}$$

$$= \frac{1}{2(2\pi)^3} \int_{k_0>0} \frac{d^3k}{k_0} \left[e^{-ik\cdot(x-y)} - e^{ik\cdot(x-y)}\right] \equiv i\Delta(x-y), \tag{6.39}$$

where $\Delta(x-y)$ is the *Klein–Gordon function* (or the *Pauli–Jordan function*). Thus, the Klein–Gordon function (which is a commutator and **not** a propagator [cf. the discussion on propagators in Section 5.1]) is given by

$$\Delta(x-y) = -\frac{1}{(2\pi)^3} \int_{k_0>0} \frac{d^3k}{k_0} e^{ik\cdot(x-y)} \sin\left(k_0(x^0 - y^0)\right). \tag{6.40}$$

Furthermore, one can compute the following commutators

$$[\phi_+(x), \phi_-(y)] = \frac{1}{2(2\pi)^3} \int_{k_0>0} \frac{d^3k}{k_0} e^{-ik\cdot(x-y)} \equiv i\Delta_+(x-y), \tag{6.41}$$

$$[\phi_-(x), \phi_+(y)] = -\frac{1}{2(2\pi)^3} \int_{k_0>0} \frac{d^3k}{k_0} e^{ik\cdot(x-y)} \equiv i\Delta_-(x-y), \tag{6.42}$$

$$[\phi_+(x), \phi_+(y)] = [\phi_-(x), \phi_-(y)] = 0, \tag{6.43}$$

which means that $\Delta(x-y) = \Delta_+(x-y) + \Delta_-(x-y)$. In order to compute the integral in the Klein–Gordon function (6.40) introduce spherical coordinates, i.e. $r \equiv |\mathbf{x} - \mathbf{y}|$ and $k = |\mathbf{k}|$. Performing the angular integrations yields

$$\Delta(x-y) = -\frac{1}{2\pi^2} \int_0^\infty \frac{k^2 dk}{\sqrt{m^2+k^2}} \frac{\sin kr \sin\left(\sqrt{m^2+k^2}(x^0 - y^0)\right)}{kr}$$

$$= \frac{1}{2\pi r}\frac{\partial}{\partial r} F(r, x^0 - y^0), \tag{6.44}$$

where

$$F(r, x^0 - y^0) \equiv \frac{1}{\pi} \int_0^\infty \frac{dk}{\sqrt{m^2+k^2}} \cos kr \sin\left(\sqrt{m^2+k^2}(x^0 - y^0)\right). \tag{6.45}$$

Then, introduce the variable substitution $k = m \sinh z$, which implies that the function F can be written as

$$F(r, x^0 - y^0) = \begin{cases} J_0(m\sqrt{(x-y)^2}), & x^0 - y^0 > r \\ 0, & |x^0 - y^0| < r \\ -J_0(m\sqrt{(x-y)^2}), & x^0 - y^0 < -r \end{cases}, \tag{6.46}$$

where J_0 is the zeroth Bessel function of the first kind. Now, we have

$$\Delta(x) = -\frac{1}{2\pi}\text{sgn}(x^0)\left[\delta(x^2) - \frac{m^2}{2}\theta(x^2)\frac{J_1(m\sqrt{x^2})}{m\sqrt{x^2}}\right], \tag{6.47}$$

where J_1 is the first Bessel function of the first kind. Therefore, $\Delta(x) = 0$ if $x^2 < 0$, i.e. if x is space-like. Thus, two Klein–Gordon fields commute if their spacetime coordinates are space-like separated, i.e. two fields can only influence each other if they can be connected by a light signal. This shows that our construction obeys so-called *microscopic causality*. The two physical requirements of a quantized theory, i.e. microscopic causality and positive-(semi)definite energy, make it possible to have a connection between spin and statistics, which was shown by Pauli (see Chapter 7).

Next, note that the functions $\Delta(x)$, $\Delta_+(x)$, and $\Delta_-(x)$ are solutions to the **free** Klein–Gordon equation $(\Box + m^2)\Delta(x) = 0$. In addition, it holds that $\Delta_-(x) = -\Delta_+(-x)$.

Finally, calculating the vacuum expectation value of the commutator (6.41) and using $\phi_+(x)|0\rangle = 0$, we obtain

$$i\Delta_+(x-y) = \langle 0|\left[\phi_+(x), \phi_-(y)\right]|0\rangle = \langle 0|\phi_+(x)\phi_-(y)|0\rangle = \langle 0|\phi(x)\phi(y)|0\rangle. \tag{6.48}$$

Similarly, we have

$$i\Delta_-(x-y) = \langle 0|\left[\phi_-(x), \phi_+(y)\right]|0\rangle = -\langle 0|\phi_+(y)\phi_-(x)|0\rangle = -\langle 0|\phi(y)\phi(x)|0\rangle. \tag{6.49}$$

Thus, using the definition of the time-ordered product for bosonic fields (5.4), we can write the Klein–Gordon Feynman propagator as

$$\begin{aligned} i\Delta_F(x-y) &\equiv \langle 0|T\phi(x)\phi(y)|0\rangle \\ &= i\left[\theta(x^0 - y^0)\Delta_+(x-y) - \theta(y^0 - x^0)\Delta_-(x-y)\right], \end{aligned} \tag{6.50}$$

where the minus sign between the two terms comes from the time-ordered product for bosonic fields together with Eqs. (6.48) and (6.49).

6.3 Green's functions and propagators

In order to find the Green's functions for the Klein–Gordon field, we need to investigate the inhomogeneous Klein–Gordon equation, which is given by

$$\left(\Box + m^2\right)G(x-y) = -\delta(x-y), \tag{6.51}$$

where $G(x-y)$ is the Green's function. The Fourier transform of $G(x-y)$ is defined as

$$G(x-y) = \frac{1}{(2\pi)^4}\int e^{-ik\cdot(x-y)}\tilde{G}(k)\,d^4k. \tag{6.52}$$

Note that the definition of the δ-function is

$$\delta(x-y) = \frac{1}{(2\pi)^4}\int e^{-ik\cdot(x-y)}\,d^4k. \tag{6.53}$$

Thus, inserting the Fourier transforms of the Green's function and the δ-function into the inhomogeneous Klein–Gordon equation, we obtain the equation

$$\left(-k^2 + m^2\right)\tilde{G}(k) = -1. \tag{6.54}$$

Solving for the Fourier transform $\tilde{G}$ implies that

$$\tilde{G}(k) = \frac{1}{k^2 - m^2}. \tag{6.55}$$

Thus, inserting the expression for $\tilde{G}$ into the Fourier transform of $G(x - y)$, we obtain

$$G(x - y) = \frac{1}{(2\pi)^4} \int \frac{e^{-ik\cdot(x-y)}}{k^2 - m^2} \, d^4k. \tag{6.56}$$

In the case that $x^0 - y^0 > 0$, we have

$$G(x - y)|_{x^0-y^0>0} = \frac{1}{(2\pi)^4} \int e^{i\mathbf{k}\cdot(\mathbf{x}-\mathbf{y})} \, d^3k \underbrace{\int_{-\infty}^{\infty} \frac{e^{-ik_0(x^0-y^0)}}{(k_0)^2 - \omega(\mathbf{k})^2} \, dk_0}_{\equiv I}, \tag{6.57}$$

where $\omega(\mathbf{k}) \equiv \sqrt{m^2 + \mathbf{k}^2} > 0$. We observe that the integrand of the integral I has poles at $k_0 = \pm\omega(\mathbf{k})$. Therefore, we shift the poles by adding $-i\epsilon$ to ω, which leads to $k_0 = \pm(\omega - i\epsilon)$ and is called the *Feynman prescription*, i.e. one pole is shifted upwards and one pole is shifted downwards, and eventually, we perform the limit $\epsilon \to 0$. Thus, we obtain

$$I = \lim_{\epsilon \to 0} \int_{-\infty}^{\infty} \frac{e^{-ik_0(x^0-y^0)}}{(k_0)^2 - [\omega(\mathbf{k}) - i\epsilon]^2} \, dk_0. \tag{6.58}$$

Note that

$$[k_0 + (\omega - i\epsilon)] \, [k_0 - (\omega - i\epsilon)] = (k_0)^2 - (\omega - i\epsilon)^2$$
$$= (k_0)^2 - \omega^2 + \underbrace{2i\omega\epsilon}_{\equiv i\epsilon, \text{ since } \omega > 0} + \underbrace{\epsilon^2}_{\approx 0}. \tag{6.59}$$

The integral I can be performed by closing the line $(-R, R)$ with a large half-circle (radius R) in the lower half-plane (see Fig. 6.1), where the part of the integral on the half-circle will vanish in the limit $R \to \infty$. Now, the residue theorem gives

$$I = -2\pi i \frac{e^{-i\omega(x^0-y^0)}}{2\omega} = -2\pi i \frac{e^{-ik_0(x^0-y^0)}}{2k_0}, \tag{6.60}$$

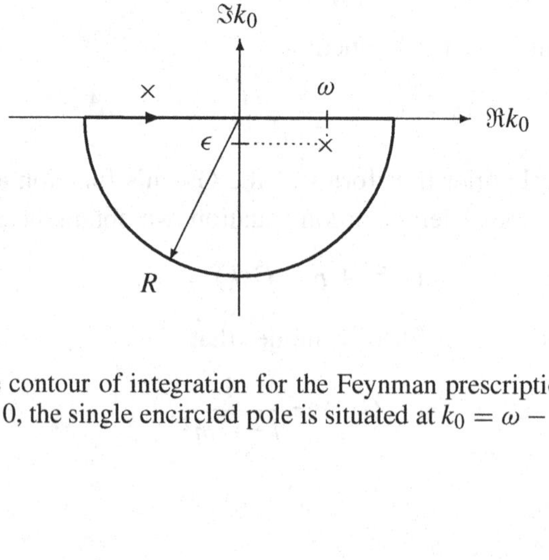

Figure 6.1 The contour of integration for the Feynman prescription. In the case that $x^0 - y^0 > 0$, the single encircled pole is situated at $k_0 = \omega - i\epsilon$.

which means that we obtain

$$G(x-y)|_{x^0-y^0>0} = -\frac{i}{(2\pi)^3}\int\frac{d^3k}{2k_0}e^{-ik\cdot(x-y)} = \Delta_+(x-y). \qquad (6.61)$$

Similarly, for $x^0 - y^0 < 0$, by interchanging x and y in the integral in Eq. (6.61), we find that

$$G(x-y)|_{x^0-y^0<0} = -\frac{i}{(2\pi)^3}\int\frac{d^3k}{2k_0}e^{-ik\cdot(y-x)}$$

$$= -\frac{i}{(2\pi)^3}\int\frac{d^3k}{2k_0}e^{ik\cdot(x-y)} = -\Delta_-(x-y). \qquad (6.62)$$

Thus, in total, using Eq. (6.50) as well as Eqs. (6.61) and (6.62), we have the *Klein–Gordon Feynman propagator*

$$i\Delta_F(x-y) = i\theta(x^0-y^0)\Delta_+(x-y) - i\theta(y^0-x^0)\Delta_-(x-y)$$

$$= \frac{1}{(2\pi)^3}\int\frac{d^3k}{2k_0}\left[\theta(x^0-y^0)e^{-ik\cdot(x-y)} + \theta(y^0-x^0)e^{ik\cdot(x-y)}\right]$$

$$= i\theta(x^0-y^0)\,G(x-y)|_{x^0-y^0>0} + i\theta(y^0-x^0)\,G(x-y)|_{x^0-y^0<0}$$

$$= iG(x-y), \qquad (6.63)$$

which, using Eqs. (6.56)–(6.59), can also be written as [cf. Eq. (5.8)]

$$i\Delta_F(x-y) = \int\frac{d^4k}{(2\pi)^4}e^{-ik\cdot(x-y)}\frac{i}{k^2-m^2+i\epsilon}, \qquad (6.64)$$

where

$$i\Delta_F(k) = \frac{i}{k^2-m^2+i\epsilon} \qquad (6.65)$$

is the Klein–Gordon Feynman propagator in momentum space.

Finally, one usually defines two other propagators in addition to the Feynman propagator. Depending on the sign of $x^0 - y^0$, we can shift **both** poles to either the lower complex plane for $x^0 - y^0 > 0$ or the upper complex plane for $x^0 - y^0 < 0$, and then, close the contour of integration in the respective complex half-planes. In these ways, we obtain the *retarded propagator* and the *advanced propagator*, respectively. Thus, we have the two cases:

(1) $G(x-y) = 0$ if $x^0 - y^0 > 0$, which gives $\Delta_R(x-y)$, i.e. the retarded propagator and
(2) $G(x-y) = 0$ if $x^0 - y^0 < 0$, which gives $\Delta_A(x-y)$, i.e. the advanced propagator.

In these two cases, we will find that $\Delta_R(x-y) = \theta(x^0-y^0)\Delta(x-y)$ and $\Delta_A(x-y) = -\theta(y^0-x^0)\Delta(x-y)$. Especially, it holds that $\Delta(x-y) =$

$\Delta_R(x-y) - \Delta_A(x-y)$, since $\theta(x^0 - y^0) + \theta(y^0 - x^0) = 1$. Note that **all** the three propagators $\Delta_F(x)$, $\Delta_R(x)$, and $\Delta_A(x)$ satisfy the inhomogeneous Klein–Gordon equation $(\Box + m^2)\Delta_i(x) = -\delta(x)$, where $i = F, R, A$.

6.4 The energy–momentum tensor

In order to obtain the Hamiltonian and momentum operators, we choose the set $\{u_n\}$ introduced in Section 6.1 as the solutions to the classical Klein–Gordon equation (cf. Chapter 2). Next, we choose to quantize the fields $\phi(x)$ and $\pi(x)$ such that the commutation relation in Eq. (6.28) is fulfilled. Then, we expand the Klein–Gordon field as was performed in Eqs. (6.35)–(6.38).

Now, according to Noether's theorem (5.69), the energy–momentum tensor is defined as

$$T^{\mu\nu} = \frac{\partial \mathcal{L}_{\text{KG}}}{\partial\left(\partial_\mu \phi\right)} \partial^\nu \phi - g^{\mu\nu} \mathcal{L}_{\text{KG}}. \tag{6.66}$$

For the free neutral Klein–Gordon field described by the Lagrangian (6.1), we find that

$$T^{\mu\nu} = \partial^\mu \phi \partial^\nu \phi + \frac{1}{2} g^{\mu\nu} \left(m^2 \phi^2 - \partial_\lambda \phi \partial^\lambda \phi\right). \tag{6.67}$$

Especially, we have the components

$$T^{00} = \left(\partial^0 \phi\right)^2 + \frac{1}{2} m^2 \phi^2 - \frac{1}{2} \partial_\lambda \phi \partial^\lambda \phi = \frac{1}{2}\dot{\phi}^2 + \frac{1}{2}(\nabla\phi)^2 + \frac{1}{2}m^2\phi^2, \tag{6.68}$$

$$T^{0i} = \left(\partial^0 \phi\right)\left(\partial^i \phi\right) = \dot{\phi}\left(\partial^i \phi\right), \tag{6.69}$$

where the component $T^{00} = \pi\dot{\phi} - \mathcal{L}_{\text{KG}} = \mathcal{H}$ is the Hamiltonian density, which is related to the Lagrangian density $\mathcal{L}_{\text{KG}}$ through a Legendre transformation [cf. the discussion on classical mechanics, Eq. (4.27)]. Note that, in this case, the energy–momentum tensor $T^{\mu\nu}$ is symmetric and that it is conserved, i.e. $\partial_\mu T^{\mu\nu} = \partial_\nu T^{\mu\nu} = 0$. Thus, we can define the (conserved) quantity

$$P^\mu = \int T^{0\mu}\, \mathrm{d}^3 x = \int T^{\mu 0}\, \mathrm{d}^3 x, \tag{6.70}$$

since one can show that it holds that

$$\frac{\mathrm{d}}{\mathrm{d}x^0} P^\mu = 0. \tag{6.71}$$

Thus, we have the components

$$P^0 = \int T^{00}\,\mathrm{d}^3x = \int \left[\frac{1}{2}\pi^2 + \frac{1}{2}(\nabla\phi)^2 + \frac{1}{2}m^2\phi^2\right]\mathrm{d}^3x = \int \mathcal{H}\,\mathrm{d}^3x = H, \quad (6.72)$$

$$P^i = \int T^{0i}\,\mathrm{d}^3x = \int \frac{\partial\mathcal{L}_{\mathrm{KG}}}{\partial\dot\phi}(\partial^i\phi)\,\mathrm{d}^3x = \int \pi(x)\left[\partial^i\phi(x)\right]\mathrm{d}^3x, \quad (6.73)$$

where the component P^0 is interpreted as the total Hamiltonian. Hence, the energy–momentum tensor yields the Hamiltonian and the 3-momentum operator.

Now, normal-ordering the 3-momentum operator in Eq. (6.73) and inserting the expansions for the fields $\phi(x)$ and $\pi(x)$ into this result as well as performing the x-integration, we have

$$P^i = \int\, :\pi(x)\left[\partial^i\phi(x)\right]:\,\mathrm{d}^3x$$

$$= -\frac{1}{2(2\pi)^3}\int_{p^0,q^0>0}\frac{\mathrm{d}^3p\,\mathrm{d}^3q}{p^0\,q^0}\,p^0q^i$$

$$\times :\left\{\left[a(\mathbf{q})a(\mathbf{p})\mathrm{e}^{-\mathrm{i}(q^0+p^0)x^0} + a^\dagger(\mathbf{q})a^\dagger(\mathbf{p})\mathrm{e}^{\mathrm{i}(q^0+p^0)x^0}\right]\right.$$

$$\left.\times\,\delta(\mathbf{q}+\mathbf{p}) - \left[a(\mathbf{q})a^\dagger(\mathbf{p}) + a^\dagger(\mathbf{p})a(\mathbf{q})\right]\delta(\mathbf{q}-\mathbf{p})\right\}:, \quad (6.74)$$

where it of course holds that $p^0 = \omega(\mathbf{p})$ and $q^0 = \omega(\mathbf{q})$. Then, using the properties of normal-ordering and performing the p-integration, we find that

$$P^i = -\frac{1}{2}\int_{q^0>0}\frac{\mathrm{d}^3q}{q^0}q^i :\left[a(\mathbf{q})a(-\mathbf{q})\mathrm{e}^{-2\mathrm{i}q^0x^0} + a^\dagger(\mathbf{q})a^\dagger(-\mathbf{q})\mathrm{e}^{2\mathrm{i}q^0x^0} - 2a^\dagger(\mathbf{q})a(\mathbf{q})\right]: . \tag{6.75}$$

Next, observing that the first two terms within the brackets in Eq. (6.75) are even with respect to the replacement $\mathbf{q} \mapsto -\mathbf{q}$, but the factor q^i is odd, it means that the corresponding integrals are equal to zero, and finally, we obtain

$$P^i = -\frac{1}{2}\int_{q^0>0}\frac{\mathrm{d}^3q}{q^0}q^i\left[-2:a^\dagger(\mathbf{q})a(\mathbf{q}):\right] = \int_{q^0>0}\frac{\mathrm{d}^3q}{q^0}q^i :a^\dagger(\mathbf{q})a(\mathbf{q}):$$

$$= \int_{q^0>0}\frac{\mathrm{d}^3q}{q^0}q^i a^\dagger(\mathbf{q})a(\mathbf{q}), \tag{6.76}$$

where in the last step we have used that the operators are already normal-ordered. In fact, similar to the 3-momentum operator in Eq. (6.76), we can write the quantity P^μ as

$$P^\mu = \int_{q^0>0}\frac{\mathrm{d}^3q}{q^0}q^\mu a^\dagger(\mathbf{q})a(\mathbf{q}), \tag{6.77}$$

which is the 4-momentum operator, i.e. four out of the ten generators of the Poincaré group.

Now, consider the state $|\mathbf{p}\rangle = a^\dagger(\mathbf{p})|0\rangle$. Applying the operator P^μ on this state gives

$$P^\mu|\mathbf{p}\rangle = \int_{q^0>0} \frac{d^3q}{q^0} q^\mu a^\dagger(\mathbf{q})a(\mathbf{q})a^\dagger(\mathbf{p})|0\rangle = \left\{[a(\mathbf{p}), a^\dagger(\mathbf{q})] = p^0\delta(\mathbf{p} - \mathbf{q})\right\}$$

$$= \int_{q^0>0} \frac{d^3q}{q^0} q^\mu a^\dagger(\mathbf{q})\left[a^\dagger(\mathbf{p})a(\mathbf{q}) + q^0\delta(\mathbf{q} - \mathbf{p})\right]|0\rangle = p^\mu a^\dagger(\mathbf{p})|0\rangle = p^\mu|\mathbf{p}\rangle,$$

$$(6.78)$$

where $(p^\mu) = (\omega(\mathbf{p}), \mathbf{p})$. Thus, the state $|\mathbf{p}\rangle = a^\dagger(\mathbf{p})|0\rangle$ is a one-particle state with 3-momentum $\mathbf{p}$ and energy $E = \omega(\mathbf{p})$.

Finally, we can express the six other generators of the Poincaré group in terms of the energy–momentum tensor as [cf. Eqs. (5.70) and (5.71)]

$$M^{\mu\nu} = \int \left(x^\mu T^{0\nu} - x^\nu T^{0\mu}\right) d^3x, \qquad (6.79)$$

which are the generators for rotations and boosts. The three generators for rotations can be interpreted as operators for the three components of the angular momentum, whereas the three generators for boosts correspond to operators for the three boosts in the Cartesian coordinate directions x, y, and z. Note that in the case of the Klein–Gordon field, the angular momentum does not contain spin, since Klein–Gordon particles are spinless.

6.5 Classical external sources

The Lagrangian for a neutral Klein–Gordon field with a classical external source is given by

$$\mathcal{L} = \frac{1}{2}\left(\partial_\mu\phi\partial^\mu\phi - m^2\phi^2\right) + g\rho\phi, \qquad (6.80)$$

where g is a coupling constant and ρ is a given c-number[4] function (a smooth function) and represents the classical external source. From this Lagrangian, using Euler–Lagrange field equations, it follows that the equation of motion is

$$\left(\Box + m^2\right)\phi(x) = g\rho(x), \qquad (6.81)$$

which constitutes an inhomogeneous Klein–Gordon equation. In addition, the conjugate momentum is again $\pi(x) = \dot{\phi}(x)$. Using the Green's functions, we can write the solution to Eq. (6.81) as

---

[4] The abbreviation c-number stands for complex (or classical or commuting) number.

$$\phi(x) = \phi_0(x) - g \int \Delta_{R,A}(x - y)\rho(y)\, d^4 y, \tag{6.82}$$

where $\phi_0(x)$ is the solution to the homogeneous Klein–Gordon equation $(\Box + m^2)$ $\phi_0(x) = 0$. Note that Eq. (6.82) is valid for Δ_R and Δ_A, since both are propagators that satisfy the same inhomogeneous Klein–Gordon equation (cf. the discussion in Section 6.3).

6.6 The charged Klein–Gordon field

The Lagrangian for a free charged (complex) Klein–Gordon field is given by

$$\mathcal{L} = \partial_\mu \phi^\dagger \partial^\mu \phi - m^2 \phi^\dagger \phi. \tag{6.83}$$

Using the Euler–Lagrange field equations for **both** the field ϕ and the field $\phi^\dagger$, we obtain the two following equations

$$(\Box + m^2)\,\phi = 0 \quad \text{and} \quad (\Box + m^2)\,\phi^\dagger = 0, \tag{6.84}$$

which are the equations of motion for the two fields, respectively. The conjugate momenta for the two fields are given by

$$\pi = \pi_\phi = \frac{\partial \mathcal{L}}{\partial \dot{\phi}} = \dot{\phi}^\dagger \quad \text{and} \quad \pi^\dagger = \pi_{\phi^\dagger} = \frac{\partial \mathcal{L}}{\partial \dot{\phi}^\dagger} = \dot{\phi}. \tag{6.85}$$

Now, we can split the non-Hermitian field ϕ into two independent Hermitian fields ϕ_1 and ϕ_2 as follows

$$\phi = \frac{1}{\sqrt{2}}(\phi_1 + i\phi_2) \quad \text{and} \quad \phi^\dagger = \frac{1}{\sqrt{2}}(\phi_1 - i\phi_2). \tag{6.86}$$

Since the fields ϕ_1 and ϕ_2 are Hermitian, we can independently quantize them in the same manner as before for the neutral Klein–Gordon field (cf. Section 6.1).

Problems

(1) Using the Klein–Gordon Feynman propagator

$$\Delta_F(x) = \frac{1}{(2\pi)^4} \int e^{-ik \cdot x} \frac{1}{k^2 - m^2 + i\epsilon}\, d^4 k,$$

show that it satisfies the inhomogeneous Klein–Gordon equation

$$(\Box + m^2)\Delta_F(x) = -\delta(x).$$

(2) Find the 3-momentum operator

$$P^i = \int T^{0i}\, d^3x, \quad \text{where } T^{\mu\nu} = \frac{\partial \mathcal{L}}{\partial(\partial_\mu \phi)} \partial^\nu \phi - g^{\mu\nu}\mathcal{L},$$

for the Klein–Gordon field and prove that

$$[P^i, \phi] = i\partial^i \phi.$$

Note that it is enough to consider the case of a free neutral Klein–Gordon field.

(3) (a) If the charged field $\phi = (\phi_1 + i\phi_2)/\sqrt{2}$, where ϕ_1 and ϕ_2 are commuting Hermitian fields, and if the charged field ϕ satisfies the canonical commutation relations, i.e. $[\phi(x), \pi(y)] = i\delta(\mathbf{x}-\mathbf{y})$, which is valid only for equal times t in the spacetime points x and y, and all other commutators are zero, show that

$$[\phi_i(x), \partial_t \phi_i(y)] = i\delta(\mathbf{x} - \mathbf{y}),$$

where ϕ_i is either ϕ_1 or ϕ_2.

(b) Show that the Lagrangian density for the charged field ϕ,

$$\mathcal{L} = \partial_\mu \phi^\dagger \partial^\mu \phi - m^2 \phi^\dagger \phi,$$

can be written as the sum of two *independent* Lagrangian densities

$$\mathcal{L} = \mathcal{L}_1 + \mathcal{L}_2,$$

where each density $\mathcal{L}_i$ is multiplied by an overall factor of 1/2 compared with the Lagrangian density for its charged counterpart.

(c) Using the density $\mathcal{L}_i$, find the conjugate momentum density π_i, for a neutral theory.

(d) Discuss the significance of your results. What is the Lagrangian density for a neutral scalar theory?

(4) By making the minimal coupling

$$\partial_\mu \phi(x) \to D_\mu \phi(x) = [\partial_\mu + ieA_\mu(x)]\phi(x),$$

$$\partial_\mu \phi^\dagger(x) \to [D_\mu \phi(x)]^\dagger = [\partial_\mu - ieA_\mu(x)]\phi^\dagger(x)$$

in the Lagrangian density of the complex Klein–Gordon field $\phi(x)$, derive the Lagrangian density $\mathcal{L}_I$ for the interaction of the charged bosons, described by the field $\phi(x)$, with the electromagnetic field $A_\mu(x)$.

(5) (a) Compute the vacuum expectation value, i.e. $\langle 0|H|0\rangle$, of the scalar field Hamiltonian

$$H = \frac{1}{2}\int \left[(\partial_0 \phi)^2 + (\nabla \phi)^2 + m^2 \phi^2\right] d^3x.$$

Hints: 1. The quantized version of the scalar field Hamiltonian can be written as

$$H = \frac{1}{2}\int_{k_0>0} \frac{\omega(\mathbf{k})}{k_0}\left[a^\dagger(\mathbf{k})a(\mathbf{k}) + a(\mathbf{k})a^\dagger(\mathbf{k})\right] d^3k,$$

where $\omega(\mathbf{k}) = \sqrt{m^2 + \mathbf{k}^2}$.

2. The result should be a product of two divergent factors.

(b) How should one interpret this result and how should one get rid of this divergent term?

Guide to additional recommended reading

The following books (see the indicated pages) and their authors have similar treatments of the content in the present chapter.

- A. Z. Capri, *Relativistic Quantum Mechanics and Introduction to Quantum Field Theory*, World Scientific (2002), pp. 80–111.
- F. Gross, *Relativistic Quantum Mechanics and Field Theory*, Wiley (1993), pp. 194–197.
- F. Mandl and G. Shaw, *Quantum Field Theory*, rev. edn, Wiley (1994), pp. 43–59.
- M. E. Peskin and D. V. Schroeder, *An Introduction to Quantum Field Theory*, Addison-Wesley (1995), pp. 13–34.
- G. Scharf, *Finite Quantum Electrodynamics*, Springer (1989), pp. 64–69.
- F. Schwabl, *Advanced Quantum Mechanics*, Springer (1999), pp. 277–287, 297–298.
- S. S. Schweber, *An Introduction to Relativistic Quantum Field Theory*, Dover (2005), pp. 156–217.
- F. J. Ynduráin, *Relativistic Quantum Mechanics and Introduction to Field Theory*, Springer (1996), pp. 193–198.

7

Quantization of the Dirac field

In this chapter, we will quantize the Dirac field using canonical quantization, i.e. with a similiar method as was used for the Klein–Gordon field (cf. Chapter 6). However, we will observe that we need to replace the canonical commutation relations with canonical anticommutation relations in order to obtain positive energy for the quantized Dirac field and to obey the Pauli exclusion principle. In addition, we will study the transformations of parity, time reversal, and charge conjugation as well as the CPT symmetry for this field. Especially, we will investigate the Majorana field, which is a special case of the Dirac field. We will also derive Green's functions and propagators and briefly investigate interactions.

The Dirac equation (cf. Chapter 3) establishes a single-particle theory (usually known as the Dirac theory), since it cannot take into account creation and annihilation of particles. Therefore, the Dirac equation for wave functions has to be replaced by the Dirac equation for quantum fields, which means that the problems of Dirac theory are circumvented by introducing a quantum field theory reformulation of this theory; thus abandoning wave functions in favour of quantum fields. In fact, adding to the theory of the Dirac equation for quantum fields the quantized version of the electromagnetic field (see Chapter 8), we will end up with the theory of QED, which will be investigated in detail in Chapters 11–13.

7.1 The free Dirac field

The Lagrangian density for a free Dirac field is given by

$$\mathcal{L}_D = \bar{\psi} \left(\frac{i}{2} \overleftrightarrow{\partial} - m \mathbb{1}_4 \right) \psi = -\frac{1}{2} \bar{\psi} \left(-i\partial\!\!\!/ + m \mathbb{1}_4 \right) \psi - \frac{1}{2} \left(i \partial_\mu \bar{\psi} \gamma^\mu + m \bar{\psi} \right) \psi,$$

$$(7.1)$$

where $a \overleftrightarrow{\partial} b = a\partial b - (\partial a)b$. Note that the fields ψ and $\bar{\psi}$ should be independently varied, since they are treated as dynamically independent fields. Using the Euler–Lagrange field equations for $\bar{\psi}$, i.e. varying $\bar{\psi}$, leads to the Dirac equation

$$(-i\partial\!\!\!/ + m\mathbb{1}_4)\,\psi = 0. \tag{7.2}$$

Similarly, one can use the Euler–Lagrange field equations for ψ to obtain the corresponding Dirac equation for $\bar{\psi}$.

Exercise 7.1 Using Euler–Lagrange field equations for the Lagrangian (7.1) and the field $\bar{\psi}$, show that the Dirac equation (7.2) is obtained.

In the case of the free Dirac field, again using Noether's theorem (5.69) and remembering that quantities for a Dirac field do not necessarily commute, the energy–momentum tensor is defined as

$$T^{\mu\nu} = \frac{\partial \mathcal{L}_D}{\partial(\partial_\mu \psi)}\partial^\nu \psi + \partial^\nu \bar{\psi}\frac{\partial \mathcal{L}_D}{\partial(\partial_\mu \bar{\psi})} - g^{\mu\nu}\mathcal{L}_D \tag{7.3}$$

and inserting Eq. (7.1) into Eq. (7.3), we find that

$$T^{\mu\nu} = \frac{i}{2}\left[\bar{\psi}\gamma^\mu\partial^\nu\psi - (\partial^\nu\bar{\psi})\gamma^\mu\psi\right], \tag{7.4}$$

which is not symmetric, but could be made symmetric. However, similar to the case of the free neutral Klein–Gordon field, the energy–momentum tensor $T^{\mu\nu}$ is conserved, i.e. $\partial_\mu T^{\mu\nu} = 0$. Now, one can compute

$$P^\mu = \int T^{0\mu}\,d^3x = \frac{i}{2}\int\left[\bar{\psi}\gamma^0\partial^\mu\psi - (\partial^\mu\bar{\psi})\gamma^0\psi\right]d^3x = i\int\bar{\psi}\gamma^0\partial^\mu\psi\,d^3x, \tag{7.5}$$

where we have used integration by parts in the last step as well as the fact that a 4-divergence does not contribute to the integral, and especially, using the definitions of derivatives (2.8) and the Dirac equation (7.2), the total Hamiltonian and 3-momentum operator become

$$H = P^0 = \int\bar{\psi}\left(-i\gamma\cdot\nabla + m\mathbb{1}_4\right)\psi\,d^3x = \int\psi^\dagger\left(-i\alpha\cdot\nabla + \beta m\right)\psi\,d^3x, \tag{7.6}$$

$$\mathbf{P} = -i\int\bar{\psi}\gamma^0\nabla\psi\,d^3x = -i\int\psi^\dagger\nabla\psi\,d^3x, \tag{7.7}$$

respectively. In addition, one has the operators for the components of the total angular momentum and the boosts

$$M^{\mu\nu} = \int\psi^\dagger\left[-i\,(x^\mu\partial^\nu - x^\nu\partial^\mu) - \frac{1}{2}\sigma^{\mu\nu}\right]\psi\,d^3x. \tag{7.8}$$

Note that in the case of the Dirac field, the total angular momentum contains spin, since Dirac particles are spin-1/2 particles, which is represented by the term $\sigma^{\mu\nu}/2$ in Eq. (7.8). Furthermore, the conjugate momenta are given by

$$\pi = \frac{\partial \mathcal{L}_D}{\partial\dot{\psi}} = \frac{i}{2}\psi^\dagger \quad \text{and} \quad \bar{\pi} = \frac{\partial \mathcal{L}_D}{\partial\dot{\bar{\psi}}} = -\frac{i}{2}\gamma^0\psi. \tag{7.9}$$

However, from this result, we observe that the conjugate momenta π and $\bar{\pi}$ are not independent, i.e. they are dependent. Thus, instead of Eq. (7.1), we consider the Lagrangian density

$$\mathcal{L}_D = \bar{\psi} \left(i\partial\!\!\!/ - m\mathbb{1}_4 \right) \psi, \tag{7.10}$$

which differs from Eq. (7.1) by the fact that we now impose that the derivative operator solely acts on ψ (and not on $\bar{\psi}$) and we are only varying the field $\bar{\psi}$ [which again leads to the Dirac equation (7.2)]. Therefore, in this case, we obtain the conjugate momenta

$$\pi = i\psi^\dagger \quad \text{and} \quad \bar{\pi} = 0. \tag{7.11}$$

Note that the difference between the Lagrangian densities (7.1) and (7.10) is a **total** divergence, which means that the physics described by the two Lagrangian densities is equivalent (cf. the discussion in Section 5.2).

Exercise 7.2 Using Eq. (7.5), show that the total Hamiltonian can be written as in Eq. (7.6).

7.2 Quantization

There are two possibilities for the quantization of the Dirac fields. Either the fields can satisfy canonical commutation or anticommutation relations, which are given by[1]

$$[\psi(x), \pi(y)]_\pm = \psi(x)\pi(y) \pm \pi(y)\psi(x) = i\delta(\mathbf{x} - \mathbf{y}), \tag{7.12}$$

$$[\psi(x), \psi(y)]_\pm = \psi(x)\psi(y) \pm \psi(y)\psi(x) = 0, \tag{7.13}$$

$$[\pi(x), \pi(y)]_\pm = \pi(x)\pi(y) \pm \pi(y)\pi(x) = 0. \tag{7.14}$$

Note that Eqs. (7.12)–(7.14) are only valid for equal times t in the spacetime points x and y, i.e. $x = (t, \mathbf{x})$ and $y = (t, \mathbf{y})$. Since $\pi = i\psi^\dagger$, we have that

$$[\psi(x), \psi^\dagger(y)]_\pm = \delta(\mathbf{x} - \mathbf{y}), \tag{7.15}$$

$$[\psi(x), \psi(y)] = [\psi^\dagger(x), \psi^\dagger(y)]_\pm = 0. \tag{7.16}$$

Now, in order to have positive energy for the free Dirac particle, we need to choose the anticommutation relations (see the discussion in Section 7.3). Expanding the Dirac field yields

---

[1] In the following, we will sometimes suppress Dirac spinor indices, which means that some expressions will not formally make sense. For example, in Eqs. (7.12)–(7.14) such a situation occurs, where products of Dirac fields [e.g. $\psi(x)\psi(y)$] have not been properly defined. However, the hope is that this sloppiness will not create any problems.

$$\psi(x) = \sum_s \int \frac{d^3p}{(2\pi)^{3/2}} \sqrt{\frac{m}{E(\mathbf{p})}} \left[b(\mathbf{p}, s)u(\mathbf{p}, s)e^{-ip\cdot x} + d^\dagger(\mathbf{p}, s)v(\mathbf{p}, s)e^{ip\cdot x} \right],$$

(7.17)

$$\psi^\dagger(x) = \sum_s \int \frac{d^3p}{(2\pi)^{3/2}} \sqrt{\frac{m}{E(\mathbf{p})}} \left[b^\dagger(\mathbf{p}, s)u^\dagger(\mathbf{p}, s)e^{ip\cdot x} + d(\mathbf{p}, s)v^\dagger(\mathbf{p}, s)e^{-ip\cdot x} \right],$$

(7.18)

where the normalization factor $\sqrt{m/E(\mathbf{p})}$ is a convention as well as $b(\mathbf{p}, s)$, $b^\dagger(\mathbf{p}, s)$, $d(\mathbf{p}, s)$, and $d^\dagger(\mathbf{p}, s)$ are creation and annihilation operators, which are different from the operators $a(\mathbf{p})$ and $a^\dagger(\mathbf{p})$ (cf. Chapter 6). Similarly, for the Dirac adjoint field, using Eq. (7.18), we obtain

$$\bar{\psi}(x) \equiv \psi^\dagger(x)\gamma^0 = \sum_s \int \frac{d^3p}{(2\pi)^{3/2}} \sqrt{\frac{m}{E(\mathbf{p})}}$$
$$\times \left[b^\dagger(\mathbf{p}, s)\bar{u}(\mathbf{p}, s)e^{ip\cdot x} + d(\mathbf{p}, s)\bar{v}(\mathbf{p}, s)e^{-ip\cdot x} \right], \quad (7.19)$$

since $u^\dagger(\mathbf{p}, s)\gamma^0 = \bar{u}(\mathbf{p}, s)$ and $v^\dagger(\mathbf{p}, s)\gamma^0 = \bar{v}(\mathbf{p}, s)$. Note that the creation and annihilation operators are interpreted as follows: The operator $b(\mathbf{p}, s)$ annihilates a particle (having positive energy) with 3-momentum $\mathbf{p}$ and spin s, the operator $b^\dagger(\mathbf{p}, s)$ creates a particle with 3-momentum $\mathbf{p}$ and spin s, the operator $d(\mathbf{p}, s)$ annhilates an antiparticle (having positive energy) with 3-momentum $\mathbf{p}$ and spin s, and finally, the operator $d^\dagger(\mathbf{p}, s)$ creates an antiparticle with 3-momentum $\mathbf{p}$ and spin s. In principle, the operators $d(\mathbf{p}, s)$ and $d^\dagger(\mathbf{p}, s)$ can instead be interpreted as creation and annihilation operators for a particle with negative energy, respectively.

7.3 Positive energy

As we have seen in Section 7.1, the Hamiltonian for a free Dirac field is given by

$$H = \int \psi^\dagger(x) \left(-i\boldsymbol{\alpha} \cdot \nabla + \beta m \right) \psi(x) \, d^3x. \quad (7.20)$$

First, inserting the expansions of the fields ψ and $\psi^\dagger$, i.e. Eqs. (7.17) and (7.18), into the Hamiltonian (7.20), we have

$$H = \int \frac{d^3p}{\sqrt{(2\pi)^3}} \frac{d^3q}{\sqrt{(2\pi)^3}} \frac{m}{\sqrt{E(\mathbf{p})E(\mathbf{q})}}$$
$$\times \sum_s \sum_{s'} \left[b^\dagger(\mathbf{p}, s)u^\dagger(\mathbf{p}, s)e^{ip\cdot x} + d(\mathbf{p}, s)v^\dagger(\mathbf{p}, s)e^{-ip\cdot x} \right]$$
$$\times \left(-i\boldsymbol{\alpha} \cdot \nabla + \beta m \right) \left[b(\mathbf{q}, s')u(\mathbf{q}, s')e^{-iq\cdot x} + d^\dagger(\mathbf{q}, s')v(\mathbf{q}, s')e^{iq\cdot x} \right] d^3x.$$

(7.21)

Second, using the relations

$$(-i\boldsymbol{\alpha} \cdot \nabla + \beta m)\, u(\mathbf{p}, s)e^{-ip\cdot x} = E(\mathbf{p})u(\mathbf{p}, s)e^{-ip\cdot x}, \qquad (7.22)$$

$$(-i\boldsymbol{\alpha} \cdot \nabla + \beta m)\, v(\mathbf{p}, s)e^{ip\cdot x} = -E(\mathbf{p})v(\mathbf{p}, s)e^{ip\cdot x}, \qquad (7.23)$$

we find that

$$
\begin{aligned}
H = \frac{1}{(2\pi)^3} \int d^3p\, d^3q\, & \frac{m}{\sqrt{E(\mathbf{p})E(\mathbf{q})}} \\
\times \sum_s \sum_{s'} \Big[& E(\mathbf{q})b^\dagger(\mathbf{p}, s)b(\mathbf{q}, s')u^\dagger(\mathbf{p}, s)u(\mathbf{q}, s')e^{i(p-q)\cdot x} \\
& - E(\mathbf{q})d(\mathbf{p}, s)d^\dagger(\mathbf{q}, s')v^\dagger(\mathbf{p}, s)v(\mathbf{q}, s')e^{-i(p-q)\cdot x} \\
& + E(\mathbf{q})d(\mathbf{p}, s)b(\mathbf{q}, s')v^\dagger(\mathbf{p}, s)u(\mathbf{q}, s')e^{-i(p+q)\cdot x} \\
& - E(\mathbf{q})b^\dagger(\mathbf{p}, s)d^\dagger(\mathbf{q}, s')u^\dagger(\mathbf{p}, s)v(\mathbf{q}, s')e^{i(p+q)\cdot x} \Big]\, d^3x. \quad (7.24)
\end{aligned}
$$

Third, using Eq. (3.129) and summing over s in the first two terms as well as performing the q- and x-integrations yields

$$
\begin{aligned}
H = \int d^3p\, \frac{m}{E(\mathbf{p})} \sum_s & E(\mathbf{p}) \left[b^\dagger(\mathbf{p}, s)b(\mathbf{p}, s) - d(\mathbf{p}, s)d^\dagger(\mathbf{p}, s) \right] \frac{E(\mathbf{p})}{m} \\
+ \int d^3p\, \frac{m}{E(\mathbf{p})} \sum_s \sum_{s'} & E(\mathbf{p}) \Big[d(\mathbf{p}, s)b(\mathbf{p}, s')v^\dagger(\mathbf{p}, s)u(-\mathbf{p}, s')e^{-2ip^0x^0} \\
& - b^\dagger(\mathbf{p}, s)d^\dagger(\mathbf{p}, s')u^\dagger(\mathbf{p}, s)v(-\mathbf{p}, s')e^{2ip^0x^0} \Big], \quad (7.25)
\end{aligned}
$$

where the last integral is equal to zero due to Eq. (3.129). Finally, we can write the Hamiltonian in momentum space as

$$H = \int E(\mathbf{p}) \sum_s \left[b^\dagger(\mathbf{p}, s)b(\mathbf{p}, s) - d(\mathbf{p}, s)d^\dagger(\mathbf{p}, s) \right] d^3p. \qquad (7.26)$$

Here we are forced to choose an anticommutation relation for the d and $d^\dagger$ operators, since otherwise the energy will not be positive (semi)definite, i.e. not bounded from below. The anticommutation relation for the d and $d^\dagger$ operators is

$$d(\mathbf{p}, s)d^\dagger(\mathbf{q}, s') = -d^\dagger(\mathbf{q}, s')d(\mathbf{p}, s) + \delta_{ss'}\delta(\mathbf{p} - \mathbf{q}). \qquad (7.27)$$

Note that Eq. (7.27) will normal-order the operators $d(\mathbf{p}, s)$ and $d^\dagger(\mathbf{p}, s)$ in Eq. (7.26), i.e. it will place the creation operator $d^\dagger(\mathbf{p}, s)$ to the left and the annihilation operator $d(\mathbf{p}, s)$ to the right, which is the same order as for the operators $b(\mathbf{p}, s)$ and $b^\dagger(\mathbf{p}, s)$ in the first term of Eq. (7.26). Thus, we obtain (modulo an irrelevant infinite constant) the quantized version of the Hamiltonian for a free Dirac field as

$$H \equiv \; : H : = \int E(\mathbf{p}) \sum_s \left[b^\dagger(\mathbf{p}, s)b(\mathbf{p}, s) + d^\dagger(\mathbf{p}, s)d(\mathbf{p}, s) \right] d^3 p \geq 0, \quad (7.28)$$

which leads to energies that are positive (semi)definite, despite the fact that there are negative energy solutions to the Dirac equation. Similarly, using Eq. (7.7), the 3-momentum operator of the quantized Dirac field is given by

$$\mathbf{P} \equiv \; : \mathbf{P} : \; = \int \mathbf{p} \sum_s \left[b^\dagger(\mathbf{p}, s)b(\mathbf{p}, s) + d^\dagger(\mathbf{p}, s)d(\mathbf{p}, s) \right] d^3 p. \quad (7.29)$$

In summary, using the anticommutation relations in Eqs. (7.15) and (7.16) as well as the expansions of the Dirac field (7.17) and (7.18), we find the corresponding canonical anticommutation relations for the creation and annihilation operators

$$\{b(\mathbf{p}, s), b^\dagger(\mathbf{q}, s')\} = \delta_{ss'}\delta(\mathbf{p} - \mathbf{q}), \quad (7.30)$$

$$\{d(\mathbf{p}, s), d^\dagger(\mathbf{q}, s')\} = \delta_{ss'}\delta(\mathbf{p} - \mathbf{q}). \quad (7.31)$$

In addition, all the other anticommutators between the different creation and annihilation operators are equal to zero. In order to conclude, we compare with the *spin statistics theorem*,[2] which states that

- integer spin fields satisfy Bose–Einstein statistics and commutation relations and
- half-integer spin fields satisfy Fermi–Dirac statistics and anticommutation relations.

In order to construct states for Dirac particles, which are fermions (or antifermions), we observe that the Dirac particles have to satisfy Fermi–Dirac statistics according to the spin statistics theorem, which means that the Pauli exclusion principle has to be obeyed. This leads to the fact that a state cannot contain more than one particle of the same given momentum and spin. In addition, a two-particle state must be antisymmetric under the exchange of the two particles. In the description of Klein–Gordon particles, we obtain a state with m_n particles with the same energy and 3-momentum by acting m_n times with $a_n^\dagger(t)$ [or $a^\dagger(\mathbf{p})$], which makes sense for bosons, but not for fermions, since this violates the Pauli exclusion principle. Thus, if one describes a state with two Klein–Gordon particles with different 3-momenta $\mathbf{p}$ and $\mathbf{q}$ by acting on the vacuum state $|0\rangle$, i.e.

$$|0, \ldots, 0, 1(\mathbf{p}), 0, \ldots, 0, 1(\mathbf{q}), 0, \ldots, 0, \ldots\rangle = a^\dagger(\mathbf{p})a^\dagger(\mathbf{q})|0\rangle, \quad (7.32)$$

then one observes that this state is symmetric with respect to the exchange $a^\dagger(\mathbf{p})$ $a^\dagger(\mathbf{q}) = a^\dagger(\mathbf{q})a^\dagger(\mathbf{p})$ due to the commutation relation $[a^\dagger(\mathbf{p}), a^\dagger(\mathbf{q})] = 0$. However, if one describes fermions with creation operators $b^\dagger(\mathbf{p}, s)$ and $b^\dagger(\mathbf{q}, s')$

---

[2] For the interested reader, please see: W. Pauli, The connection between spin and statistics, *Phys. Rev.* **58**, 716–722 (1940); R. F. Streater and A. S. Wightman, *PCT, Spin and Statistics, and All That*, Princeton (2000).

[or antifermions with creation operators $d^\dagger(\mathbf{p}, s)$ and $d^\dagger(\mathbf{q}, s')$], then one needs antisymmetry, i.e.

$$b^\dagger(\mathbf{p}, s)b^\dagger(\mathbf{q}, s')|0\rangle = -b^\dagger(\mathbf{q}, s')b^\dagger(\mathbf{p}, s)|0\rangle, \tag{7.33}$$

due to the anticommutation relation $\{b^\dagger(\mathbf{p}, s), b^\dagger(\mathbf{q}, s')\} = 0$. In addition, it immediately follows that

$$[b^\dagger(\mathbf{p}, s)]^2 = b^\dagger(\mathbf{p}, s)b^\dagger(\mathbf{p}, s) = 0 \tag{7.34}$$

for all 3-momenta $\mathbf{p}$ and spins s, since one can only have one Dirac particle in each state, if one postulates anticommutation relations instead of commutation relations. In the case of Dirac particles and antiparticles, the number operators are still given by $N(\mathbf{p}, s) = b^\dagger(\mathbf{p}, s)b(\mathbf{p}, s)$ and $\bar{N}(\mathbf{p}, s) = d^\dagger(\mathbf{p}, s)d(\mathbf{p}, s)$ [cf. Eq. (6.23)], but now, using the anticommutation relations, it holds that

$$N(\mathbf{p}, s)^2 = b^\dagger(\mathbf{p}, s)b(\mathbf{p}, s)b^\dagger(\mathbf{p}, s)b(\mathbf{p}, s) = b^\dagger(\mathbf{p}, s)\left[\mathbb{1} - b^\dagger(\mathbf{p}, s)b(\mathbf{p}, s)\right]b(\mathbf{p}, s)$$
$$= b^\dagger(\mathbf{p}, s)b(\mathbf{p}, s) = N(\mathbf{p}, s), \tag{7.35}$$

and similarly for the number operator $\bar{N}$, which means that the possible excitations can only be one or zero, in accordance with the Pauli exclusion principle.

7.4 The charge operator

What happens when the electromagnetic field is coupled to the Dirac field? Again, we start with the Lagrangian for a free Dirac field (7.1) [or Eq. (7.10)]. If we introduce the electromagnetic field through the minimal coupling replacement, i.e. $i\partial_\mu \mapsto i\partial_\mu - eA_\mu$ [cf. Eqs. (3.185) and/or (5.26)], then we observe that the interaction term in the quantized Hamiltonian (7.28) becomes of the form $-eA_\mu j^\mu$, where the electromagnetic 4-current density is given by

$$j^\mu(x) = \bar{\psi}(x)\gamma^\mu\psi(x). \tag{7.36}$$

Thus, using the technique in Section 7.3, the total electric charge operator is found to be

$$Q = q\int j^0(x)\,\mathrm{d}^3x = -e\int \bar{\psi}(x)\gamma^0\psi(x)\,\mathrm{d}^3x$$
$$= -e\int \sum_s \left[b^\dagger(\mathbf{p}, s)b(\mathbf{p}, s) - d^\dagger(\mathbf{p}, s)d(\mathbf{p}, s)\right]\mathrm{d}^3p, \tag{7.37}$$

where we have assumed the quantity q to be the charge of the electron, i.e. $q = -e < 0$. Note the difference in signs of the $b^\dagger b$ and $d^\dagger d$ terms in Eq. (7.37). The reason is the following. The energies contributed by fermions (described by the b and $b^\dagger$ operators) and antifermions (described by the d and $d^\dagger$ operators) are both positive [cf. Eq. (7.28)], but the charges have opposite signs. Thus, the natural interpretation of the quantized Dirac field is that the particles described by the b operators are positive-energy electrons and the particles described by the d operators are positive-energy positrons. In conclusion, quantum field theory solves the problems with the Dirac equation that were discussed in Section 3.2.

7.5 Parity, time reversal, and charge conjugation

7.5.1 Parity

The *parity operator* P reverses the 3-momentum of a particle without flipping its spin (cf. the discussion on parity transformation in Sections 1.2 and 2.1). Mathematically, P is a unitary operator $U(P)$, which we will simply denote by P. For Dirac particles, P transforms the state $b^\dagger(\mathbf{p}, s)|0\rangle$ into the state $b^\dagger(-\mathbf{p}, s)|0\rangle$, and similarly for the d states. Thus, we have

$$Pb(\mathbf{p}, s)P^{-1} = \eta_b b(-\mathbf{p}, s) \quad \text{and} \quad Pd(\mathbf{p}, s)P^{-1} = \eta_d d(-\mathbf{p}, s), \qquad (7.38)$$

where η_b and η_d are possible phases, which are restricted by the condition that two applications of the parity operator should return the observables to their original values. Since all observables are constructed from an even number of fermion operators, we have that $\eta_b^2 = \pm 1$ and $\eta_d^2 = \pm 1$. Using Eq. (7.38) together with the Dirac field (7.17), we find that

$$P\psi(x)P^{-1} = \sum_s \int \frac{\mathrm{d}^3 p}{(2\pi)^{3/2}} \sqrt{\frac{m}{E(\mathbf{p})}}$$
$$\times \left[\eta_b b(-\mathbf{p}, s)u(\mathbf{p}, s)e^{-ip\cdot x} + \eta_d^* d^\dagger(-\mathbf{p}, s)v(\mathbf{p}, s)e^{ip\cdot x} \right]. \qquad (7.39)$$

Changing variables to $\tilde{p} = (p^0, -\mathbf{p})$ and noting that $p \cdot x = \tilde{p} \cdot \tilde{x}_p$, where $\tilde{x}_p = (x^0, -\mathbf{x})$, as well as $u(p) = \gamma^0 u(\tilde{p})$ and $v(p) = -\gamma^0 v(\tilde{p})$, we have that

$$P\psi(x)P^{-1} = \sum_s \int \frac{\mathrm{d}^3 \tilde{p}}{(2\pi)^{3/2}} \sqrt{\frac{m}{E(\tilde{\mathbf{p}})}}$$
$$\times \left[\eta_b b(\tilde{\mathbf{p}}, s)\gamma^0 u(\tilde{\mathbf{p}}, s)e^{-i\tilde{p}\cdot \tilde{x}_p} - \eta_d^* d^\dagger(\tilde{\mathbf{p}}, s)\gamma^0 v(\tilde{\mathbf{p}}, s)e^{i\tilde{p}\cdot \tilde{x}_p} \right] \propto \psi(\tilde{x}_p), \qquad (7.40)$$

which can be realized if $\eta_d^* = -\eta_b$. Therefore, we obtain $\eta_b \eta_d = -\eta_b \eta_b^* = -|\eta_b|^2 = -1$, and thus, we have the parity transformation for the Dirac field $\psi(x)$

$$P\psi(x)P^{-1} = \eta_b \gamma^0 \psi(\tilde{x}_p). \qquad (7.41)$$

Similarly, for the Dirac adjoint field $\bar{\psi}(x)$, we obtain

$$P\bar{\psi}(x)P^{-1} = P\psi^\dagger(x)P^{-1}\gamma^0 = \left[P\psi(x)P^{-1}\right]^\dagger \gamma^0 = \eta_b^* \bar{\psi}(\tilde{x}_p)\gamma^0. \qquad (7.42)$$

Exercise 7.3 Show the relations $u(p) = \gamma^0 u(\tilde{p})$ and $v(p) = -\gamma^0 v(\tilde{p})$.

7.5.2 Time reversal

The *time-reversal operator* T reverses the 3-momentum of a particle as well as flipping the spin. Thus, we must find a mathematical operation that flips the spin of a spinor. Defining again $\tilde{p} = (p^0, -\mathbf{p})$, this is achieved by

$$u(\tilde{\mathbf{p}}, -s) = -i \begin{pmatrix} \sigma^2 & 0 \\ 0 & \sigma^2 \end{pmatrix} \left[u(\mathbf{p}, s) \right]^* = -\gamma^1 \gamma^3 \left[u(\mathbf{p}, s) \right]^*, \tag{7.43}$$

$$v(\tilde{\mathbf{p}}, -s) = -\gamma^1 \gamma^3 \left[v(\mathbf{p}, s) \right]^*. \tag{7.44}$$

Note that the operation of complex conjugation is antilinear. Thus, T is **not** a unitary operator as P, but referred to as an antiunitary operator, which means that $T(\text{c-number}) = (\text{c-number})^* T$. Therefore, we define the time reversal transformation of Dirac fermion annihilation operators as

$$T b(\mathbf{p}, s) T^{-1} = b(-\mathbf{p}, -s) \quad \text{and} \quad T d(\mathbf{p}, s) T^{-1} = d(-\mathbf{p}, -s). \tag{7.45}$$

Using Eqs. (7.43)–(7.45) together with the Dirac field (7.17), we find that

$$T \psi(x) T^{-1} = \sum_s \int \frac{d^3 p}{(2\pi)^{3/2}} \sqrt{\frac{m}{E(\mathbf{p})}}$$
$$\times T \left[b(\mathbf{p}, s) u(\mathbf{p}, s) e^{-ip \cdot x} + d^\dagger(\mathbf{p}, s) v(\mathbf{p}, s) e^{ip \cdot x} \right] T^{-1}$$
$$= \sum_s \int \frac{d^3 p}{(2\pi)^{3/2}} \sqrt{\frac{m}{E(\mathbf{p})}}$$
$$\times \left\{ b(-\mathbf{p}, -s) \left[u(\mathbf{p}, s) \right]^* e^{ip \cdot x} + d^\dagger(-\mathbf{p}, -s) \left[v(\mathbf{p}, s) \right]^* e^{-ip \cdot x} \right\}$$
$$= \gamma^1 \gamma^3 \sum_s \int \frac{d^3 \tilde{p}}{(2\pi)^{3/2}} \sqrt{\frac{m}{E(\tilde{\mathbf{p}})}}$$
$$\times \left[b(\tilde{\mathbf{p}}, -s) u(\tilde{\mathbf{p}}, -s) e^{-i\tilde{p} \cdot \tilde{x}_t} + d^\dagger(\tilde{\mathbf{p}}, -s) v(\tilde{\mathbf{p}}, -s) e^{i\tilde{p} \cdot \tilde{x}_t} \right]$$
$$= \gamma^1 \gamma^3 \psi(\tilde{x}_t), \tag{7.46}$$

where $\tilde{p} \cdot (x^0, -\mathbf{x}) = -\tilde{p} \cdot (-x^0, \mathbf{x}) = -\tilde{p} \cdot \tilde{x}_t$ and the complex conjugates in the second line of Eq. (7.46) are due to the fact that T is antiunitary. Thus, we have the time reversal transformation for the Dirac field $\psi(x)$

$$T \psi(x) T^{-1} = \gamma^1 \gamma^3 \psi(\tilde{x}_t). \tag{7.47}$$

Similarly, for the Dirac adjoint field $\bar{\psi}(x)$, we obtain

$$T \bar{\psi}(x) T^{-1} = T \psi^\dagger(x) T^{-1} \gamma^0 = \left[T \psi(x) T^{-1} \right]^\dagger \gamma^0$$
$$= \psi^\dagger(\tilde{x}_t) \left(\gamma^1 \gamma^3 \right)^\dagger \gamma^0 = -\bar{\psi}(\tilde{x}_t) \gamma^1 \gamma^3. \tag{7.48}$$

Exercise 7.4 Show the relations in Eqs. (7.43) and (7.44).

7.5.3 Charge conjugation

Finally, we investigate the *charge-conjugation operator C*, which is the particle–antiparticle symmetry operator (cf. the discussion on charge conjugation in Section 3.8). As in the case of P, there is no problem in defining C as a unitary operator. Normally, charge conjugation is defined as transforming a fermion with spin into an antifermion with the same spin. Thus, we define the charge conjugation transformation of Dirac annihilation operators as

$$Cb(\mathbf{p}, s)C^{-1} = d(\mathbf{p}, s) \quad \text{and} \quad Cd(\mathbf{p}, s)C^{-1} = b(\mathbf{p}, s). \tag{7.49}$$

In addition, the relations between the positive- and negative-energy Dirac spinors are

$$u(\mathbf{p}, s) = -i\gamma^2 \left[v(\mathbf{p}, s) \right]^*, \tag{7.50}$$

$$v(\mathbf{p}, s) = -i\gamma^2 \left[u(\mathbf{p}, s) \right]^*. \tag{7.51}$$

Using Eqs. (7.49)–(7.51) together with the expansion of the Dirac field (7.17), we find that

$$C\psi(x)C^{-1} = \sum_s \int \frac{d^3 p}{(2\pi)^{3/2}} \sqrt{\frac{m}{E(\mathbf{p})}}$$

$$\times \left\{ -i\gamma^2 d(\mathbf{p}, s) \left[v(\mathbf{p}, s) \right]^* e^{-ip\cdot x} - i\gamma^2 b^\dagger(\mathbf{p}, s) \left[u(\mathbf{p}, s) \right]^* e^{ip\cdot x} \right\}$$

$$= -i\gamma^2 \psi^*(x) = -i\gamma^2 \left[\psi^\dagger(x) \right]^T = -i \left[\bar{\psi}(x)\gamma^0\gamma^2 \right]^T. \tag{7.52}$$

Thus, we have the charge conjugation transformation for the Dirac field $\psi(x)$

$$C\psi(x)C^{-1} = -i \left[\bar{\psi}(x)\gamma^0\gamma^2 \right]^T. \tag{7.53}$$

Note that C is a unitary operator, i.e. $C^{-1} = C^\dagger$, even though it transforms ψ into ψ^*.[3] Similarly, for the Dirac adjoint field, we obtain

$$C\bar{\psi}(x)C^{-1} = C\psi^\dagger(x)C^{-1}\gamma^0 = \left[C\psi(x)C^{-1} \right]^\dagger \gamma^0$$

$$= \left[-i\gamma^2\psi(x) \right]^T \gamma^0 = \left[-i\gamma^0\gamma^2\psi(x) \right]^T. \tag{7.54}$$

Exercise 7.5 Show the relations in Eqs. (7.50) and (7.51).

---

[3] For fields and states, the charge-conjugation operator C is unitary. However, for wave functions, it is antiunitary (cf. Section 3.8).

7.5.4 The charged Klein–Gordon field

As in the case of the Dirac field, we can find the transformations of the charged (complex-valued) Klein–Gordon field (cf. Section 6.6) for the unitary operators P and C as well as the antiunitary operator T. The transformations are the following:

$$P\phi(x)P^{-1} = \phi(x^0, -\mathbf{x}), \tag{7.55}$$

$$T\phi(x)T^{-1} = \phi(-x^0, \mathbf{x}), \tag{7.56}$$

$$C\phi(x)C^{-1} = \phi^\dagger(x), \tag{7.57}$$

where $\phi(x)$ is a charged Klein–Gordon field and potential phase factors have been suppressed.

7.5.5 The CPT symmetry operator

In addition, defining the operator

$$\Theta = CPT, \tag{7.58}$$

which is the CPT *symmetry operator* and an antiunitary operator, since the charge-conjugation operator C and the parity operator P are unitary operators and the time-reversal operator T is an antiunitary operator, the Θ conjugated field is given by

$$\psi^\Theta(x) = \Theta\psi(x)\Theta^{-1} = -\eta_\Theta^* \gamma^5 \psi^*(-x), \tag{7.59}$$

where η_Θ is a CPT phase factor. The appearance of γ^5 in Eq. (7.59) can be understood as follows. The T operation gives a product of a matrix $\gamma^1\gamma^3$ and the complex conjugation operator, whereas the CP operation gives a matrix $i\gamma^2\gamma^0$. Thus, we have $(i\gamma^2\gamma^0)(\gamma^1\gamma^3) = i\gamma^0\gamma^1\gamma^2\gamma^3 = \gamma^5$. The negative sign in the argument of ψ^* is obvious, since both the P and T operations are involved.

7.6 The Majorana field

We now discuss the quantization of a special case when the fermion is a self-charge-conjugate particle, i.e. a *Majorana particle*. In other words, such a fermion is its own antiparticle. This case was first studied by E. Majorana in 1937. A *Majorana field* ψ_M (as opposed to a Dirac field) is defined as a field with the following property:

$$\psi_M = \psi_M^c \equiv C\psi_M C^{-1}, \tag{7.60}$$

where C is again the charge-conjugation operator, which acts on states as

$$C|\psi(\mathbf{p}, s)\rangle = \eta_C |\bar\psi(\mathbf{p}, s)\rangle \tag{7.61}$$

with the quantity η_C being an arbitrary unobservable charge-conjugation phase factor that is normally set equal to unity. In general, we could also add a phase factor η_M in Eq. (7.60) such that $\psi_M = \eta_M \psi_M^c$. However, we have set $\eta_M = 1$ for simplicity. In addition, since $|\psi(\mathbf{p}, s)\rangle = b^\dagger(\mathbf{p}, s)|0\rangle$ and $|\bar{\psi}(\mathbf{p}, s)\rangle = d^\dagger(\mathbf{p}, s)|0\rangle$ are states, where $|0\rangle$ is the vacuum state with $C|0\rangle = |0\rangle$, we find, using Eq. (7.61), that

$$C|\psi(\mathbf{p}, s)\rangle = Cb^\dagger(\mathbf{p}, s)|0\rangle = Cb^\dagger(\mathbf{p}, s)C^{-1}C|0\rangle = \eta_C|\bar{\psi}(\mathbf{p}, s)\rangle = \eta_C d^\dagger(\mathbf{p}, s)|0\rangle,$$
(7.62)

which means that

$$Cb^\dagger(\mathbf{p}, s)C^{-1} = \eta_C d^\dagger(\mathbf{p}, s)$$
(7.63)

or

$$Cd^\dagger(\mathbf{p}, s)C^{-1} = \eta_C^* b^\dagger(\mathbf{p}, s),$$
(7.64)

where the last equation is obtained by a similar consideration of $C|\bar{\psi}(\mathbf{p}, s)\rangle$ as was performed in Eq. (7.62). Taking the Hermitian conjugate of Eq. (7.63), we have

$$Cb(\mathbf{p}, s)C^{-1} = \eta_C^* d(\mathbf{p}, s),$$
(7.65)

where we have used that $\eta_C = e^{i\alpha}$ ($\alpha \in \mathbb{R}$), since $|\eta_C| = 1$. Thus, using the expansion of the Dirac field in Eq. (7.17) as well as Eqs. (7.64) and (7.65), we obtain

$$\psi^c(x) = C\psi(x)C^{-1}$$
$$= \sum_s \int \frac{d^3 p}{(2\pi)^{3/2}} \sqrt{\frac{m}{E(\mathbf{p})}}$$
$$\times \left[Cb(\mathbf{p}, s)C^{-1}u(\mathbf{p}, s)e^{-ip\cdot x} + Cd^\dagger(\mathbf{p}, s)C^{-1}v(\mathbf{p}, s)e^{ip\cdot x} \right]$$
$$= \sum_s \int \frac{d^3 p}{(2\pi)^{3/2}} \sqrt{\frac{m}{E(\mathbf{p})}} \left[\eta_C^* d(\mathbf{p}, s)u(\mathbf{p}, s)e^{-ip\cdot x} + \eta_C^* b^\dagger(\mathbf{p}, s)v(\mathbf{p}, s)e^{ip\cdot x} \right].$$
(7.66)

Now, using the definition of the Majorana field (7.60) together with Eqs. (7.17) and (7.66), implies that

$$b(\mathbf{p}, s) = \eta_C^* d(\mathbf{p}, s) \quad \text{and} \quad d^\dagger(\mathbf{p}, s) = \eta_C^* b^\dagger(\mathbf{p}, s).$$
(7.67)

Therefore, the Majorana field can be written as

$$\psi_M(x) = \sum_s \int \frac{d^3 p}{(2\pi)^{3/2}} \sqrt{\frac{m}{E(\mathbf{p})}} \left[b(\mathbf{p}, s)u(\mathbf{p}, s)e^{-ip\cdot x} + \lambda b^\dagger(\mathbf{p}, s)v(\mathbf{p}, s)e^{ip\cdot x} \right],$$
(7.68)

where λ is an additional phase. Since we set $\eta_M = 1$, we have that $\lambda = \eta_C^*$. The phase λ is called the *creation phase factor*.

In Eq. (7.68), we only have one set of creation and annihilation operators, whereas in the case of a Dirac field, we have two sets. Thus, a Majorana field has only two degrees of freedom, corresponding to two spin states, whereas a Dirac field has four degrees of freedom.

For example, neutrinos are electrically neutral and almost massless, i.e. they have very small masses compared to other fermions. However, neutrinos have fundamental properties that are not yet known. In fact, it is not even known if neutrinos are Dirac or Majorana particles. Since neutrinos are uncharged, they could be Dirac or Majorana particles. Thus, in the case that neutrinos are Majorana particles, it means that neutrinos are their own antiparticles.

7.7 Green's functions and propagators

In order to find the Green's functions for the Dirac field, we need to investigate the inhomogeneous Dirac equation, which is given by

$$\sum_{\rho} \left(-i\partial_x + m\mathbb{1}_4\right)_{\alpha\rho} S_{F\rho\beta}(x-y) = -\delta(x-y)\delta_{\alpha\beta}, \qquad (7.69)$$

where α and β are indices for components of Dirac spinors. Thus, we need to look for the matrix function $S_{F\alpha\beta}(x-y)$. Suppressing Dirac spinor indices, we have

$$\left(-i\partial_x + m\mathbb{1}_4\right) S_F(x-y) = -\delta(x-y)\mathbb{1}_4. \qquad (7.70)$$

Now, introduce the function $F(x-y)$ such that

$$S_F(x-y) = \left(i\partial_x + m\mathbb{1}_4\right) F(x-y), \qquad (7.71)$$

which implies that we have the equation

$$\left(\Box_x + m^2\right) F(x-y) = -\delta(x-y), \qquad (7.72)$$

which is an inhomogeneous Klein–Gordon equation and means that F is one of the Klein–Gordon Green's functions, e.g. Δ_F, Δ_R, or Δ_A (cf. Section 6.3). Therefore, we have the Dirac Feynman propagator

$$S_F(x-y) = \left(i\partial_x + m\mathbb{1}_4\right) \Delta_F(x-y). \qquad (7.73)$$

Of course, we also have the corresponding retarded and advanced propagators

$$S_R(x-y) = \left(i\partial_x + m\mathbb{1}_4\right) \Delta_R(x-y), \qquad (7.74)$$

$$S_A(x-y) = \left(i\partial_x + m\mathbb{1}_4\right) \Delta_A(x-y). \qquad (7.75)$$

Especially, note that $S_R - S_A$ is **not** a propagator, since $S_R(x) - S_A(x) = (i\partial + m\mathbb{1}_4)\Delta(x)$, where $\Delta(x)$ is the Klein–Gordon function (cf. again Section 6.3). Furthermore, in analogy with the commutators for the bosonic fields (6.39)–(6.43), we have the anticommutators for the fermionic fields

$$\{\psi(x), \bar{\psi}(y)\} = iS(x - y) \tag{7.76}$$

and

$$\{\psi_\pm(x), \bar{\psi}_\mp(y)\} = iS_\pm(x - y), \tag{7.77}$$

where $\psi_\pm(x)$ are the positive- and negative-frequency parts of the Dirac field operator $\psi(x)$, respectively (cf. the discussion in Section 6.2).[4] Actually, we have

$$S_\pm(x) = (i\partial_x + m\mathbb{1}_4)\Delta_\pm(x), \tag{7.78}$$

and thus, we find that

$$S(x) = S_+(x) + S_-(x) = (i\partial_x + m\mathbb{1}_4)[\Delta_+(x) + \Delta_-(x)] = (i\partial_x + m\mathbb{1}_4)\Delta(x). \tag{7.79}$$

Again, note that $S(x)$ and $S_\pm(x)$ are **not** propagators (or Green's functions), but anticommutators. In other words, $S(x - y)$ does not satisfy Eq. (7.70). Finally, calculating the vacuum expectation value of the anticommutator (7.77) and using $\psi_+(x)|0\rangle = 0$, we obtain

$$iS_+(x - y) = \langle 0|\{\psi_+(x), \bar{\psi}_-(y)\}|0\rangle = \langle 0|\psi_+(x)\bar{\psi}_-(y)|0\rangle = \langle 0|\psi(x)\bar{\psi}(y)|0\rangle. \tag{7.80}$$

Similarly, using $\bar{\psi}_+(x)|0\rangle = 0$, we have

$$iS_-(x - y) = \langle 0|\{\psi_-(x), \bar{\psi}_+(y)\}|0\rangle = \langle 0|\bar{\psi}_+(y)\psi_-(x)|0\rangle = \langle 0|\bar{\psi}(y)\psi(x)|0\rangle. \tag{7.81}$$

Note that the signs on the right-hand sides of Eqs. (7.80) and (7.81) are the same, since two Dirac fields evaluated at two different spacetime points anticommute [cf. Eqs. (6.48) and (6.49)]. Thus, using the definition of the time-ordered product for fermionic fields (5.5), we find the *Dirac Feynman propagator*

$$iS_F(x - y) = i[\theta(x^0 - y^0)S_+(x - y) - \theta(y^0 - x^0)S_-(x - y)]$$
$$= (i\partial_x + m\mathbb{1}_4)i\Delta_F(x - y), \tag{7.82}$$

---

[4] Note that the notation $\bar{\psi}_\pm(x)$ means $(\bar{\psi})_\pm(x)$ and **not** $\overline{\psi_\pm}(x)$, i.e. $\bar{\psi}_+(x)$ [$\bar{\psi}_-(x)$] is the positive-frequency (negative-frequency) part of the Dirac adjoint field operator $\bar{\psi}(x)$.

which can also be written as [cf. Eq. (5.10)]

$$iS_F(x - y) = \int \frac{d^4k}{(2\pi)^4} e^{-ik\cdot(x-y)} \frac{i(\not{k} + m\mathbb{1}_4)}{k^2 - m^2 + i\epsilon}$$

$$= i\left(\frac{\not{x} - \not{y}}{|x - y|^5} + \frac{m\mathbb{1}_4}{|x - y|^3}\right) J_1(m|x - y|), \quad (7.83)$$

where

$$iS_F(k) = \frac{i(\not{k} + m\mathbb{1}_4)}{k^2 - m^2 + i\epsilon} \quad (7.84)$$

is the Dirac Feynman propagator in momentum space. Note that we obtain the same sign between the terms in Eq. (7.82) as in Eq. (6.50), which is due to the definition of the time-ordered product of Dirac fields together with Eqs. (7.80) and (7.81). In fact, this is actually the reason that we can express the Green's functions (or propagators) for the Dirac field according to Eq. (7.71).

7.8 Perturbation of electromagnetic interaction

The dynamics of a Dirac field coupled to an external electromagnetic field is provided by the inhomogeneous Dirac equation

$$(i\not{\partial} - m\mathbb{1}_4)\,\psi = e\not{A}\psi, \quad (7.85)$$

where the source term on the right-hand side is the minimal coupling term (cf. Section 3.8) and e acts as the electromagnetic coupling constant. Using the Green's functions of the Dirac field, we obtain the so-called *Källén–Yang–Feldmann equations*, which are given by

$$\psi(x) = \psi_{\text{in}}(x) + e \int S_R(x - y)A(y)\psi(y)\,d^4y, \quad (7.86)$$

$$\psi(x) = \psi_{\text{out}}(x) + e \int S_A(x - y)A(y)\psi(y)\,d^4y. \quad (7.87)$$

Note that the *free fields* ψ_{in} and ψ_{out} will be investigated in detail in what follows. Now, using perturbation theory, we expand these equations in powers of e such that

$$\psi(x) = \psi^{(0)}(x) + e\psi^{(1)}(x) + e^2\psi^{(2)}(x) + \cdots. \quad (7.88)$$

Thus, we obtain

$$\psi^{(0)}(x) = \psi_{\text{in}}(x), \quad (7.89)$$

$$\psi^{(1)}(x) = \int S_R(x - y)A(y)\psi_{\text{in}}(y)\,d^4y \quad (7.90)$$

and similar expressions instead using the field ψ_{out}.

7.9 Expansion of the S operator

Next, we will investigate the S operator, which relates the fields ψ_{in} and ψ_{out} as

$$\psi_{\text{out}} = S^\dagger \psi_{\text{in}} S. \tag{7.91}$$

Expanding the operator S in powers of e yields

$$S = 1 + eS^{(1)} + e^2 S^{(2)} + \cdots. \tag{7.92}$$

Since S is a unitary operator, which means that $SS^\dagger = S^\dagger S = 1$, we have to first order in e

$$\left(1 + eS^{(1)}\right)\left(1 + eS^{(1)\dagger}\right) = \left(1 + eS^{(1)\dagger}\right)\left(1 + eS^{(1)}\right) = 1, \tag{7.93}$$

which can be simplified to

$$1 + e\left(S^{(1)\dagger} + S^{(1)}\right) = 1. \tag{7.94}$$

Thus, we obtain

$$S^{(1)\dagger} = -S^{(1)}. \tag{7.95}$$

Using the Källén–Yang–Feldmann equations (7.86) and (7.87), we can relate the fields ψ_{in} and ψ_{out} in the following way

$$\psi_{\text{out}}(x) = \psi_{\text{in}}(x) + e \int S(x - y) A(y)\psi(y)\, d^4 y, \tag{7.96}$$

where we have used the fact that $S(x) = S_R(x) - S_A(x)$. To first order in e, i.e. the Born approximation, we have

$$\psi_{\text{out}}(x) = \psi_{\text{in}}(x) + e \int S(x - y) A(y)\psi_{\text{in}}(y)\, d^4 y$$
$$= S^\dagger \psi_{\text{in}}(x) S = \psi_{\text{in}}(x) + e\left(S^{(1)\dagger}\psi_{\text{in}}(x) + \psi_{\text{in}}(x)S^{(1)}\right)$$
$$= \psi_{\text{in}}(x) + e\left(S^{(1)\dagger}\psi_{\text{in}}(x) - \psi_{\text{in}}(x)S^{(1)\dagger}\right). \tag{7.97}$$

Thus, using Eqs. (7.96) and (7.97) to first order in e, we find that

$$\left[S^{(1)\dagger}, \psi_{\text{in}}(x)\right] = \int S(x - y) A(y)\psi_{\text{in}}(y)\, d^4 y. \tag{7.98}$$

Computing the commutator and assuming normal-ordering, we finally obtain

$$S^{(1)\dagger} = i \int\, : \bar{\psi}_{\text{in}}(z) A(z)\psi_{\text{in}}(z) :\, d^4 z. \tag{7.99}$$

Exercise 7.6 Show that the expression in Eq. (7.99), ignoring the normal-ordering, is a solution to Eq. (7.98).

Problems

(1) Show that

$$\gamma^0 S_R^\dagger(x - y)\gamma^0 = S_A(y - x).$$

(2) Show that the fermion Feynman propagator $S_F(x)$ fulfils the relation

$$S_F(x - y) = \left(i\partial\!\!\!/_x + m\mathbb{1}_4\right)\Delta_F(x - y),$$

where the propagator $\Delta_F(x)$ is a Klein–Gordon Green's function, which satisfies the inhomogeneous Klein–Gordon equation $\left(\Box + m^2\right)\Delta_F(x) = -\delta(x)$.

(3) Use the anticommutation relations for the Dirac annihilation and creation operators to show that

$$\{\psi_\alpha(x), \bar{\psi}_\beta(y)\} = iS_{\alpha\beta}(x - y).$$

Guide to additional recommended reading

The following books (see the indicated pages) and their authors have similar treatments of the content in the present chapter.

- A. Z. Capri, *Relativistic Quantum Mechanics and Introduction to Quantum Field Theory*, World Scientific (2002), pp. 112–141.
- F. Gross, *Relativistic Quantum Mechanics and Field Theory*, Wiley (1993), pp. 198–201.
- C. W. Kim and A. Pevsner, *Neutrinos in Physics and Astrophysics*, Harwood Academic Publishers (1993), pp. 17–27.
- F. Mandl and G. Shaw, *Quantum Field Theory*, rev. edn, Wiley (1994), pp. 61–80.
- M. E. Peskin and D. V. Schroeder, *An Introduction to Quantum Field Theory*, Addison-Wesley (1995), pp. 35–76.
- L. H. Ryder, *Quantum Field Theory*, 2nd edn., Cambridge (1996), pp. 137–140.
- F. Schwabl, *Advanced Quantum Mechanics*, Springer (1999), pp. 287–302.
- S. S. Schweber, *An Introduction to Relativistic Quantum Field Theory*, Dover (2005), pp. 218–239.
- F. J. Ynduráin, *Relativistic Quantum Mechanics and Introduction to Field Theory*, Springer (1996), pp. 198–203.
- For the interested reader: W. Pauli, The connection between spin and statistics, *Phys. Rev.* **58**, 716–722 (1940); R. F. Streater and A. S. Wightman, *PCT, Spin and Statistics, and All That*, Princeton (2000).

8

Maxwell's equations and quantization of the electromagnetic field

In 1861, J. C. Maxwell published a theoretical and important work on electromagnetism. Indeed, this work contains essentially the equations that today are known as *Maxwell's equations*. These equations, together with the Lorentz force law $\mathbf{F} = q(\mathbf{E} + \mathbf{v} \times \mathbf{B})$, constitute the complete set of laws for classical electromagnetism. An important property of Maxwell's equations is that they are Lorentz covariant. However, classical electromagnetism is, of course, non-quantized, as the prefix 'classical' suggests. Later on, in 1900, M. Planck[1] introduced the concept of quantization of radiation that can be studied by using emission of radiation from heated bodies, which has become known as *black-body radiation*. Thus, the idea of quantization was born by Planck's discovery. Then, in 1905, Einstein introduced the concept of 'quanta of light', since the quantum nature of light was revealed by the *photoelectric effect* that was claimed by himself. In addition, the quanta of light were dubbed *photons*.

In this chapter, we will first present Maxwell's equations, we will then discuss different gauges and quantization of the electromagnetic field, we will next consider the Casimir effect that arises when quantizing the electromagnetic field, and finally, we will try to obtain a covariant quantization of the electromagnetic field.

8.1 Maxwell's equations

Classical (i.e. non-quantized) electromagnetism is described by Maxwell's equations, which are given by the four vector expressions

$$\nabla \cdot \mathbf{E} = \rho, \tag{8.1}$$

$$\nabla \times \mathbf{E} = -\frac{\partial \mathbf{B}}{\partial t}, \tag{8.2}$$

---

[1] In 1918, Planck was awarded the Nobel prize in physics 'in recognition of the services he rendered to the advancement of Physics by his discovery of energy quanta'.

$$\nabla \cdot \mathbf{B} = 0, \tag{8.3}$$

$$\nabla \times \mathbf{B} = \mathbf{j} + \frac{\partial \mathbf{E}}{\partial t}, \tag{8.4}$$

where $\mathbf{E} = (E_x, E_y, E_z)$ and $\mathbf{B} = (B_x, B_y, B_z)$ are the electric and magnetic field strength vectors, respectively.[2] Introducing the scalar (electric) potential ϕ and the 3-vector (magnetic) potential $\mathbf{A}$ as

$$\mathbf{E} = -\nabla\phi - \frac{\partial \mathbf{A}}{\partial t} \quad \text{and} \quad \mathbf{B} = \nabla \times \mathbf{A}, \tag{8.5}$$

these potentials solve Eqs. (8.2) and (8.3) exactly. In order to be able to write these equations in relativistic form (i.e. in a Lorentz covariant form), we introduce the 4-vector potential A and the 4-vector current j such that

$$A = (A^\mu) = (\phi, \mathbf{A}) \quad \text{and} \quad j = (j^\mu) = (\rho, \mathbf{j}). \tag{8.6}$$

Furthermore, we define the *electromagnetic field strength tensor* as [cf. Eq. (3.197)]

$$F^{\mu\nu} = \partial^\mu A^\nu - \partial^\nu A^\mu. \tag{8.7}$$

This tensor is antisymmetric, i.e. $F^{\mu\nu} = -F^{\nu\mu}$, which means that the diagonal components are equal to zero, and its other off-diagonal components are given by $F^{i0} = E^i$ and $F^{ij} = -\epsilon^{ijk}B^k$.[3] Therefore, we have

$$F = (F^{\mu\nu}) = \begin{pmatrix} 0 & -E^1 & -E^2 & -E^3 \\ E^1 & 0 & -B^3 & B^2 \\ E^2 & B^3 & 0 & -B^1 \\ E^3 & -B^2 & B^1 & 0 \end{pmatrix} = \begin{pmatrix} 0 & -E_x & -E_y & -E_z \\ E_x & 0 & -B_z & B_y \\ E_y & B_z & 0 & -B_x \\ E_z & -B_y & B_x & 0 \end{pmatrix}. \tag{8.8}$$

Thus, we observe that the electromagnetic field strength tensor combines both the electric and magnetic fields. Now, Maxwell's equations in relativistic form are given by

$$\partial_\mu F^{\mu\nu} = j^\nu, \tag{8.9}$$

$$\partial^\mu F^{\nu\lambda} + \partial^\nu F^{\lambda\mu} + \partial^\lambda F^{\mu\nu} = 0, \tag{8.10}$$

where the first equation is an expression for the two inhomogeneous Maxwell's equations, i.e. Gauß' law (8.1) and Ampere's law with Maxwell's correction (8.4), whereas the second equation is an expression for the two homogeneous Maxwell's equations, i.e. Faraday's law of induction (8.2) and Gauß' law for magnetism (8.3). Note that the second equation (8.10) is equivalent to[4]

---

[2] Usually, the electric and magnetic field strength vectors are simply referred to as the electric and magnetic fields.

[3] Note that it, of course, holds that $F_{\mu\nu} = g_{\mu\lambda}g_{\nu\omega}F^{\lambda\omega}$.

[4] In addition, note that Eq. (8.10) is also called the *Bianchi identity* and is a kinematical condition.

$$\epsilon_{\omega\mu\nu\lambda} \partial^{\mu} F^{\nu\lambda} = 0. \tag{8.11}$$

The basic postulate in the theory of classical electromagnetism is that the electromagnetic field strength tensor is really transforming as a second-rank tensor under Lorentz transformations, i.e.

$$F'^{\mu\nu}(x') = \Lambda^{\mu}{}_{\lambda} \Lambda^{\nu}{}_{\omega} F^{\lambda\omega}(x), \tag{8.12}$$

where $x' = \Lambda x$ and Λ is a Lorentz transformation, or equivalently, in matrix form, we have

$$F' = \Lambda F \Lambda^{T}. \tag{8.13}$$

Therefore, using this postulate as well as the fact that $\partial'_{\mu} = \Lambda_{\mu}{}^{\nu} \partial_{\nu}$ [cf. Eqs. (3.148) and (3.149)], we find that

$$\begin{aligned} \partial'_{\mu} F'^{\mu\nu}(x') &= \Lambda_{\mu}{}^{\alpha} \Lambda^{\mu}{}_{\beta} \Lambda^{\nu}{}_{\gamma} \partial_{\alpha} F^{\beta\gamma}(x) = \delta^{\alpha}_{\beta} \Lambda^{\nu}{}_{\gamma} \partial_{\alpha} F^{\beta\gamma}(x) \\ &= \Lambda^{\nu}{}_{\gamma} \partial_{\alpha} F^{\alpha\gamma}(x) = \Lambda^{\nu}{}_{\gamma} j^{\gamma}(x). \end{aligned} \tag{8.14}$$

Now, assuming that the 4-vector current transforms as a vector under Lorentz transformations, i.e.

$$j'^{\mu}(x') = \Lambda^{\mu}{}_{\nu} j^{\nu}(x), \tag{8.15}$$

we obtain

$$\partial'_{\mu} F'^{\mu\nu} = j'^{\nu}, \tag{8.16}$$

which shows that the first set of inhomogeneous Maxwell's equations in relativistic form, i.e. Eq. (8.9), is indeed Lorentz covariant. In a similar way, the second set of homogeneous Maxwell's equations can be shown to be Lorentz covariant.

Exercise 8.1   Show that Eq. (8.10) is Lorentz covariant.

8.2 Quantization of the electromagnetic field

There exist various methods for quantization of the electromagnetic field, which indeed show that there are some difficulties with the quantization thereof. Such difficulties are that photons are massless and the photons' vector character.

Now, the electric and magnetic fields, **E** and **B**, respectively, are measurable quantities. In quantized theory, they should be replaced by operators $\hat{\mathbf{E}}$ and $\hat{\mathbf{B}}$ such that $\mathbf{E}_{cl} \simeq \langle \Psi | \hat{\mathbf{E}} | \Psi \rangle$ and $\mathbf{B}_{cl} \simeq \langle \Psi | \hat{\mathbf{B}} | \Psi \rangle$. However, **E** and **B** are **not** the basic quantities in quantum electromagnetism, but the 4-vector potential A (cf. the Aharonov–Bohm effect and bremsstrahlung). Thus, we have to construct an operator for A^{μ}. One problem is what gauge to choose.

Since we want to recover the standard theory of electromagnetism in the classical limit,[5] we start with the first set of Maxwell's equations (8.9) and the definition of the electromagnetic field strength tensor (8.7), i.e.

$$\partial_\mu F^{\mu\nu} = \partial_\mu(\partial^\mu A^\nu - \partial^\nu A^\mu) = j^\nu, \tag{8.17}$$

which can be written as

$$\Box A^\nu - \partial^\nu(\partial_\mu A^\mu) = j^\nu. \tag{8.18}$$

Note that the electromagnetic field strength tensor $F^{\mu\nu}$ does not uniquely define the 4-vector potential A^μ. If we introduce the following transformation

$$A^\mu \mapsto A'^\mu = A^\mu + \partial^\mu \chi, \tag{8.19}$$

which is again a gauge transformation (cf. the discussion in Section 5.3), where χ is any scalar field, i.e. an arbitrary function of the spacetime point x, then we obtain

$$F'^{\mu\nu} = \partial^\mu A'^\nu - \partial^\nu A'^\mu = \partial^\mu (A^\nu + \partial^\nu\chi) - \partial^\nu (A^\mu + \partial^\mu\chi) = \partial^\mu A^\nu - \partial^\nu A^\mu = F^{\mu\nu}, \tag{8.20}$$

which means that (i) $\mathbf{E}$ and $\mathbf{B}$ given in Eq. (8.5) are unchanged by applying the transformation (8.19) and (ii) $F^{\mu\nu}$ is invariant under such gauge transformations. On the other hand, it is only $\mathbf{E}$ and $\mathbf{B}$ which contain measurable information of the system. Thus, the 4-vector potential A^μ contains redundant degrees of freedom, corresponding to the transformation (8.19). One can actually take advantage of the gauge degrees of freedom. Inserting the gauge transformation (8.19) into Eq. (8.18), we find that

$$\Box (A^\nu + \partial^\nu\chi) - \partial^\nu \left(\partial_\mu A^\mu + \partial_\mu\partial^\mu\chi\right) = j^\nu, \tag{8.21}$$

which can be rewritten as

$$\Box A^\nu - \partial^\nu \left(\partial_\mu A^\mu\right) - j^\nu = -\Box\partial^\nu\chi + \partial^\nu\Box\chi \equiv 0. \tag{8.22}$$

Thus, we explicitly observe that Eq. (8.18) is invariant under gauge transformations in the form of Eq. (8.19). Indeed, the fundamental question arises: what gauge could we choose? Or equivalently, what condition on the function χ could we impose? Anyway, the effect of choosing a gauge is to reduce the number of degrees of freedom from four to two for the electromagnetic field.

---

[5] Here, the classical limit refers to the non-quantized limit, and **not** the non-relativistic limit.

Now, we will discuss gauges in detail. Some possible gauges to choose from are:

- $\partial_\mu A^\mu = 0$ (Lorenz gauge[6]),
- $\nabla \cdot \mathbf{A} = 0$ (Coulomb gauge),
- $A^0 = 0$ (temporal gauge),
- $A^3 = 0$ (axial gauge).

If we choose the function χ to be the solution of the linear partial differential equation $\Box \chi = -\partial_\mu A^\mu$, then for the definition of the gauge transformation (8.19), we obtain

$$\partial_\mu A'^\mu = 0 \tag{8.23}$$

and the first set of Maxwell's equations can be written in the simple form

$$\Box A'^\nu = j^\nu. \tag{8.24}$$

The choice of the 4-vector potential A'^μ satisfying Eq. (8.23) is known as the Lorenz gauge, i.e. the first gauge in the bullet list above. Note that this gauge has the advantage that the gauge condition itself is Lorentz covariant. In addition, note that the reason that several different gauges are possible is that the components of the 4-vector potential A'^μ are not coupled in the inhomogeneous relativistic wave equation (8.24). This is not the case in Eq. (8.18).

Exercise 8.2 Show that the Lorenz gauge condition is Lorentz covariant.

Although choosing the Lorenz gauge, there is still some extra freedom in the choice of the 4-vector potential A'^μ. In fact, we can perform yet another gauge transformation of the type given in Eq. (8.19), i.e.

$$A'^\mu \mapsto A''^\mu = A'^\mu + \partial^\mu \xi, \tag{8.25}$$

where ξ is a function that fulfils the equation

$$\Box \xi(x) = 0, \tag{8.26}$$

which ensures that the Lorenz gauge still holds. For a free electromagnetic field (i.e. $j = 0$), using Eq. (8.24), the 4-vector potential A'^μ satisfies the equation

$$\Box A'^\mu(x) = 0, \tag{8.27}$$

which is the wave equation for each component A'^μ and has the solutions

$$A'^\mu(x) = \varepsilon^\mu(\mathbf{k}) e^{-ik \cdot x}, \tag{8.28}$$

---

[6] The Lorenz gauge was first published by the Danish physicist L. Lorenz. It is often misspelled as the 'Lorentz gauge', since many people believe that the Dutch physicist H. Lorentz was the first to publish this gauge. Anyway, in 1902, Lorentz was awarded the Nobel Prize in physics together with P. Zeeman 'in recognition of the extraordinary service they rendered by their researches into the influence of magnetism upon radiation phenomena'.

where the 4-vector ε^μ is the polarization vector of the photon and k is the 4-momentum of the photon. Equation (8.27) is again the relativistic wave equation, which we also encountered for the Klein–Gordon wave function in the massless case. See the discussion at the end of Section 4.1. However, in this case, the situation is more complicated, since this wave function is a 4-vector, i.e. it has vector character. Anyway, inserting the solutions (8.28) into Eq. (8.27), we find that $k^2 = 0$, which implies that $m_\gamma = 0$, i.e. the photon is indeed massless. Thus, it is clear that the relativistic wave equation describes massless particles. In general, note that the equations for the electromagnetic field are much simpler in the Lorenz gauge [cf. Eq. (8.24)] than in a general gauge [cf. Eq. (8.18)]; in particular, this is the case for a free electromagnetic field [cf. Eq. (8.27)].

Since the polarization vector ε^μ is a 4-vector, it has four components, but describes a spin-1 particle. How should this be interpreted? First, the Lorenz gauge leads to the condition $k_\mu \varepsilon^\mu = 0$, which means that the number of independent components of ε^μ is reduced from four to three. Second, investigating the extra gauge degrees of freedom given by the additional gauge transformation (8.25), we introduce a gauge parameter

$$\xi(x) = \mathrm{i}a e^{-\mathrm{i}k\cdot x}, \tag{8.29}$$

where a is a constant such that Eq. (8.26) is fulfilled. Inserting Eqs. (8.28) and (8.29) into Eq. (8.25), we obtain the transformation rule for the polarization vector

$$\varepsilon^\mu \mapsto \varepsilon'^\mu = \varepsilon^\mu + ak^\mu, \tag{8.30}$$

which shows that the polarization vectors ε^μ and ε'^μ differ by a multiple of the 4-momentum of the photon. Thus, the physics is the same before and after gauge transformations, and therefore, the two polarization vectors describe the same photon. Finally, we can use this extra gauge freedom to impose the 'temporal gauge' $\varepsilon^0 = 0$, which means that (i) the 'Lorenz gauge' $k_\mu \varepsilon^\mu = 0$ is replaced by the 'Coulomb gauge' $\mathbf{k} \cdot \boldsymbol{\varepsilon} = 0$ and (ii) the number of independent components of ε^μ is reduced from three to two. Thus, in the Coulomb gauge, there are only two independent polarization vectors, which are both transversal to the 3-momentum of the photon.[7] For example, for a photon moving in the z direction, we can choose

$$\boldsymbol{\varepsilon}_1 = (1, 0, 0) \quad \text{and} \quad \boldsymbol{\varepsilon}_2 = (0, 1, 0), \tag{8.31}$$

since these two polarization vectors are orthogonal to each other and to the z direction, i.e. $\boldsymbol{\varepsilon}_1 \cdot \boldsymbol{\varepsilon}_2 = 0$, $\boldsymbol{\varepsilon}_1 \cdot \mathbf{e}_z = 0$, and $\boldsymbol{\varepsilon}_2 \cdot \mathbf{e}_z = 0$.[8] This means that the free photon

---

[7] Note that the Coulomb gauge is non-covariant and decomposes fields into transversal and longitudinal components and therefore is manifestly dependent on the inertial frame.

[8] It also holds that $\boldsymbol{\varepsilon}_1 \times \boldsymbol{\varepsilon}_2 = \mathbf{e}_z$, $\boldsymbol{\varepsilon}_2 \times \mathbf{e}_z = \boldsymbol{\varepsilon}_1$, and $\mathbf{e}_z \times \boldsymbol{\varepsilon}_1 = \boldsymbol{\varepsilon}_2$.

is described by its 4-momentum k and its polarization vector that is a combination of ε_1 and ε_2.

In addition, for a free electromagnetic field without imposing the Lorenz gauge, the second and third gauges in the bullet list can be simultaneously satisfied. This particular gauge is called the *radiation gauge* (or sometimes just the Coulomb gauge), and using Eq. (8.18), it also leads to the relativistic wave equation $\Box A^\nu(x) = 0$ [cf. Eq. (8.27)]. In the radiation gauge, using Eq. (8.5), we have

$$\mathbf{E} = -\frac{\partial \mathbf{A}}{\partial t} \quad \text{and} \quad \mathbf{B} = \nabla \times \mathbf{A}. \tag{8.32}$$

Normally, the Coulomb gauge condition $\nabla \cdot \mathbf{A} = 0$ is written as $\mathbf{k} \cdot \hat{\mathbf{A}} = 0$ in momentum space, which means that the electromagnetic field is **transversal**. Since we also have the temporal gauge condition $A^0 = 0$, there are no **scalar** or **longitudinal** photons in the radiation gauge.

Since we want $\mathbf{E}$ and $\mathbf{B}$ to be real fields, we assume A^μ to be a self-adjoint (or Hermitian) operator, i.e. $A^{\mu\dagger} = A^\mu$. In the radiation gauge, the solution to the relativistic wave equation [cf. Eq. (8.27)] for the electromagnetic field is a field operator for the 3-vector potential depending on time t and position $\mathbf{x}$

$$\begin{aligned}
\mathbf{A}(t, \mathbf{x}) = \sum_{\mathbf{k}} \frac{1}{\sqrt{2V\omega}} \sum_{\eta=\pm 1} & \left[e^{i(\mathbf{k}\cdot\mathbf{x}-\omega t)} \varepsilon(\mathbf{k}, \eta) \hat{a}_{\text{nr}}(\mathbf{k}, \eta) + e^{-i(\mathbf{k}\cdot\mathbf{x}-\omega t)} \varepsilon^*(\mathbf{k}, \eta) \hat{a}_{\text{nr}}^\dagger(\mathbf{k}, \eta) \right] \\
\rightarrow \frac{1}{\sqrt{2(2\pi)^3}} \int \frac{d^3 k}{\sqrt{\omega}} \sum_{\eta=\pm 1} & \left[e^{i(\mathbf{k}\cdot\mathbf{x}-\omega t)} \varepsilon(\mathbf{k}, \eta) \hat{a}_{\text{nr}}(\mathbf{k}, \eta) \right. \\
& \left. + e^{-i(\mathbf{k}\cdot\mathbf{x}-\omega t)} \varepsilon^*(\mathbf{k}, \eta) \hat{a}_{\text{nr}}^\dagger(\mathbf{k}, \eta) \right], \tag{8.33}
\end{aligned}$$

when $V \rightarrow \infty$, where $\omega \equiv \omega(\mathbf{k}) = |\mathbf{k}|$, $\varepsilon(\mathbf{k}, \eta)$ are the polarization vectors, and $\eta = \pm 1$ label the two transversal helicity states (cf. Section 1.7.2). For the polarization vectors, the radiation gauge condition $\nabla \cdot \mathbf{A} = 0$ implies $\mathbf{k} \cdot \varepsilon(\mathbf{k}, \eta) = 0$ for $\eta = \pm 1$ and it also holds that $\varepsilon(\mathbf{k}, \eta) \cdot \varepsilon(\mathbf{k}, \eta') = \delta_{\eta\eta'}$ for $\eta, \eta' = \pm 1$. In addition to Eq. (8.33), we have $A^0(t, \mathbf{x}) = 0$, since we are working in the radiation gauge. The classical interpretation of the electromagnetic field is a set of harmonic oscillators (cf. Section 4.4). Therefore, we postulate harmonic oscillator commutation relations for the operators $\hat{a}_{\text{nr}}$ and $\hat{a}_{\text{nr}}^\dagger$ such that

$$\left[\hat{a}_{\text{nr}}(\mathbf{k}, \eta), \hat{a}_{\text{nr}}^\dagger(\mathbf{k}', \eta') \right] = \delta_{\eta\eta'} \delta(\mathbf{k} - \mathbf{k}') f(k), \tag{8.34}$$

where f is a normalization function and the abbreviation nr stands for non-relativistic. In order for Eqs. (8.33) and (8.34) to be consistent with each other implies that $f(k) \equiv 1$. Thus, we have

$$\left[\hat{a}_{\mathrm{nr}}(\mathbf{k}, \eta), \hat{a}_{\mathrm{nr}}^{\dagger}(\mathbf{k}', \eta')\right] = \delta_{\eta\eta'}\delta(\mathbf{k} - \mathbf{k}'), \tag{8.35}$$

$$\left[\hat{a}_{\mathrm{nr}}(\mathbf{k}, \eta), \hat{a}_{\mathrm{nr}}(\mathbf{k}', \eta')\right] = \left[\hat{a}_{\mathrm{nr}}^{\dagger}(\mathbf{k}, \eta), \hat{a}_{\mathrm{nr}}^{\dagger}(\mathbf{k}', \eta')\right] = 0. \tag{8.36}$$

Now, it turns out that the energy from a continuous set of oscillators is

$$H_{\mathrm{rad}}^{\mathrm{osc}} = \sum_{\eta} \int \omega \hat{a}_{\mathrm{nr}}^{\dagger}(\mathbf{k}, \eta)\hat{a}_{\mathrm{nr}}(\mathbf{k}, \eta)\, \mathrm{d}^3 k + C^{\mathrm{osc}}, \tag{8.37}$$

whereas the energy from the correspondence principle is

$$H_{\mathrm{rad}}^{\mathrm{cp}} = \frac{1}{2} \int \left(\mathbf{E}^2 + \mathbf{B}^2\right) \mathrm{d}^3 x + C^{\mathrm{cp}}, \tag{8.38}$$

which is the total energy of the electromagnetic field. In general, the constants C^{osc} and C^{cp} are not equal to each other, i.e. $C^{\mathrm{osc}} \neq C^{\mathrm{cp}}$. In fact, the constant C^{osc} is divergent, since the integral

$$C^{\mathrm{osc}} = \sum_{\eta} \int \frac{1}{2}\omega\, \mathrm{d}^3 k \tag{8.39}$$

is divergent. In the following section, we will investigate the constant C^{osc} in great detail.

Next, we introduce the operator

$$\hat{a}(k, \eta) \equiv \sqrt{k_0}\,\hat{a}_{\mathrm{nr}}(\mathbf{k}, \eta), \tag{8.40}$$

where $k_0 = |\mathbf{k}|$, which implies that we have the following field operator for the 3-vector potential depending on the spacetime point x [instead of Eq. (8.33)]

$$\mathbf{A}(x) = \frac{1}{\sqrt{2(2\pi)^3}} \int_{k_0 > 0} \frac{\mathrm{d}^3 k}{k_0} \sum_{\eta = \pm 1} \left[\mathrm{e}^{-\mathrm{i}k \cdot x}\boldsymbol{\varepsilon}(\mathbf{k}, \eta)\hat{a}(k, \eta) + \mathrm{e}^{\mathrm{i}k \cdot x}\boldsymbol{\varepsilon}^*(\mathbf{k}, \eta)\hat{a}^{\dagger}(k, \eta)\right], \tag{8.41}$$

where the operators $\hat{a}$ and $\hat{a}^{\dagger}$ fulfil the commutation relation

$$\left[\hat{a}(k, \eta), \hat{a}^{\dagger}(k', \eta')\right] = k_0\delta_{\eta\eta'}\delta(\mathbf{k} - \mathbf{k}'), \tag{8.42}$$

where $k_0 = |\mathbf{k}|$. Note that, introducing the operator in Eq. (8.40), the situation has become more relativistically apparent, since $\hat{a}(k, \eta)$ depends on the 4-momentum $k = (k_0, \mathbf{k})$, whereas $\hat{a}_{\mathrm{nr}}(\mathbf{k}, \eta)$ depends only on the 3-momentum $\mathbf{k}$. The system of oscillators, i.e. the operators $\hat{a}$ and $\hat{a}^{\dagger}$, can be interpreted as

- $\hat{a}(k, \eta)$ annihilates an energy excitation with 4-momentum k and helicity state η,
- $\hat{a}^{\dagger}(k, \eta)$ creates energy excitations with 4-momentum k and helicity state η.

In addition, photons are particles, which are the quanta (or energy excitations) of the electromagnetic field. The vacuum state, i.e. the state without photons, is denoted $|0\rangle$, which is the state of minimum energy and $\hat{a}(k, \eta)|0\rangle = 0$. Thus, the state of n photons is given by $\hat{a}^\dagger(k_1, \eta_1)\hat{a}^\dagger(k_2, \eta_2)\ldots\hat{a}^\dagger(k_n, \eta_n)|0\rangle$, where $k_1, k_2, \ldots, k_n$ are 4-momenta and $\eta_1, \eta_2, \ldots, \eta_n$ are helicity states.

8.3 The Casimir effect

In this section, we investigate the so-called *Casimir effect*, which is due to Eq. (8.39) that gives the vacuum energy (or zero-point energy). Since the expression for this energy is divergent, one would like to subtract it from all other energies (i.e. to renormalize the energy) in order to obtain energy differences, which are the only experimentally measurable and physically meaningful energies. However, in 1948, H. Casimir showed that this procedure was too trivial by studying the quantization of the electromagnetic field in the presence of two conducting planes, but in the absence of any charges.[9] The Casimir effect says that there is a force between the two planes, even though no charges are involved. Therefore, in order to show the Casimir effect, consider, in vacuum, two parallel and perfectly conducting plates, which are squares with side length L that is large (in principle, $L \to \infty$), located at $z = 0$ and $z = a$, respectively, and thus separated by a distance a from each other (see Fig. 8.1). Note that $a \ll L$, but a is much larger than interatomic distances anyway. A typical order of magnitude for the separation distance is $a \sim 1\,\mu\text{m}$, which means that the plates can be considered to be smooth.

In the case that the two planes are present, the electromagnetic field must fulfil certain boundary conditions. The boundary conditions on the electric and magnetic fields at the surface of a perfect conductor, i.e. the parallel component of the electric field and the normal component of the magnetic field vanish on the surface, give rise to the following wave vectors

$$\mathbf{k}(a) = \left(k_x, k_y, \frac{n\pi}{a}\right),\tag{8.43}$$

where n is an integer, if the planes are situated at $z = 0$ and $z = a$, respectively. Thus, in the radiation gauge, using Eq. (8.33), the 3-vector potential (in the presence of the planes) is given by

$$\mathbf{A}_a(t, \mathbf{x}) = \frac{1}{2\pi}\iint dk_x dk_y \frac{\pi}{a}\sum_{n=-\infty}^{\infty}\frac{1}{\sqrt{2|\mathbf{k}(a)|}}$$
$$\times\sum_{\eta=\pm\eta}\left[e^{-i\omega t}e^{i\mathbf{k}(a)\cdot\mathbf{x}}\boldsymbol{\varepsilon}(\mathbf{k}(a), \eta)\hat{a}_{nr}(\mathbf{k}(a), \eta)\right.$$
$$\left. + e^{i\omega t}e^{-i\mathbf{k}(a)\cdot\mathbf{x}}\boldsymbol{\varepsilon}^*(\mathbf{k}(a), \eta)\hat{a}_{nr}^\dagger(\mathbf{k}(a), \eta)\right].\tag{8.44}$$

In addition, we have the following modified commutation relation for the $\hat{a}_{nr}$ and $\hat{a}_{nr}^\dagger$ operators [cf. Eq. (8.35)]

---

[9] For the interested reader, please see: H. B. G. Casimir, On the attraction between two perfectly conducting plates, *Proc. Kon. Nederland. Akad. Wetenschap.* **51**, 793–795 (1948).

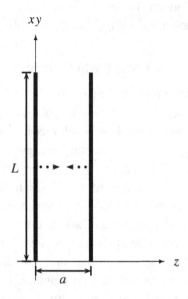

Figure 8.1 The setup of the Casimir effect. This schematic picture shows two parallel, perfectly conducting plates (thick solid lines), which are squares with side length L, placed in vacuum. The plates are separated by a distance a. It should hold that $a \ll L$. In addition, the picture displays that the plates are attracting each other with a force, i.e. the Casimir effect, which is indicated by the dotted arrows.

$$\left[\hat{a}_{\mathrm{nr}}(\mathbf{k}(a), \eta), \hat{a}^{\dagger}_{\mathrm{nr}}(\mathbf{k}'(a), \eta') \right] = \delta_{\eta\eta'} \delta \left(k_x - k'_x \right) \delta \left(k_y - k'_y \right) \frac{a}{\pi} \delta_{nn'}. \tag{8.45}$$

Now, we can compute the energy in the presence of the planes as well as in the absence of the planes. Using Eq. (8.38), the energy with the planes is given by

$$H_a = \frac{1}{2} \int \left[(\partial_0 \mathbf{A}_a)^2 - \mathbf{A}_a \Delta \mathbf{A}_a \right] \mathrm{d}^3 x + C, \tag{8.46}$$

whereas the energy without the planes is given by

$$H = \frac{1}{2} \int \left[(\partial_0 \mathbf{A})^2 - \mathbf{A} \Delta \mathbf{A} \right] \mathrm{d}^3 x + C, \tag{8.47}$$

where the constant C is the same in both cases. The difference between the cases with and without planes introduces an energy shift. The vacuum expectation value of this energy shift is

$$\delta E(a) = \langle 0| H_a - H |0\rangle = \frac{1}{2} \int \langle 0| (\partial_0 \mathbf{A}_a)^2 - (\partial_0 \mathbf{A})^2 - \mathbf{A}_a \Delta \mathbf{A}_a + \mathbf{A} \Delta \mathbf{A} |0\rangle \, \mathrm{d}^3 x. \tag{8.48}$$

Inserting Eqs. (8.33) and (8.44) into Eq. (8.48), we obtain after some tedious computations

$$\delta E(a) = \frac{L^2}{8\pi^2} \int_{-\infty}^{\infty} \int_{-\infty}^{\infty} \left(\sum_{n=-\infty}^{\infty} \sqrt{k_x^2 + k_y^2 + \frac{\pi^2 n^2}{a^2}} \right.$$

$$\left. - \int_{-\infty}^{\infty} \sqrt{k_x^2 + k_y^2 + \frac{\pi^2 v^2}{a^2}} \, dv \right) dk_x dk_y$$

$$= \left\{ \text{introducing the polar coordinate } \rho = \sqrt{k_x^2 + k_y^2} \right\}$$

$$= \left\{ \text{using the symmetry of the integrand with respect to } n \to -n \text{ and} \right.$$

$$\left. v \to -v \right\}$$

$$= \frac{L^2}{2\pi} \int_0^{\infty} \left(\sum_{n=0}^{\infty} \sqrt{\rho^2 + \frac{\pi^2 n^2}{a^2}} - \frac{1}{2}\rho - \int_0^{\infty} \sqrt{\rho^2 + \frac{\pi^2 v^2}{a^2}} \, dv \right) \rho \, d\rho,$$

$$(8.49)$$

and performing the variable substitution $u = a^2 \rho^2 / \pi^2$, the corresponding energy shift per unit area is

$$\delta\epsilon(a) \equiv \frac{\delta E(a)}{L^2} = \frac{\pi^2}{4a^3} \int_0^{\infty} \left(\sum_{n=0}^{\infty} \sqrt{u + n^2} - \frac{1}{2}\sqrt{u} - \int_0^{\infty} \sqrt{u + v^2} \, dv \right) du,$$

$$(8.50)$$

which is ill-defined, since it diverges. The reason is the following. For large wave vectors $\mathbf{k}$ (or in other words, for large values of $|\mathbf{k}| = k$), the values of k are of similar order as the interatomic distances, i.e. $k \sim (1\,\text{Å})^{-1} \gg 1/a$, where $a \sim$ 1 μm. In order to handle the situation $u \propto k^2 \gtrsim (1\,\text{Å})^{-2}$, we have to introduce a so-called *cut-off* in Eq. (8.50), which is done by multiplying the integrand of the u integration by a cut-off function. In fact, the cut-off dependent contributions will cancel, and therefore, we will end up with a finite and cut-off independent result. Introducing the function

$$f(v) \equiv \int_0^{\infty} \sqrt{u + v^2} g\left(\sqrt{u + v^2} \frac{\pi}{aK}\right) du, \qquad (8.51)$$

where the cut-off function g is defined as

$$g(k/K) = \begin{cases} 1, & k \ll K \\ \frac{1}{2}, & k = K \\ 0, & k \gg K \end{cases} \qquad (8.52)$$

and the quantity K is given through the relation $g(1) = 1/2$, we can rewrite Eq. (8.50) as

$$\delta\epsilon(a) = \frac{\pi^2}{4a^3} \left[\sum_{n=0}^{\infty} f(n) - \frac{1}{2}f(0) - \int_0^{\infty} f(v) \, dv \right]. \qquad (8.53)$$

Using the Euler–Maclaurin formula

$$\int_0^\infty f(v)\, dv = \sum_{n=0}^\infty f(n) - \frac{1}{2} f(0) + \sum_{n=1}^\infty \frac{B_{2n}}{(2n)!} f^{(2n-1)}(0), \qquad (8.54)$$

where the coefficients B_{2n} $(n = 1, 2, \ldots)$ are *Bernoulli numbers*,[10] we find that

$$\delta\epsilon(a) = -\frac{\pi^2}{4a^3} \sum_{n=1}^\infty \frac{B_{2n}}{(2n)!} f^{2n-1}(0) = -\frac{\pi^2}{4a^3} \left[\frac{B_2}{2!} f'(0) + \frac{B_4}{4!} f'''(0) + \cdots \right].$$
$$(8.55)$$

In addition, using the defintion (8.51), performing the variable substitution $v = u + v^2$, and applying the general form of Leibniz integral rule for differentiation of an integral, we have

$$f(v) = \int_{v^2}^\infty \sqrt{v} g\left(\sqrt{v}\,\tfrac{\pi}{aK}\right) dv,$$
$$f'(v) = -2v^2 g\left(v\tfrac{\pi}{aK}\right), \qquad f'(0) = 0,$$
$$f''(v) = -4vg\left(v\tfrac{\pi}{aK}\right) - 2v^2 g'\left(v\tfrac{\pi}{aK}\right)\tfrac{\pi}{aK}, \qquad f''(0) = 0,$$
$$f'''(v) = -4g\left(v\tfrac{\pi}{aK}\right) - 8vg'\left(v\tfrac{\pi}{aK}\right)\tfrac{\pi}{aK} - 2v^2 g''\left(v\tfrac{\pi}{aK}\right)\left(\tfrac{\pi}{aK}\right)^2, \qquad f'''(0) = -4.$$

Note that 'higher' derivatives are finite. Thus, inserting $f'(0) = 0$ and $f'''(0) = -4$ as well as $B_2 = 1/6$ and $B_4 = -1/30$ into Eq. (8.55), we obtain

$$\delta\epsilon(a) = -\frac{\pi^2}{4a^3} \frac{-1/30}{4!} (-4) = -\frac{\pi^2}{720} \frac{1}{a^3}, \qquad (8.56)$$

which shows that the energy difference is dependent on the separation distance a between the two planes. Finally, deriving the force using the energy shift (8.56), we find that

$$f = -\frac{\partial \delta\epsilon(a)}{\partial a} = -\frac{\pi^2}{240} \frac{1}{a^4} < 0, \qquad (8.57)$$

which means that the force between the two planes is attractive, since f is negative.[11] This is the Casimir effect in the case of two planes. However, note that this effect is small. For example, inserting $a = 1\,\mu\text{m}$ into Eq. (8.57), we obtain $F = fL^2 \simeq -1.3 \cdot 10^{-7}$ N, where $L = 1$ cm was used for the side length of the planes.[12] The Casimir effect was first measured in 1958 by M. J. Sparnaay, but with large experimental errors.[13]

---

[10] The values of the first few non-zero Bernoulli numbers are: $B_0 = 1$, $B_1 = -1/2$, $B_2 = 1/6$, $B_4 = -1/30, \ldots$

[11] Note that there seems to be some other setups of the Casimir effect that would lead to a force between uncharged objects that is repulsive.

[12] When numerically calculating the Casimir force, we need to reinsert the fundamental constants $\hbar$ and c, since we have used natural units where $\hbar = 1$ and $c = 1$, and we have $F = -\hbar c\pi^2 L^2/(240a^4)$. Indeed, the presence of $\hbar$ shows that this force has a quantum-mechanical origin.

[13] For the interested reader, please, see: M. J. Sparnaay, Attractive forces between flat plates, *Nature* **180**,

In fact, the Casimir effect is our first example of regularization, which we used to renormalize the energy. In Chapter 13, we will elaborate on regularization and renormalization. Nevertheless, note that the energy shift per unit area $\delta\epsilon(a)$ given in Eq. (8.50) can also be regularized using other methods than the one we used here. One such method is Riemann's ζ-function regularization and another method is performed by introducing the thermodynamical partition function corresponding to the vacuum energy.

In conclusion, it would be wrong not to consider the vacuum energy in Eq. (8.39), since in this case the Casimir effect would not appear. Thus, it seems that the vacuum is not as empty as one might think, but filled with vacuum fluctuations of the electromagnetic field. In addition, in the derivation of the Casimir effect, we have included all wave vectors **k** with values of k up to infinity, which made the expression in Eq. (8.50) undefined, but solved by a cut-off function (8.51), which contributions cancel, and finally, Eq. (8.55) led to a finite and cut-off independent result. The fact that this result does not depend on the cut-off implies that it is not sensitive to large wave vectors, and thus, the Casimir effect is indeed a low-energy phenomenon.

Exercise 8.3 Perform the tedious computations that lead from Eq. (8.48) to Eq. (8.49).

8.4 Covariant quantization of the electromagnetic field

So far, the quantization of the electromagnetic field has **not** been (Lorentz) covariant. Now, we want to obtain a quantization that is covariant, which means that the electromagnetic field operator A^μ should transform covariantly. This is more convenient for fully relativistic situations. However, in 1967, F. Strocchi stated the following theorem.[14]

Theorem 8.1 *At the same time, one* **cannot** *have a manifestly covariant quantization of the electromagnetic field and a positive-definite metric in Hilbert space.*

In fact, the covariant electromagnetic field operator A^μ is given by

$$A^\mu(x) = \frac{1}{\sqrt{2(2\pi)^3}} \int_{k_0>0} \frac{d^3k}{k_0} \sum_{\ell=1}^{2} \left[e^{-ik\cdot x} e^\mu(k, \ell)\hat{a}(k, \ell) + e^{ik\cdot x} e^\mu(k, \ell)\hat{a}^\dagger(k, \ell) \right]$$

$$(8.58)$$

334–335 (1957) and M. J. Sparnaay, Measurement of attractive forces between flat plates, *Physica* **24**, 751 (1958).

[14] For the interested reader, please see: F. Strocchi, Gauge problem in quantum field theory, *Phys. Rev.* **162**, 1429–1438 (1967).

and does not have a positive-definite metric. Comparing Eqs. (8.41) and (8.58), we have changed the labels of the helicity states from $\eta = \pm 1$ to $\ell = 1, 2$ and replaced the polarization vectors $\boldsymbol{\varepsilon}(\mathbf{k}, \eta)$ with the quantities $e^{\mu}(k, \ell)$, which denote the *covariant polarization vectors*. Furthermore, the operators $\hat{a}(k, 1)$ and $\hat{a}(k, 2)$ annihilate transversal photons with 4-momentum k and the operators $\hat{a}^{\dagger}(k, 1)$ and $\hat{a}^{\dagger}(k, 2)$ create the same type of photons. The idea is now to extend and 'complete' the covariant electromagnetic field operator A^{μ} in Eq. (8.58) in order to have a Minkowski product of e^{μ}s and $\hat{a}$s such that[15]

$$A^{\mu}(x) = \frac{1}{\sqrt{2(2\pi)^3}} \int_{k_0 > 0} \frac{d^3 k}{k_0} \sum_{\lambda=0}^{3} (-g_{\lambda\lambda})$$
$$\times \left[e^{-ik \cdot x} e^{\mu}(k, \lambda) \hat{a}(k, \lambda) + e^{ik \cdot x} e^{\mu}(k, \lambda) \hat{a}^{\dagger}(k, \lambda) \right], \quad (8.59)$$

where we have again changed the labels of the helicity states from ℓ to λ and the covariant polarization vectors, which consist of four linearly independent polarization vectors $e^{\mu}(k, \lambda)$ ($\lambda = 0, 1, 2, 3$), fulfil the products

$$e^{\mu}(k, \lambda) e_{\mu}(k, \lambda') = g_{\lambda\lambda'}, \quad (8.60)$$

$$\sum_{\lambda} (-g_{\lambda\lambda}) e^{\mu}(k, \lambda) e^{\nu}(k, \lambda) = -g^{\mu\nu}. \quad (8.61)$$

Here, the quantity $g_{\lambda\lambda'}$ denotes an element of the metric tensor as usual. Note that the product (8.60) can be written explicitly as

$$e^{\mu}(k, \lambda) e_{\mu}(k, \lambda') = e^{\mu}(k, \lambda) g_{\mu\nu} e^{\nu}(k, \lambda') = e^0(k, \lambda) e^0(k, \lambda') - e^1(k, \lambda) e^1(k, \lambda')$$
$$- e^2(k, \lambda) e^2(k, \lambda') - e^3(k, \lambda) e^3(k, \lambda') = g_{\lambda\lambda'}. \quad (8.62)$$

Investigating Eq. (8.62), we observe that the polarization vector $e(k, 0)$ is time-like (since $g_{00} = 1$), whereas the polarization vectors $e(k, 1)$, $e(k, 2)$, and $e(k, 3)$ are space-like (since $g_{11} = g_{22} = g_{33} = -1$). In general, it is enough to require the properties (8.60) and (8.61) for the covariant polarization vectors. However, in particular without loss of generality, one can choose a specific inertial frame in which a photon is moving along the z-axis, i.e. the normalized 4-momentum of the photon is given by $\bar{k} = (1, 0, 0, 1)$, and we have the unit covariant polarization vectors

$$e^{\mu}(\bar{k}, \lambda) = g^{\mu}_{\ \lambda} = \delta^{\mu}_{\lambda}, \quad \lambda = 0, 1, 2, 3, \quad (8.63)$$

which explicitly read

$$e(\bar{k}, 0) = (1, 0, 0, 0), \quad e(\bar{k}, 1) = (0, 1, 0, 0),$$
$$e(\bar{k}, 2) = (0, 0, 1, 0), \quad e(\bar{k}, 3) = (0, 0, 0, 1). \quad (8.64)$$

---

[15] Note that the covariant electromagnetic field operator A^{μ} in Eq. (8.59) contains four linearly independent plane-wave solutions to Eq. (8.27), i.e. one solution for each value of μ ($\mu = 0, 1, 2, 3$).

Using Eq. (8.63) [or Eq. (8.64)] and taking the inner product of $\bar{k}$ and the unit covariant polarization vectors, it of course holds that

$$\bar{k}_\mu e^\mu(\bar{k}, 1) = \bar{k}_\mu e^\mu(\bar{k}, 2) = 0, \tag{8.65}$$
$$\bar{k}_\mu e^\mu(\bar{k}, 0) = -\bar{k}_\mu e^\mu(\bar{k}, 3) = 1. \tag{8.66}$$

Note that photons with the polarization vectors $e(\bar{k}, 1)$ and/or $e(\bar{k}, 2)$ are called *transversal photons*, whereas photons with the polarization vector $e(\bar{k}, 0)$ are called *scalar* (or *time-like*) *photons* and photons with the polarization vector $e(\bar{k}, 3)$ are called *longitudinal photons*.

Exercise 8.4 Check explicitly that the unit covariant polarization vectors given in Eq. (8.63) fulfil the product (8.60).

Now, since the sum in Eq. (8.59) runs from 0 to 3 instead of 1 and 2 as in Eq. (8.58), we have two new types of creation and annihilation operators

- $\hat{a}(k, 0)$ annihilates a (non-existent) scalar photon with 4-momentum k,
- $\hat{a}^\dagger(k, 0)$ creates a scalar photon with 4-momentum k,
- $\hat{a}(k, 3)$ annihilates a (non-existent) longitudinal photon with 4-momentum k,
- $\hat{a}^\dagger(k, 3)$ creates a longitudinal photon with 4-momentum k.

The scalar and longitudinal photon states have not been observed in Nature, and therefore, they are considered to be unphysical. Hence, we have too many states, and therefore, we need more conditions imposed on the creation and annihilation operators in order to be able to eliminate the superfluous states.

If Eq. (8.59) should be covariant, then we also want the commutation relations for the creation and annihilation operators to be covariant. The covariant commutation relations are given by[16]

$$\left[\hat{a}^\mu(k), \hat{a}^{\nu\dagger}(k')\right] = -g^{\mu\nu} k_0 \delta(\mathbf{k} - \mathbf{k}'), \tag{8.67}$$
$$\left[\hat{a}^\mu(k), \hat{a}^\nu(k')\right] = \left[\hat{a}^{\mu\dagger}(k), \hat{a}^{\nu\dagger}(k')\right] = 0, \tag{8.68}$$

where $k_0 = |\mathbf{k}| > 0$ and we have defined

$$\hat{a}^\mu(k) \equiv -\sum_\lambda g_{\lambda\lambda} e^\mu(k, \lambda) \hat{a}(k, \lambda), \tag{8.69}$$

but Eq. (8.67) implies that

$$||\hat{a}^{0\dagger}(k)|0\rangle||^2 = \langle 0|\hat{a}^0(k)\hat{a}^{0\dagger}(k)|0\rangle = \langle 0|\left[\hat{a}^0(k), \hat{a}^{0\dagger}(k)\right]|0\rangle = -g^{00} k_0 \delta(\mathbf{0}) < 0, \tag{8.70}$$

since $g^{00} = 1 > 0$ and $k_0 > 0$, which means that the scalar (or time-like) one-photon states $\hat{a}^{0\dagger}(k)|0\rangle$ have negative norm. In addition, there exist states with zero

---

[16] Note that Eq. (8.67) replaces Eq. (8.42).

norm. Thus, we need to get rid of both the states with negative and zero norms. The difficulty with the states having negative and zero norms is that they are problematic for the probabilistic interpretation of the theory based on a covariant quantization of the electromagnetic field. Note that, according to Theorem 8.1, states cannot only have positive norms if the quantization of the electromagnetic field is covariant. In the following, we will investigate how this problem can be solved.

In 1932, E. Fermi[17] interpreted $\hat{a}(k, \ell)$, where $\ell = 1, 2, 3$, as annihilation operators and $\hat{a}^\dagger(k, \ell)$, where $\ell = 1, 2, 3$, as creation operators, but $\hat{a}(k, 0)$ as a creation operator and $\hat{a}^\dagger(k, 0)$ as an annihilation operator.[18] This interpretation means that the negative norm disappears and that the non-physical operators $\hat{a}(k, 0)$ and $\hat{a}(k, 3)$ cancel out. However, this interpretation leads to another difficulty: the Hamiltonian H is not bounded from below, since for scalar one-photon states $\hat{a}^0(k)|0\rangle$, one can show that

$$H\hat{a}^0(k)|0\rangle = \left[H, \hat{a}^0(k)\right]|0\rangle = -k_0\hat{a}^0(k)|0\rangle, \tag{8.71}$$

since $k_0 > 0$, which means that such states with arbitrarily negative energy are possible.

Another method, which is equivalent to Fermi's method, is to interpret $\hat{a}$ as annihilation operators and $\hat{a}^\dagger$ as creation operators as usual, but to keep the indefinite metric. This is called the *Gupta–Bleuler formalism*.[19] Introducing the operators [cf. Eq. (8.69)]

$$\hat{a}^\mu(k) \equiv -\sum_\lambda g_{\lambda\lambda} e^\mu(k, \lambda)\hat{a}(k, \lambda), \tag{8.72}$$

where

$$\hat{a}(k, \lambda) = -\sum_\mu g_{\mu\mu} e^\mu(k, \lambda)\hat{a}^\mu(k), \tag{8.73}$$

we can write the covariant electromagnetic field operator as

$$A^\mu(x) = A_+^\mu(x) + \left[A_+^\mu(x)\right]^\dagger, \tag{8.74}$$

where the positive-frequency part (including annihilation operators) is given by

$$A_+^\mu(x) = \frac{1}{\sqrt{2(2\pi)^3}} \int_{k_0 > 0} \frac{d^3k}{k_0} e^{-ik \cdot x} \hat{a}^\mu(k). \tag{8.75}$$

---

[17] In 1938, Fermi was awarded the Nobel Prize in physics 'for his demonstrations of the existence of new radioactive elements produced by neutron irradiation, and for his related discovery of nuclear reactions brought about by slow neutrons'.

[18] For the interested reader, please see: E. Fermi, Quantum Theory of Radiation, *Rev. Mod. Phys.* **4**, 87–132 (1932).

[19] For the interested reader, please see: S. N. Gupta, Theory of Longitudinal Photons in Quantum Electrodynamics, *Proc. Phys. Soc. (London)* A **63**, 681–691 (1950) and K. Bleuler, *Helv. Phys. Acta* **23**, 567–586 (1950).

The covariant canonical commutation relations for the covariant electromagnetic field operator and its conjugate momentum operator are given by[20]

$$[A^\mu(x), \pi^\nu(y)] = ig^{\mu\nu}\delta(\mathbf{x} - \mathbf{y}), \tag{8.76}$$

$$[A^\mu(x), A^\nu(y)] = [\pi^\mu(x), \pi^\nu(y)] = 0. \tag{8.77}$$

Note that Eqs. (8.76) and (8.77) are only valid for equal times t in the spacetime points x and y, i.e. $x = (t, \mathbf{x})$ and $y = (t, \mathbf{y})$. In addition, it holds that

$$[A^\mu(x), A^\nu(y)] = iD^{\mu\nu}(x - y), \tag{8.78}$$

where the function $D^{\mu\nu}$ is defined as

$$D^{\mu\nu}(x) = -g^{\mu\nu} \lim_{m\to 0} \Delta(x), \tag{8.79}$$

which should be compared with Eq. (5.14). Note that the function $\Delta(x)$ is the Klein–Gordon function defined in Eq. (6.39). Now, using the expansion of the covariant electromagnetic field operator (8.74), it can be shown that the covariant canonical commutation relations for the covariant electromagnetic field (8.76) and (8.77) reduce to the covariant commutation relations for the creation and annihilation operators (8.67) and (8.68).

Exercise 8.5 Verify by direct computation that Eq. (8.73) is the inverse of Eq. (8.72).

Again, the vacuum state (the state with no photons) is denoted $|0\rangle$, and using Eq. (8.72), it is defined as

$$\hat{a}^\mu(k)|0\rangle = 0 \quad \forall k, \tag{8.80}$$

which means that the vacuum is annihilated by all $\hat{a}^\mu$s ($\mu = 0, 1, 2, 3$), or equivalently,

$$A^\mu_+(x)|0\rangle = 0 \quad \forall x. \tag{8.81}$$

---

[20] Note that the conjugate momentum operator of the covariant electromagnetic field operator is defined as (see Section 9.1)

$$\pi^\mu(x) = \frac{\partial \mathcal{L}}{\partial \dot{A}_\mu} = \frac{\partial \mathcal{L}}{\partial(\partial_0 A_\mu)}.$$

Using $\mathcal{L} = \mathcal{L}_{EM,0}$ [see Eq. (9.10)], we have $\pi^\mu(x) = F^{\mu 0}$, which leads to $\pi^0(x) = 0$, since $F^{\mu\nu}$ is antisymmetric. This means that the covariant canonical commutation relations for the electromagnetic field will not resemble the canonical commutation relations for the Klein–Gordon field [cf. Eqs. (6.12) and (6.13)]. Therefore, using $\mathcal{L} = \mathcal{L}_{EM,0,g}$ instead [see Eq. (9.16) with $j = 0$], we obtain $\pi^\mu(x) = F^{\mu 0} - \frac{1}{\alpha}(\partial_\lambda A^\lambda)g^{\mu 0}$ in general and $\pi^0(x) = -\frac{1}{\alpha}(\partial_\lambda A^\lambda)$ in particular. However, in the Lorenz gauge, i.e. $\partial_\mu A^\mu = 0$, π^0 is still zero. Thus, in order to perform a covariant canonical quantization and to end up with Maxwell's equations, we have to quantize the theory using $\mathcal{L}_{EM,0,g}$, but not applying the strong Lorenz gauge condition (before quantization), and then (after quantization) impose a weaker Lorenz gauge condition. This will be discussed in detail in what follows.

In addition, using Eqs. (8.80) and (8.81), it should hold that

$$\langle 0|\hat{a}^{\mu\dagger}(k) = 0 \quad \forall k \quad \text{and} \quad \langle 0|A^{\mu}_{-}(x) = 0 \quad \forall x, \tag{8.82}$$

where $A^{\mu}_{-}(x) = \left[A^{\mu}_{+}(x)\right]^{\dagger}$. Note that the present formalism is not yet equivalent to Maxwell's equations in the classical limit, since we have ignored the Lorenz gauge condition. Therefore, we must impose this condition. However, it turns out that it cannot be imposed as the normal condition

$$\partial_{\mu}A^{\mu}(x) = 0, \tag{8.83}$$

since it is incompatible with the covariant canonical commutation relation

$$\left[\partial_{\mu}A^{\mu}(x), A^{\nu}(y)\right] = i\partial_{\mu}D^{\mu\nu}(x-y), \tag{8.84}$$

which is not equal to zero as it should be in the case that Eq. (8.83) holds. Thus, in the Gupta–Bleuler formalism, the 'strong' condition in Eq. (8.83) is replaced by the weaker condition

$$\partial_{\mu}A^{\mu}_{+}(x)|\Psi\rangle = 0 \quad \forall x, \tag{8.85}$$

where $|\Psi\rangle \in \mathcal{H}$ denotes a state that is allowed by the Gupta–Bleuler formalism and the quantity $A^{\mu}_{+}(x)$ contains annihilation operators only [cf. Eq. (8.75)]. Note that the so-called *Gupta–Bleuler space* $\mathcal{H}_{\text{GB}}$ has an indefinite metric and contains states with both negative and zero norms. If we were to impose conditions on the covariant electromagnetic field operator A^{μ}, then we will lose covariance. Therefore, we instead impose conditions on the states in $\mathcal{H}_{\text{GB}}$. Imposing the weaker Lorenz condition on the states in $\mathcal{H}_{\text{GB}}$, we obtain the subspace $\mathcal{H}_{\text{L}} \subset \mathcal{H}_{\text{GB}}$, which is called the *Lorenzian space*, but still contains states with zero norm. Finally, imposing another condition on the states in $\mathcal{H}_{\text{L}}$, which is equivalent to the radiation gauge condition, we obtain the subspace $\mathcal{H} \subset \mathcal{H}_{\text{L}}$ that contains no states with negative or zero norm, i.e. the physical states. Next, using Eq. (8.85), it naturally follows that

$$\langle\Psi|\partial_{\mu}A^{\mu}_{-}(x) = 0 \quad \forall x. \tag{8.86}$$

Thus, combining Eqs. (8.85) and (8.86), we obtain the vacuum expectation value of the Lorenz gauge condition (8.83), i.e.

$$\langle\Psi|\partial_{\mu}A^{\mu}(x)|\Psi\rangle = \langle\Psi|\partial_{\mu}A^{\mu}_{+}(x) + \partial_{\mu}A^{\mu}_{-}(x)|\Psi\rangle = 0, \tag{8.87}$$

which means that both the Lorenz gauge condition and Maxwell's equations hold in the classical limit of the Gupta–Bleuler formalism.

In the Gupta–Bleuler formalism, the 4-momentum operator is given by

$$P^{\mu} = -\sum_{\nu} g_{\nu\nu} \int_{k_0>0} \frac{d^3k}{k_0} k^{\mu} \hat{a}^{\nu\dagger}(k)\hat{a}^{\nu}(k), \tag{8.88}$$

which means that we have the Hamiltonian and 3-momentum operator

$$H = -\sum_{\mu} g_{\mu\mu} \int \hat{a}^{\mu\dagger}(k)\hat{a}^{\mu}(k)\, d^3k, \tag{8.89}$$

$$P^i = -\sum_{\mu} g_{\mu\mu} \int_{k_0>0} \frac{d^3k}{k_0} k^i \hat{a}^{\mu\dagger}(k)\hat{a}^{\mu}(k). \tag{8.90}$$

Considering the Hamiltonian, we obtain its vacuum expectation value as

$$\langle\Psi|H|\Psi\rangle = \int \sum_{\mu}(-g_{\mu\mu})\langle\Psi|\hat{a}^{\mu\dagger}(k)\hat{a}^{\mu}(k)|\Psi\rangle\, d^3k$$

$$= \int \langle\Psi| -\hat{a}^{0\dagger}(k)\hat{a}^0(k) + \hat{a}^{1\dagger}(k)\hat{a}^1(k) + \hat{a}^{2\dagger}(k)\hat{a}^2(k)$$

$$+ \hat{a}^{3\dagger}(k)\hat{a}^3(k)|\Psi\rangle\, d^3k. \tag{8.91}$$

In order to study the vacuum expectation value of the Hamiltonian in detail, we write the weaker Lorenz condition (8.85) in momentum space and use Eqs. (8.75), (8.72), (8.65), and (8.66), and finally, we obtain[21]

$$\bar{k}_{\mu}\hat{a}^{\mu}(\bar{k})|\Psi\rangle = \left[-\hat{a}(\bar{k},0) - \hat{a}(\bar{k},3)\right]|\Psi\rangle = \left[\hat{a}^0(\bar{k}) - \hat{a}^3(\bar{k})\right]|\Psi\rangle = 0, \tag{8.92}$$

which means that scalar and longitudinal photons cancel out for any physical state of the electromagnetic field. Equation (8.92) is the main idea of the Gupta–Bleuler formalism. In addition, using Eq. (8.92) and its adjoint, we have

$$\langle\Psi|\hat{a}^{0\dagger}(\bar{k})\hat{a}^0(\bar{k}) - \hat{a}^{3\dagger}(\bar{k})\hat{a}^3(\bar{k})|\Psi\rangle = \langle\Psi|\hat{a}^{0\dagger}(\bar{k})\hat{a}^0(\bar{k}) - \hat{a}^{0\dagger}(\bar{k})\hat{a}^3(\bar{k})|\Psi\rangle$$

$$= \langle\Psi|\hat{a}^{0\dagger}\left[\hat{a}^0(\bar{k}) - \hat{a}^3(\bar{k})\right]|\Psi\rangle = 0, \tag{8.93}$$

which immediately leads to

$$\langle\Psi|\hat{a}^{0\dagger}(\bar{k})\hat{a}^0(\bar{k})|\Psi\rangle = \langle\Psi|\hat{a}^{3\dagger}(\bar{k})\hat{a}^3(\bar{k})|\Psi\rangle. \tag{8.94}$$

Therefore, inserting Eq. (8.94) into Eq. (8.91), we finally obtain

$$\langle\Psi|H|\Psi\rangle = \int \langle\Psi|\hat{a}^{1\dagger}(k)\hat{a}^1(k) + \hat{a}^{2\dagger}(k)\hat{a}^2(k)|\Psi\rangle\, d^3k \geq 0, \tag{8.95}$$

which implies that (i) only physical transversal photons contribute to the vacuum expectation value of the Hamiltonian[22] and (ii) there are only states with positive norm (and no states with negative or zero norm) due to the restriction on the Gupta–Bleuler space by the weaker Lorenz condition and the condition that is equivalent to the radiation gauge condition. In conclusion, the Gupta–Bleuler formalism leads

---

[21] In the case that $k = \bar{k}$, it holds that $\hat{a}^0(\bar{k}) = -\hat{a}(\bar{k},0)$ and $\hat{a}^i(\bar{k}) = \hat{a}(\bar{k},i)$ for $i = 1,2,3$.
[22] In fact, this result holds for all other observables too.

to the fact that there exist states with scalar and longitudinal photons, but they do not contribute to physical observables.

Exercise 8.6 Verify Eq. (8.92).

Note that our analysis is only valid for **free** photons. We must respect gauge invariance in order to extend it to interactions. Otherwise, the equivalence between the radiation gauge formalism and the Gupta–Bleuler formalism will break down.

Problems

(1) Use the relation $F^{ij} = -\epsilon^{ijk} B^k$ to solve for B^i.
(2) Show that the quantities $\mathbf{E}^2 - \mathbf{B}^2$ and $\mathbf{E} \cdot \mathbf{B}$ are Lorentz invariants.
(3) (a) Show that if the electric and magnetic fields $\mathbf{E}$ and $\mathbf{B}$ are orthogonal for one observer, they are orthogonal for any observer.
   (b) Show that $\mathbf{E}$ and $\mathbf{B}$ are orthogonal for free plane waves with the 4-vector potential $A^\mu(x) = \varepsilon^\mu \mathrm{e}^{\mathrm{i}k \cdot x}$, where ε is the polarization vector.
   (c) Show for the plane waves that $\mathbf{E} \times \mathbf{B} = A\mathbf{k}$, where $\mathbf{k}$ is the wave vector and A is a non-vanishing expression.
(4) Calculate the Lorentz invariants $F^{\mu\nu} F_{\mu\nu}$ and $\epsilon^{\mu\nu\lambda\omega} F_{\mu\nu} F_{\lambda\omega}$ for a free electromagnetic plane wave $A^\mu(x) = \varepsilon^\mu \mathrm{e}^{\mathrm{i}k \cdot x}$, where ε is the polarization vector. Give a physical interpretation of your result.
(5) Maxwell's equations can be expressed by means of the 4-vector potential A. When $\partial_\mu A^\mu = 0$ (i.e. the Lorenz gauge is fulfilled), they take on a simple form. What is this form? Assuming that Maxwell's equations are of this simple form, and in addition, $j = 0$ (i.e. the setup is current free), show for a plane wave, $A^\mu(x) = \varepsilon^\mu \mathrm{e}^{\mathrm{i}k \cdot x}$, where ε is the polarization vector, that

$$\mathbf{E} \cdot \mathbf{k} = \mathbf{B} \cdot \mathbf{k} = 0,$$

i.e. the electric and magnetic fields are perpendicular to the direction of motion.

Guide to additional recommended reading

The following books (see the indicated pages) and their authors have similar treatments of the content in the present chapter.

- F. Gross, *Relativistic Quantum Mechanics and Field Theory*, Wiley (1993), pp. 28–56.
- F. Halzen and A. D. Martin, *Quarks & Leptons: An Introductory Course in Modern Particle Physics*, Wiley (1984), pp. 132–135.
- J. D. Jackson, *Classical Electrodynamics*, 3rd edn., Wiley (1999), pp. 239–242, 553–558, 612–615.
- R. H. Landau, *Quantum Mechanics II – A Second Course in Quantum Theory*, 2nd edn., Wiley-VCH (2004), pp. 176–180, 309–312.
- F. Mandl and G. Shaw, *Quantum Field Theory*, rev. edn, Wiley (1994), pp. 1–26, 81–94.
- J. Mickelsson, T. Ohlsson, and H. Snellman, *Relativity Theory*, KTH (2005), pp. 17–25.
- L. H. Ryder, *Quantum Field Theory*, 2nd edn., Cambridge (1996), pp. 64–67, 140–150.
- F. Schwabl, *Advanced Quantum Mechanics*, Springer (1999), pp. 307–320.

- S. S. Schweber, *An Introduction to Relativistic Quantum Field Theory*, Dover (2005), pp. 240–253.
- F. J. Ynduráin, *Relativistic Quantum Mechanics and Introduction to Field Theory*, Springer (1996), pp. 19–21, 160–169, 204–210.
- For the interested reader: K. Bleuler, *Helv. Phys. Acta* **23**, 567–586 (1950); H. B. G. Casimir, On the attraction between two perfectly conducting plates, *Proc. Kon. Nederland. Akad. Wetenschap.* **51**, 793–795 (1948); E. Fermi, quantum theory of radiation, *Rev. Mod. Phys.* **4**, 87–132 (1932); S. N. Gupta, Theory of longitudinal photons in quantum electrodynamics, *Proc. Phys. Soc. (London) A* **63**, 681–691 (1950); M. J. Sparnaay, Attractive forces between flat plates, *Nature* **180**, 334–335 (1957); M. J. Sparnaay, Measurement of attractive forces between flat plates, *Physica* **24**, 751 (1958); and F. Strocchi, Gauge problem in quantum field theory, *Phys. Rev.* **162**, 1429–1438 (1967).

9

The electromagnetic Lagrangian and introduction to Yang–Mills theory

In this chapter, we will discuss a famous theory known as Yang–Mills theory, which is a gauge theory based on the non-Abelian group SU(n). Note that our discussion will only be performed classically, i.e. we will not quantize Yang–Mills theory. Nevertheless, Yang–Mills theory serves as one of the possible extensions of the electromagnetic theory. First, we will recapitulate and continue to study the Lagrangian for the electromagnetic theory. Then, we will investigate possible models for extending the electromagnetic theory such as massive vector fields and other couplings. Next, we will introduce the important concept of covariant derivative. Finally, we will present the Lagrangian for Yang–Mills theory and discuss its properties.

9.1 The electromagnetic Lagrangian

In general, the Lagrangian for the electromagnetic theory is Lorentz invariant (constructed from the 4-vector potential A^μ, the 4-vector current j^μ, and the electromagnetic field strength tensor $F^{\mu\nu}$) and contains scalar products. It is given by

$$\mathcal{L}_{\text{EM}} = -\frac{1}{4} F^{\mu\nu} F_{\mu\nu} - j^\mu A_\mu, \tag{9.1}$$

where the first term is the free kinetic part for the electromagnetic field [cf. Eq. (5.24)] and the second term is an external source term. Again, $F_{\mu\nu} = \partial_\mu A_\nu - \partial_\nu A_\mu$ [cf. Eqs. (3.197) and (8.7)]. Writing the free kinetic part $-\frac{1}{4} F^{\mu\nu} F_{\mu\nu}$ as

$$-\frac{1}{4} F^{\mu\nu} F_{\mu\nu} = -\frac{1}{2} \partial^\mu A^\nu \left(\partial_\mu A_\nu - \partial_\nu A_\mu \right) = -\frac{1}{2} g^{\mu\mu'} g^{\nu\nu'} \partial_{\mu'} A_{\nu'} \left(\partial_\mu A_\nu - \partial_\nu A_\mu \right)$$

$$= -\frac{1}{2} g^{\mu\mu'} g^{\nu\nu'} \left(\partial_{\mu'} A_{\nu'} \partial_\mu A_\nu - \partial_{\mu'} A_{\nu'} \partial_\nu A_\mu \right) \tag{9.2}$$

and using Euler–Lagrange field equations with the Lagrangian (9.1), we find that

$$\frac{\partial \mathcal{L}_{EM}}{\partial(\partial_\alpha A_\beta)} = -\partial^\alpha A^\beta + \partial^\beta A^\alpha = -F^{\alpha\beta}, \tag{9.3}$$

$$\partial_\alpha \left(\frac{\partial \mathcal{L}_{EM}}{\partial(\partial_\alpha A_\beta)} \right) = -\partial_\alpha F^{\alpha\beta}, \tag{9.4}$$

$$\frac{\partial \mathcal{L}_{EM}}{\partial A_\beta} = -j^\beta, \tag{9.5}$$

which lead to the conjugate momentum

$$\pi^\beta = \frac{\partial \mathcal{L}_{EM}}{\partial \dot{A}_\beta} = \frac{\partial \mathcal{L}_{EM}}{\partial(\partial_0 A_\beta)} = F^{\beta 0} \tag{9.6}$$

and Maxwell's equations

$$\partial_\alpha F^{\alpha\beta} = j^\beta. \tag{9.7}$$

Note that $\partial_\beta j^\beta = \partial_\beta \partial_\alpha F^{\alpha\beta} \equiv 0$, because of the antisymmetric property of $F^{\alpha\beta}$, which implies that the 4-vector current j is conserved.

If j is conserved, then the Lagrangian function $L = \int \mathcal{L}_{EM} \, d^3x$ and the action $A = \int \mathcal{L}_{EM} \, d^4x$ for the electromagnetic theory are invariant under local gauge transformations of the following type (modulo an irrelevant boundary term)

$$A_\mu \mapsto A'_\mu = A_\mu + \partial_\mu \chi, \tag{9.8}$$

where χ is a scalar function. However, the Lagrangian for the electromagnetic theory $\mathcal{L}_{EM}$ is not invariant under local gauge transformations (9.8), since for the external source term, we have

$$-j^\mu A_\mu \mapsto -j^\mu A'_\mu = -j^\mu A_\mu - j^\mu \partial_\mu \chi, \tag{9.9}$$

where the last term is only equal to zero if $\chi = \text{const.}$, i.e. for global gauge transformations. Nevertheless, the Lagrangian for the free electromagnetic field, i.e.

$$\mathcal{L}_{EM,0} \equiv \mathcal{L}_{EM}(j=0) = -\frac{1}{4}F^{\mu\nu}F_{\mu\nu} = \frac{1}{2}\left(\mathbf{E}^2 - \mathbf{B}^2\right), \tag{9.10}$$

is invariant under local gauge transformations (9.8), since the electromagnetic field strength tensor is invariant under such transformations [cf. Eq. (8.20)]. As stated above, the action A is indeed invariant under local gauge transformations, since using integration by parts and the fact that j is conserved, we find that

$$-\int j^\mu \partial_\mu \chi \, d^4x = \int (\partial_\mu j^\mu) \chi \, d^4x = 0. \tag{9.11}$$

The electromagnetic field has a special property for the 0-component, i.e. A_0. Using Eq. (9.3), we observe that the conjugate momentum corresponding to the 0-component is given by

$$\pi^0 = \frac{\partial \mathcal{L}_{\text{EM}}}{\partial \dot{A}_0} = \frac{\partial \mathcal{L}_{\text{EM}}}{\partial (\partial_0 A_0)} = 0, \qquad (9.12)$$

since the time derivative of A_0 does not naturally occur in the Lagrangian (9.1). This means that $[A^0, \pi^0] = 0$. Therefore, quantizing A^0, it will commute with all operators. Thus, the 0-component of the electromagnetic field A^0 is indeed special. In the Coulomb gauge, A^0 can be replaced by the other components of the electromagnetic field. Using Maxwell's equations (9.7), the definition of $F^{\mu\nu}$, and the Coulomb gauge condition $\nabla \cdot \mathbf{A} = 0$, we obtain

$$\partial_\mu F^{\mu 0} = \partial_\mu \left(\partial^\mu A^0 - \partial^0 A^\mu \right) = -\Delta A^0 - \partial^0 \nabla \cdot \mathbf{A} = -\Delta A^0 = \rho = j^0, \quad (9.13)$$

which is the Poisson equation. The solution to this equation is Coulomb's law. If $\rho = 0$, then $A^0 = 0$. On the other hand, in the Lorenz gauge, we have

$$\Box A^0 = \rho = j^0, \qquad (9.14)$$

which is Lorentz covariant. However, the solution for A^0 is not zero, even though $\rho = 0$. Thus, in the Lorenz gauge, we need another approach to study the electromagnetic field and its Lagrangian (9.1). Introducing a so-called *gauge-fixing term*

$$\mathcal{L}_{\text{gauge}} = -\frac{1}{2\alpha} \left(\partial_\mu A^\mu \right) \left(\partial_\nu A^\nu \right), \qquad (9.15)$$

where α is the corresponding *gauge parameter* (and, in principle, an arbitrary real number), and adding it to the Lagrangian (9.1), we obtain

$$\mathcal{L}_{\text{EM,g}} = \mathcal{L}_{\text{EM}} + \mathcal{L}_{\text{gauge}} = -\frac{1}{4} F^{\mu\nu} F_{\mu\nu} - j^\mu A_\mu - \frac{1}{2\alpha} \left(\partial_\mu A^\mu \right) \left(\partial_\nu A^\nu \right), \qquad (9.16)$$

where the gauge-fixing term $\mathcal{L}_{\text{gauge}}$ is, in principle, zero due to the Lorenz gauge condition $\partial_\mu A^\mu = 0$. Three popular and obvious choices of the gauge parameter exist:

- $\alpha \to 0$ (Landau gauge parameter),
- $\alpha = 1$ (Feynman gauge parameter),
- $\alpha \to \infty$ (unitary gauge parameter).

Again, using Euler–Lagrange field equations, but this time for the Lagrangian (9.16), we find that

$$\frac{\partial \mathcal{L}_{\text{EM,g}}}{\partial (\partial_\mu A_\nu)} = -F^{\mu\nu} - \frac{1}{\alpha} (\partial_\lambda A^\lambda) g^{\mu\nu}, \qquad (9.17)$$

$$\partial_\mu \frac{\partial \mathcal{L}_{\text{EM,g}}}{\partial (\partial_\mu A_\nu)} = -\partial_\mu F^{\mu\nu} - \frac{1}{\alpha} \partial_\mu (\partial_\lambda A^\lambda) g^{\mu\nu}, \qquad (9.18)$$

$$\frac{\partial \mathcal{L}_{\text{EM,g}}}{\partial A_\nu} = -j^\nu, \qquad (9.19)$$

which lead to the conjugate momentum

$$\pi^\mu = \frac{\partial \mathcal{L}_{EM,g}}{\partial \dot{A}_\mu} = \frac{\partial \mathcal{L}_{EM,g}}{\partial(\partial_0 A_\mu)} = F^{\mu 0} - \frac{1}{\alpha}(\partial_\lambda A^\lambda)g^{\mu 0} \qquad (9.20)$$

and the equations of motion

$$\Box A^\mu + \left(1 - \frac{1}{\alpha}\right)\partial^\mu(\partial_\lambda A^\lambda) = j^\mu. \qquad (9.21)$$

Especially, we have $\pi^0 = -\frac{1}{\alpha}(\partial_\lambda A^\lambda)$. For example, in the Lorenz gauge, we have $\pi^\mu = F^{\mu 0}$ and $\Box A^\mu = j^\mu$, since $\partial_\mu A^\mu = 0$, whereas using the Feynman gauge parameter (not assuming the Lorenz gauge), we have $\pi^\mu = F^{\mu 0} - (\partial_\lambda A^\lambda)g^{\mu 0}$ and again $\Box A^\mu = j^\mu$, since $\alpha = 1$. Note that the electromagnetic theory, i.e. the form of the Lagrangian (9.16), is affected by the addition of $\mathcal{L}_{gauge}$. However, the physical observables are not affected by this addition, since $\mathcal{L}_{gauge}$ is zero due to the assumption of the Lorenz gauge condition $\partial_\mu A^\mu = 0$. Indeed, physics must be unaffected by the choice of gauge and the final result for any physical observable must be independent of the gauge (cf. the discussion on gauge interactions in Section 5.3).

Finally, in the case of a free electromagnetic field, i.e. $j = 0$, the energy–momentum tensor is defined as [once again using Noether's theorem (5.69)]

$$T^{\mu\nu} = \frac{\partial \mathcal{L}_{EM,0}}{\partial(\partial_\mu A^\lambda)}\partial^\nu A^\lambda - g^{\mu\nu}\mathcal{L}_{EM,0}, \qquad (9.22)$$

and inserting Eq. (9.10) into Eq. (9.22), we have

$$T^{\mu\nu} = -F^\mu{}_\lambda \partial^\nu A^\lambda + \frac{1}{4}g^{\mu\nu}F^{\lambda\omega}F_{\lambda\omega}. \qquad (9.23)$$

However, note that the first term of $T^{\mu\nu}$ in Eq. (9.23) is not symmetric, but can be made symmetric. In addition, Eq. (9.23) is not gauge invariant, since it contains an explicit factor of the 4-vector potential and not only factors of the electromagnetic field strength tensor. Thus, symmetrizing $T^{\mu\nu}$ in Eq. (9.23), we obtain the symmetric energy–momentum tensor for the free electromagnetic field[1]

$$T^{\mu\nu} := \frac{1}{2}(T^{\mu\nu} + T^{\nu\mu}) = F^\mu{}_\lambda F^{\lambda\nu} + \frac{1}{4}g^{\mu\nu}F^{\lambda\omega}F_{\lambda\omega}, \qquad (9.24)$$

which is gauge invariant. In addition, note that the trace of $T^{\mu\nu}$ vanishes

$$T^\mu{}_\mu = g_{\mu\nu}T^{\mu\nu} = F^\mu{}_\lambda F^\lambda{}_\mu + \frac{1}{4}\delta^\mu{}_\mu F^{\lambda\omega}F_{\lambda\omega} = -F^{\mu\lambda}F_{\mu\lambda} + F^{\lambda\omega}F_{\lambda\omega} = 0. \quad (9.25)$$

---

[1] Instead of symmetrizing $T^{\mu\nu}$ in Eq. (9.23), we can perform the transformation $T^{\mu\nu} \mapsto T^{\mu\nu} + \partial^\lambda\left(F^\mu{}_\lambda A^\nu\right)$, which will also lead to the symmetric $T^{\mu\nu}$ given in Eq. (9.24).

Without sources ($j = 0$), $T^{\mu\nu}$ is conserved, i.e. $\partial_\mu T^{\mu\nu} = 0$. Especially, investigating the components T^{00} and $T^{0i} = T^{i0}$ and using the definition of $F^{\mu\nu}$, we can compute

$$\mathcal{H} \equiv T^{00} = \frac{1}{2}\left(\mathbf{E}^2 + \mathbf{B}^2\right), \tag{9.26}$$

$$\mathcal{P}^i \equiv T^{0i} = (\mathbf{E} \times \mathbf{B})^i. \tag{9.27}$$

Hence, the T^{00} component is interpreted as the Hamiltonian density of the free electromagnetic field, whereas the spatial components T^{0i} are interpreted as the components of the 3-momentum density. The vector $\mathcal{P} = \mathbf{E} \times \mathbf{B}$ is sometimes called the *Poynting vector*. Note that both the total energy and the 3-momentum are conserved when there are no sources. Furthermore, it holds that the energy–momentum tensor $T^{\mu\nu}$ transforms as a second-rank tensor under Lorentz transformations, i.e.

$$T'^{\mu\nu}(x') = \Lambda^\mu{}_\lambda \Lambda^\nu{}_\omega T^{\lambda\omega}(x). \tag{9.28}$$

Exercise 9.1 Compute the components T^{00} and T^{0i} of the symmetric energy–momentum tensor $T^{\mu\nu}$ for the free electromagnetic field.

Exercise 9.2 Show that $T^{\mu\nu}$ (the symmetric version) transforms as a second-rank tensor with respect to Lorentz transformations.

Exercise 9.3 Show that $T^{\mu\nu}$ (the symmetric version) is gauge invariant.

We should now be set to study generalizations of the electromagnetic Lagrangian.

9.2 Massive vector fields

In order to investigate massive vector fields, we add a 'mass' term to the Lagrangian for the electromagnetic theory

$$\mathcal{L}_M = -\frac{1}{4}F^{\mu\nu}F_{\mu\nu} + \frac{1}{2}M^2 A^\mu A_\mu - j^\mu A_\mu, \tag{9.29}$$

where M is assumed to be real. Massive vector fields play an important role in physics. For example, there are the $W^\pm$ and Z^0 bosons for electroweak interactions.[2] Using Euler–Lagrange field equations with the Lagrangian (9.29), we obtain

$$\partial_\mu F^{\mu\nu} + M^2 A^\nu = j^\nu, \tag{9.30}$$

---

[2] In 1979, S. Glashow, A. Salam and S. Weinberg were awarded the Nobel Prize in physics 'for their contributions to the theory of the unified weak and electromagnetic interaction between elementary particles, including, inter alia, the prediction of the weak neutral current', in 1984, C. Rubbia and S. van der Meer 'for their decisive contributions to the large project, which led to the discovery of the gauge field particles W and Z, communicators of weak interaction', and in 1999, G. 't Hooft and M. Veltman 'for elucidating the quantum structure of electroweak interactions in physics'.

which are the *Proca equations*. Taking the derivative ∂_ν of both sides of these equations, we find

$$M^2 \partial_\nu A^\nu = \partial_\nu j^\nu = 0, \qquad (9.31)$$

since j is assumed to be conserved. However, since $M \neq 0$, the Lorenz gauge condition $\partial_\nu A^\nu = 0$ appears as a necessary constraint. Thus, we **cannot** choose gauge. The reason is that the mass term is **not** gauge invariant. This can be seen as follows. Applying a gauge transformation to the mass term $M^2 A^\mu A_\mu$, we obtain in general

$$M^2 A^\mu A_\mu \mapsto M^2 A'^\mu A'_\mu = M^2 \left(A^\mu A_\mu + \partial^\mu \chi A_\mu + A^\mu \partial_\mu \chi + \partial^\mu \chi \partial_\mu \chi \right) \neq M^2 A^\mu A_\mu. \qquad (9.32)$$

However, using the Lorenz gauge, we have the following simplified version of the Proca equations

$$\left(\Box + M^2 \right) A^\nu = j^\nu. \qquad (9.33)$$

Note the similarity of these equations with the Klein–Gordon equation [cf. Eq. (2.10)], the wave equation for the non-relativistic string [cf. Eq. (4.15)], and the wave equation for the electromagnetic field [cf. Eq. (8.27)]. In principle, the Proca equations (9.33) are generalizations of the above mentioned equations. If $j = 0$, i.e. the situation is source free, then the solutions to the Proca equations are given by plane-wave solutions

$$A^\nu(x) \sim \varepsilon^\nu(k) e^{-i k \cdot x}, \qquad (9.34)$$

where k satisfies the relation $k^2 = M^2$, which has the solutions $\pm k_0 = E_k = \sqrt{M^2 + \mathbf{k}^2}$. Indeed, M is a mass. Furthermore, using the plane-wave solutions, the Lorenz gauge implies that $k_\mu \varepsilon^\mu(k) = 0$, which means that there are **three** independent polarization states (cf. the discussion on the two polarization states of the electromagnetic field in Section 8.2).

Next, consider the following Lagrangian for the field A_μ

$$\mathcal{L}_f = \lambda_1 F^{\mu\nu} F_{\mu\nu} + \lambda_2 G^{\mu\nu} G_{\mu\nu} + \lambda_3 A^\mu A_\mu, \qquad (9.35)$$

where $G_{\mu\nu} = \partial_\mu A_\nu + \partial_\nu A_\mu$ is symmetric. Performing a gauge transformation, we find that $F'_{\mu\nu} = F_{\mu\nu}$ is invariant as before, whereas we obtain $G'_{\mu\nu} = G_{\mu\nu} + 2\partial_\mu \partial_\nu \chi$, which means that $G_{\mu\nu}$ does not transform in a gauge invariant way. Thus, in order for the Lagrangian $\mathcal{L}_f$ to be gauge invariant, i.e. $\mathcal{L}'_f = \mathcal{L}_f$, we have to require that $\lambda_2 = 0$ and $\lambda_3 = 0$. Furthermore, if $\mathcal{L}_f$ should be gauge invariant, then A_μ needs to be **massless** and have a free Lagrangian of the form $\lambda_1 F^{\mu\nu} F_{\mu\nu}$, where $\lambda_1 = -1/4$ in the electromagnetic theory. Thus, (local) gauge invariance dictates the form of the electromagnetic theory and its Lagrangian in particular and of QED in general.

Exercise 9.4 Show that $G_{\mu\nu} = \partial_\mu A_\nu + \partial_\nu A_\mu$ transforms as $G_{\mu\nu} \mapsto G'_{\mu\nu} = G_{\mu\nu} + 2\partial_\mu\partial_\nu\chi$ with respect to gauge transformations of the form $A_\mu \mapsto A'_\mu = A_\mu + \partial_\mu\chi$.

9.3 Gauge transformations and the covariant derivative

A generalization of QED is Yang–Mills theory.[3] In order to investigate Yang–Mills theory, we first need to study how different fields transform under gauge transformations and to introduce the concept of covariant derivative. Under local gauge transformations, it is assumed that the vector field A_μ and the Dirac field transform in the following way

$$A_\mu(x) \mapsto A'_\mu(x) = A_\mu(x) + \partial_\mu\chi(x), \tag{9.36}$$

$$\psi(x) \mapsto \psi'(x) = e^{i\alpha(x)}\psi(x), \tag{9.37}$$

where $\chi(x) \equiv -\alpha(x)/e$.

Now, introduce the generalization of the normal derivative, i.e. the *covariant derivative* as (cf. the discussion in Section 5.3)

$$D_\mu\psi(x) = \partial_\mu\psi(x) + ieA_\mu(x)\psi(x). \tag{9.38}$$

Using local gauge transformations, we find the transformation for the covariant derivative

$$D_\mu\psi(x) \mapsto D'_\mu\psi'(x) = \left[\partial_\mu + ie\left(A_\mu(x) - \frac{1}{e}\partial_\mu\alpha(x)\right)\right]e^{i\alpha(x)}\psi(x)$$

$$= e^{i\alpha(x)}\left[\partial_\mu + ieA_\mu(x)\right]\psi(x) = e^{i\alpha(x)}D_\mu\psi(x). \tag{9.39}$$

Thus, the covariant derivative of the Dirac field $D_\mu\psi$ transforms in the same way as the Dirac field ψ, but this is **not** the case for the normal derivative of the Dirac field $\partial_\mu\psi$. In addition, note that the commutator of two covariant derivatives transforms as

$$\left[D_\mu, D_\nu\right]\psi(x) \mapsto \left[D'_\mu, D'_\nu\right]\psi'(x) = e^{i\alpha(x)}\left[D_\mu, D_\nu\right]\psi(x), \tag{9.40}$$

which follows from Eq. (9.39). However, the commutator is 'not' a derivative, since

$$\left[D_\mu, D_\nu\right]\psi = \left[\partial_\mu, \partial_\nu\right]\psi + ie\left(\left[\partial_\mu, A_\nu\right] - \left[\partial_\nu, A_\mu\right]\right)\psi - e^2\left[A_\mu, A_\nu\right]\psi$$

$$= ie\left(\partial_\mu A_\nu - \partial_\nu A_\mu\right)\psi = ieF_{\mu\nu}\psi. \tag{9.41}$$

Thus, we have that the commutator is proportional to the electromagnetic field strength tensor, i.e.

$$\left[D_\mu, D_\nu\right] = ieF_{\mu\nu}. \tag{9.42}$$

---

[3] For the interested reader, please see: C. N. Yang and R. L. Mills, Conservation of isotopic spin and isotopic gauge invariance, *Phys. Rev.* **96**, 191–195 (1954).

Exercise 9.5 Show that the commutator of two covariant derivatives transforms according to Eq. (9.40) with respect to gauge transformations.

9.4 The Yang–Mills Lagrangian

9.4.1 The Lie group SU(2)

If we replace the **single** Dirac field by a **doublet** of Dirac fields, then we have

$$\psi(x) = \begin{pmatrix} \psi_1(x) \\ \psi_2(x) \end{pmatrix}, \tag{9.43}$$

where $\psi_i(x)$ $(i = 1, 2)$ transform into one another under **abstract** three-dimensional rotations as

$$\psi \mapsto \psi' = \exp\left(i\alpha^j \frac{\sigma^j}{2}\right)\psi. \tag{9.44}$$

Here the σ^js are the Pauli matrices, which are the generators of the Lie group SU(2) (see Appendix A). In 1954, Yang and Mills interpreted $(\psi_1 \quad \psi_2)^T$ as the proton–neutron doublet, and in this case, the abstract three-dimensional rotation is the *isotopic spin* (or *isospin*).

We will now require that the Lagrangian should be invariant under local gauge transformation of the form (9.44), which means that the Dirac doublet and the vector fields should transform as

$$\psi(x) \mapsto \psi'(x) = V(x)\psi(x), \tag{9.45}$$

$$A_\mu^j(x)\frac{\sigma^j}{2} \mapsto A'^j_\mu(x)\frac{\sigma^j}{2} = V(x)\left[A_\mu^j(x)\frac{\sigma^j}{2} + \frac{i}{g}\partial_\mu\right]V^\dagger(x), \tag{9.46}$$

respectively, where

$$V(x) = \exp\left(i\alpha^j(x)\frac{\sigma^j}{2}\right) \quad \Leftrightarrow \quad V^\dagger(x) = \exp\left(-i\alpha^j(x)\frac{\sigma^j}{2}\right), \tag{9.47}$$

g is a coupling constant, and the A_μ^js are three vector fields, one for each generator σ^j. In this case, the covariant derivative is given by

$$D_\mu = \partial_\mu - ig A_\mu^j \frac{\sigma^j}{2}. \tag{9.48}$$

Again, expanding the commutator of covariant derivatives, we find that [cf. the result in Eq. (9.42)]

$$[D_\mu, D_\nu] = -ig F_{\mu\nu}^j \frac{\sigma^j}{2}, \tag{9.49}$$

where

$$F^j_{\mu\nu}\frac{\sigma^j}{2} = \partial_\mu A^j_\nu\frac{\sigma^j}{2} - \partial_\nu A^j_\mu\frac{\sigma^j}{2} - ig\left[A^j_\mu\frac{\sigma^j}{2}, A^k_\nu\frac{\sigma^k}{2}\right], \tag{9.50}$$

which has an extra term compared with the electromagnetic field strength tensor. Using the commutation relations for the Pauli matrices

$$\left[\frac{\sigma^j}{2}, \frac{\sigma^k}{2}\right] = i\epsilon^{jk\ell}\frac{\sigma^\ell}{2}, \tag{9.51}$$

we obtain the *non-Abelian field strength tensor*

$$F^i_{\mu\nu} = \partial_\mu A^i_\nu - \partial_\nu A^i_\mu + g\epsilon^{ijk}A^j_\mu A^k_\nu, \tag{9.52}$$

where the last term is due to the fact that the Pauli matrices do not commute. The gauge transformation rule for $F^i_{\mu\nu}$ is given by

$$F^j_{\mu\nu}\frac{\sigma^j}{2} \mapsto F'^j_{\mu\nu}\frac{\sigma^j}{2} = F^j_{\mu\nu}\frac{\sigma^j}{2} + \left[i\alpha^j\frac{\sigma^j}{2}, F^k_{\mu\nu}\frac{\sigma^k}{2}\right], \tag{9.53}$$

which is not gauge invariant. However, the Lagrangian

$$\mathcal{L}_{YM} = -\frac{1}{2}\text{tr}\left[\left(F^i_{\mu\nu}\frac{\sigma^i}{2}\right)^2\right] = -\frac{1}{2}\text{tr}\left(F^{i\,\mu\nu}\frac{\sigma^i}{2}F^j_{\mu\nu}\frac{\sigma^j}{2}\right) = -\frac{1}{8}F^{i\,\mu\nu}F^j_{\mu\nu}\underbrace{\text{tr}(\sigma^i\sigma^j)}_{2\delta^{ij}}$$

$$= -\frac{1}{4}F^{i\,\mu\nu}F^i_{\mu\nu} \tag{9.54}$$

is gauge invariant. This Lagrangian is called the *Yang–Mills Lagrangian* for the case of SU(2). Note that the Lagrangian contains **cubic** and **quartic** terms in the vector fields A^i_μ.

9.4.2 General symmetry groups

In general, for a general local symmetry group, we replace the 'Pauli matrices' $\sigma^i/2$ by general generators t^a. The commutation relations for the generators t^a are given by

$$[t^a, t^b] = if^{abc}t^c, \tag{9.55}$$

where the f^{abc}s are the so-called *structure constants*, which have to be computed for the specific symmetry group under consideration, but independently of the symmetry group they are totally antisymmetric with respect to all three indices. For example, it holds that $f^{abc} = f^{cab} = -f^{acb}$. Note that the imaginary unit i in Eq. (9.55) is a convention. Thus, we have the generalized version of the covariant derivative

$$D_\mu = \partial_\mu - ig A^a_\mu t^a \tag{9.56}$$

and the infinitesimal transformation rules for the fields

$$\psi \mapsto \psi' = (1 + i\alpha^a t^a)\psi \tag{9.57}$$

$$A_\mu^a \mapsto A_\mu'^a = A_\mu^a + \frac{1}{g}\partial_\mu\alpha^a + f^{abc}A_\mu^b\alpha^c. \tag{9.58}$$

In addition, using Eq. (9.56), the commutator of covariant derivatives is [cf. the result in Eq. (9.49)]

$$[D_\mu, D_\nu] = -ig F_{\mu\nu}^a t^a, \tag{9.59}$$

where

$$F_{\mu\nu}^a = \partial_\mu A_\nu^a - \partial_\nu A_\mu^a + g f^{abc}A_\mu^b A_\nu^c. \tag{9.60}$$

The infinitesimal transformation rule for $F_{\mu\nu}^a$ is given by

$$F_{\mu\nu}^a \mapsto F_{\mu\nu}'^a = F_{\mu\nu}^a - f^{abc}\alpha^b F_{\mu\nu}^c. \tag{9.61}$$

Note that the vector space spanned by the generators t^a (with commutation relations) is called a *Lie algebra* if bilinearity, i.e. $[x, x] = 0$, and the Jacobi identity holds (see Appendix A.3). The Jacobi identity is $[t^a, [t^b, t^c]] + \text{cycl.} = 0$, which implies that

$$f^{ade}f^{bcd} + f^{bde}f^{cad} + f^{cde}f^{abd} = 0. \tag{9.62}$$

Exercise 9.6 Derive the infinitesimal transformation rule for the commutator of covariant derivatives in Eq. (9.61).

Now, the Lagrangian for pure Yang–Mills theory is given by

$$\mathcal{L}_{YM} = -\frac{1}{4}F^{a\,\mu\nu}F_{\mu\nu}^a. \tag{9.63}$$

Expanding $\mathcal{L}_{YM}$ in terms of A_μ^a, i.e. inserting Eq. (9.60) into Eq. (9.63) as well as relabelling indices and using the antisymmetric property of the structure constants, gives

$$\mathcal{L}_{YM} = -\frac{1}{2}\partial^\mu A^{a\,\nu}\left(\partial_\mu A_\nu^a - \partial_\nu A_\mu^a\right) + \mathcal{L}_3 + \mathcal{L}_4, \tag{9.64}$$

where

$$\mathcal{L}_3 = -\frac{1}{2}g f^{abc}A^{b\,\mu}A^{c\,\nu}\left(\partial_\mu A_\nu^a - \partial_\nu A_\mu^a\right) = -g f^{abc}A^{b\,\mu}A^{c\,\nu}\partial_\mu A_\nu^a, \tag{9.65}$$

$$\mathcal{L}_4 = -\frac{1}{4}g^2 f^{abe}f^{cde}A_\mu^a A_\nu^b A^{c\,\mu}A^{d\,\nu}. \tag{9.66}$$

The terms in $\mathcal{L}_3$ and $\mathcal{L}_4$ are so-called *gauge self-couplings*, which are a **direct** consequence of non-Abelian gauge invariance and mean that the vector fields are self-interacting. Note that the first term in Eq. (9.64) corresponds to the Abelian case, which can be achieved by setting the structure constants equal to zero, and in this case, $\mathcal{L}_3 = 0$ and $\mathcal{L}_4 = 0$.

Exercise 9.7 Perform the computations that lead from Eq. (9.63) to Eqs. (9.64)–(9.66).

Finally, using $\mathcal{L}_{\text{YM}}$ and Euler–Lagrange field equations, we can derive the classical equations of motion for Yang–Mills theory

$$\partial^\mu F^a_{\mu\nu} + g f^{abc} A^{b\mu} F^c_{\mu\nu} = 0. \tag{9.67}$$

Introducing the quantity $\tilde{F}_{\mu\nu} := t^a F^a_{\mu\nu}$ as well as using Eq. (9.55) and the extended version of Eq. (9.56), i.e. $D_\mu X = \partial_\mu X - ig A^a_\mu [t^a, X]$, where the quantity X should be proportional to the generators, Eq. (9.67) can be rewritten as

$$D^\mu \tilde{F}_{\mu\nu} = 0, \tag{9.68}$$

and in addition, using Eq. (9.60), the antisymmetric property of the structure constants, and Eq. (9.62), the Bianchi identity [cf. Eq. (8.10)] is satisfied, i.e.

$$D_\mu \tilde{F}_{\nu\lambda} + D_\nu \tilde{F}_{\lambda\mu} + D_\lambda \tilde{F}_{\mu\nu} = 0. \tag{9.69}$$

To this end, using the definition of $\tilde{F}_{\mu\nu}$ and Eq. (9.59), this Bianchi identity can be generalized to

$$D_\mu [D_\nu, D_\lambda] + D_\nu [D_\lambda, D_\mu] + D_\lambda [D_\mu, D_\nu] = 0, \tag{9.70}$$

which is a Jacobi identity.

Exercise 9.8 Show by direct computation that Eq. (9.68) leads to Eq. (9.67).

Exercise 9.9 Prove the Bianchi identity in Eq. (9.69).

Problems

(1) Find the Euler–Lagrange field equations for the following Lagrangians

(a)
$$\mathcal{L} = \frac{1}{2}(\partial_\mu \phi)(\partial^\mu \phi) - \frac{1}{2}m^2\phi^2 - \frac{1}{4}\lambda\phi^4,$$

(b)
$$\mathcal{L} = -(\partial_\mu A^\nu)(\partial_\nu A^\mu) + \frac{1}{2}m^2 A_\mu A^\mu + \frac{\Lambda}{2}(\partial_\mu A^\mu)^2.$$

(2) Show that the two Lagrangians

$$\mathcal{L}_{\text{EM,0,g}} = -\frac{1}{4}F^{\mu\nu}F_{\mu\nu} - \frac{1}{2}(\partial_\mu A^\mu)(\partial_\nu A^\nu)$$

and

$$\mathcal{L}_F = -\frac{1}{2} \left(\partial_\nu A_\mu \right) \left(\partial^\nu A^\mu \right)$$

are equivalent, i.e. they lead to the same equations of motion. Note that the second Lagrangian was proposed by Fermi.

(3) Derive the Proca equations $\partial_\mu F^{\mu\nu} + M^2 A^\nu = j^\nu$, where $M > 0$, from the Lagrangian

$$\mathcal{L} = -\frac{1}{4} F^{\mu\nu} F_{\mu\nu} + \frac{1}{2} M^2 A_\mu A^\mu - j_\mu A^\mu.$$

(4) For a general local symmetry group, the covariant derivative is given by

$$D_\mu = \partial_\mu - ig A_\mu^a t^a,$$

where g is the coupling constant and t^a are generators with commutation relations $[t^a, t^b] = if^{abc} t^c$, where f^{abc} are structure constants. Find the non-Abelian field strength tensor $F_{\mu\nu}^a$ using the commutator of covariant derivatives, i.e. $[D_\mu, D_\nu] = -ig F_{\mu\nu}^a t^a$.

(5) The free Yang–Mills Lagrangian is given by

$$\mathcal{L}_{YM} = -\frac{1}{4} F^{a\,\mu\nu} F_{\mu\nu}^a,$$

where $F_{\mu\nu}^a$ is the non-Abelian field strength tensor. [It is sufficient if you consider only the case of SU(2).]

(a) Show that this Lagrangian is gauge invariant.
(b) Show that this Lagrangian contains cubic and quartic terms in the vector potentials A_μ^a.
(c) Why do the cubic and quartic terms arise?
(d) What is the physical interpretation of the cubic and quartic terms?

Guide to additional recommended reading

The following books (see the indicated pages) and their authors have similar treatments of the content in the present chapter.

- F. Gross, *Relativistic Quantum Mechanics and Field Theory*, Wiley (1993), pp. 415–427.
- J. D. Jackson, *Classical Electrodynamics*, 3rd edn., Wiley (1999), pp. 598–603, 605–610.
- M. E. Peskin and D. V. Schroeder, *An Introduction to Quantum Field Theory*, Addison-Wesley (1995), pp. 481–504.
- L. H. Ryder, *Quantum Field Theory*, 2nd edn., Cambridge (1996), pp. 67–69, 150–153.
- For the interested reader: C. N. Yang and R. L. Mills, Conservation of isotopic spin and isotopic gauge invariance, *Phys. Rev.* **96**, 191–195 (1954).

10

Asymptotic fields and the LSZ formalism

In this chapter, the Lehmann–Symanzik–Zimmermann (LSZ) formalism is presented. This was developed in 1955 and is named after its inventors, namely, the three German physicists H. Lehmann, K. Symanzik, and W. Zimmermann.[1] Especially, the LSZ formalism includes the LSZ reduction formula that provides an explicit way of expressing physical S matrix elements (i.e. scattering amplitudes) in terms of T-ordered correlation functions of an interacting field within a quantum field theory to all orders in perturbation theory. Thus, the goal is to express the S matrix elements in terms of asymptotic free fields instead of the unknown interacting field. Therefore, given the Lagrangian of some quantum field theory, it leads to predictions of measurable quantities. Note that the LSZ reduction formula cannot treat massless particles, bound states, and so-called *topological defects*. For example, in quantum chromodynamics (QCD), the asymptotic states are bound, which means that they are not free states. However, the original formula can be generalized in order to include bound states, which are states described by non-local composite fields. In addition, in statistical physics, a general formulation of the fluctuation–dissipation theorem, which is a theorem that can be used to predict the non-equilibrium behaviour of a system, can be obtained using the LSZ formalism.

10.1 Asymptotic fields and the S operator

Let us start by considering a real scalar field ϕ that fulfils the equation

$$\left(\Box + m_0^2\right)\phi(x) = j(x), \tag{10.1}$$

where j is a functional of the field ϕ, i.e. $j = j[\phi(x)] \equiv j(x)$. In addition, the field ϕ satisfies the canonical commutation relations

---

[1] For the interested reader, please see: H. Lehmann, K. Symanzik and W. Zimmermann, Zur Formulierung quantisierter Feldtheorien, *Nuovo Cimento* **1**, 205–255 (1955).

$$[\phi(x), \pi(y)] = i\delta(\mathbf{x} - \mathbf{y}), \tag{10.2}$$

$$[\phi(x), \phi(y)] = 0, \tag{10.3}$$

$$[\pi(x), \pi(y)] = 0, \tag{10.4}$$

where again $\pi \equiv \partial \mathcal{L}/\partial \dot{\phi}$ is the conjugate momentum of the field ϕ. Note that Eqs. (10.2)–(10.4) are valid only for equal times $t = x^0 = y^0$.

In general, the Lagrangian

$$\mathcal{L} = \frac{1}{2}\left(\partial_\mu \phi \partial^\mu \phi - m_0^2 \phi^2\right) + \mathcal{L}_{\text{int.}}, \tag{10.5}$$

where $\mathcal{L}_{\text{int.}}$ is the interaction Lagrangian, leads to Eq. (10.1) using Euler–Lagrange field equations. In fact, the actual form of the interaction Lagrangian is not of importance to us. However, in the case that there is no derivative coupling in the interaction Lagrangian, it follows that $\pi = \dot{\phi}$ and $j = \partial \mathcal{L}_{\text{int.}}/\partial \phi$. In particular, the Lagrangian

$$\mathcal{L} = \frac{1}{2}\left(\partial_\mu \phi \partial^\mu \phi - m_0^2 \phi^2 + \frac{\lambda}{2}\phi^4\right) \tag{10.6}$$

is such a case, since j does not depend on $\partial_\mu \phi$, where λ is a coupling constant. In this case, the functional is given by $j(x) = \lambda \phi(x)^3$. The interaction Lagrangian, $\lambda \phi^4/4$, describes self-interactions with the field ϕ itself. Other possible terms in the interaction Lagrangian could be interactions with other fields, such as the Yukawa interaction, $g \bar{\psi} \psi \phi$, where g is a coupling constant and ψ denotes a Dirac field.

Now, we want to find an asymptotic field operator $\phi_{\text{in}}(x)$ formed by the field operator $\phi(x)$ and having the same properties, i.e.

(1) the fields $\phi_{\text{in}}(x)$ and $\phi(x)$ transform in the same way under transformations of the Poincaré group, which means that

$$[P^\mu, \phi_{\text{in}}] = i\frac{\partial \phi_{\text{in}}}{\partial x_\mu} = i\partial^\mu \phi_{\text{in}}; \tag{10.7}$$

(2) the field $\phi_{\text{in}}(x)$ is a **free** Klein–Gordon field with the *physical mass* m such that

$$\left(\Box + m^2\right)\phi_{\text{in}}(x) = 0. \tag{10.8}$$

Note that the mass m_0 is called the *bare mass*. In general, it holds that $m \neq m_0$.

If the two conditions are fulfilled, then the asymptotic field $\phi_{\text{in}}(x)$ creates physical one-particle states from the *physical vacuum*, i.e. the ground state of the interacting theory, which we will denote by $|\Omega\rangle$. Note that the ground state of the interacting theory $|\Omega\rangle$ is, in general, different from the ground state of the free theory $|0\rangle$.

First, considering a state $|n\rangle$ such that $P^\mu |n\rangle = p_n^\mu |n\rangle$, implies that

$$i\partial^\mu \langle n|\phi_{\text{in}}|\Omega\rangle = \langle n|i\partial^\mu \phi_{\text{in}}|\Omega\rangle = \langle n|[P^\mu, \phi_{\text{in}}]|\Omega\rangle = p_n^\mu \langle n|\phi_{\text{in}}|\Omega\rangle. \tag{10.9}$$

In addition, we have

$$(\Box + m^2) \langle n|\phi_{in}(x)|\Omega\rangle = (-p_n^2 + m^2) \langle n|\phi_{in}(x)|\Omega\rangle = 0. \tag{10.10}$$

Thus, as we will see, it will be possible to write the asymptotic field $\phi_{in}(x)$ as 'usual' in terms of plane waves.

Second, considering the equation

$$(\Box + m_0^2) \phi(x) = j(x) \tag{10.11}$$

and adding the mass term $\delta m^2 = m^2 - m_0^2$ to both sides of the equation yields

$$(\Box + m^2) \phi(x) = j(x) + \delta m^2 \phi(x) \equiv \tilde{j}(x). \tag{10.12}$$

Thus, we can express the asymptotic field $\phi_{in}(x)$ in terms of the field $\phi(x)$, or vice versa.

In this case, we can write the integral equation [cf. Eq. (6.82)]

$$\sqrt{Z}\phi_{in}(x) = \phi(x) + \int \Delta_R(x - y; m)\tilde{j}(y)\,d^4y, \tag{10.13}$$

where $\sqrt{Z}$ normalizes the single-particle amplitude and $\Delta_R(x; m)$ is the retarded propagator for the Klein–Gordon field (cf. the definition of Δ_R in Section 6.3). This normalization procedure is normally called the *wave function renormalization*, since it normalizes the field operator. Note that Eq. (10.13) is the formal solution to Eq. (10.12), which can be checked by inserting Eq. (10.13) into Eq. (10.12). Thus, we want $\tilde{j}$ to be such that

$$\lim_{x^0 \to -\infty} \phi(x) = \sqrt{Z}\phi_{in}(x), \tag{10.14}$$

which is a *strong asymptotic condition* that means that the interacting field should smoothly approach the asymptotic free field up to a normalization factor when time goes to minus infinity. In other words, it means that the behaviour of the interacting field resembles the asymptotic field and that all interactions become negligible, i.e. the current j is negligible, when particles are far away from each other, which is called the *adiabatic hypothesis*. However, the current j contains self-interactions that, in principle, induce the mass term δm^2, and therefore, such interactions are, of course, never negligible. Thus, this means that the strong asymptotic condition is incorrect and that we have to be careful when establishing relations between interacting fields and asymptotic fields. In fact, if the strong asymptotic condition is naively assumed, then it can be shown that the S matrix is trivial and that no scattering takes place. Therefore, according to Lehmann, Symanzik, and Zimmermann, we consider two normalizable states $|\alpha\rangle$ and $|\beta\rangle$ and introduce the functional

$$\phi(t, f) = i \int f^*(t, \mathbf{x}) \overleftrightarrow{\partial_0} \phi(t, \mathbf{x})\,d^3x, \tag{10.15}$$

where f is a normalizable solution to the Klein–Gordon equation $(\Box + m^2)$
$f(x) = 0$. Thus, we have

$$\lim_{t \to -\infty} \langle \alpha | \phi(t, f) | \beta \rangle = \sqrt{Z} \lim_{t \to -\infty} \langle \alpha | \phi_{\text{in}}(t, f) | \beta \rangle, \tag{10.16}$$

which is well-defined and is called a *weak asymptotic condition*. In addition, we of
course have

$$\sqrt{Z} \phi_{\text{out}}(x) = \phi(x) + \int \Delta_A(x - y; m) \tilde{j}(y) \, d^4 y, \tag{10.17}$$

$$\lim_{x^0 \to \infty} \phi(x) = \sqrt{Z} \phi_{\text{out}}(x), \tag{10.18}$$

$$\lim_{t \to \infty} \langle \alpha | \phi(t, f) | \beta \rangle = \sqrt{Z} \lim_{t \to \infty} \langle \alpha | \phi_{\text{out}}(t, f) | \beta \rangle, \tag{10.19}$$

where ϕ_{out} is another asymptotic free field. Since the asymptotic fields $\phi_{\text{in}}(x)$ and
$\phi_{\text{out}}(x)$ are free scalar fields, they can both be expanded in plane waves as

$$\phi_{\text{in}}(x) = \int_{p_0 > 0} \frac{d^3 p}{p_0} \left[f_p(x) a_{\text{in}}(\mathbf{p}) + f_p^*(x) a_{\text{in}}^\dagger(\mathbf{p}) \right], \tag{10.20}$$

$$\phi_{\text{out}}(x) = \int_{p_0 > 0} \frac{d^3 p}{p_0} \left[f_p(x) a_{\text{out}}(\mathbf{p}) + f_p^*(x) a_{\text{out}}^\dagger(\mathbf{p}) \right], \tag{10.21}$$

where the normalized plane-wave functions are given by

$$f_p(x) = \frac{1}{\sqrt{2(2\pi)^3}} e^{-i p \cdot x}. \tag{10.22}$$

Now, using Eq. (10.15) as well as Eq. (6.38) and a similar expansion for $\pi(x)$, we
can write the operators a_{in} and a_{out} as inverse functions in terms of the asymptotic
fields ϕ_{in} and ϕ_{out}

$$a_{\text{in}}(\mathbf{p}) = i \int f_p^*(x) \overleftrightarrow{\partial_0} \phi_{\text{in}}(x) \, d^3 x, \tag{10.23}$$

$$a_{\text{out}}(\mathbf{p}) = i \int f_p^*(x) \overleftrightarrow{\partial_0} \phi_{\text{out}}(x) \, d^3 x. \tag{10.24}$$

Note that it is not immediately obvious that $a_{\text{in}}(\mathbf{p})$ and $a_{\text{out}}(\mathbf{p})$ are time-independent,
but it can be shown using the Klein–Gordon equation.

Finally, the S matrix is the matrix of the S operator that is formed between 'in'
and 'out' states, i.e.

$$S_{\alpha\beta} = \langle \beta | S | \alpha \rangle = \langle \beta; \text{out} | \alpha; \text{in} \rangle, \tag{10.25}$$

where the S matrix elements $S_{\alpha\beta}$ are transition amplitudes between 'in' and 'out' states. Here the 'in' state $|\alpha; \text{in}\rangle$ represents a state of a set of particles with, for example, definite 3-momenta $\mathbf{p}_1, \mathbf{p}_2, \ldots, \mathbf{p}_n$, moving freely in the far past $(t \to -\infty)$ before interacting, whereas the 'out' state $|\beta; \text{out}\rangle$ represents a state of a set of particles with, for example, definite 3-momenta $\mathbf{q}_1, \mathbf{q}_2, \ldots, \mathbf{q}_m$, moving freely in the far future $(t \to \infty)$ after interacting. Note that the numbers of particles in the 'in' and 'out' states, respectively, do not necessarily need to be the same. Thus, the S operator transforms 'in' states into 'out' states, and vice versa, i.e. from one asymptotic set of states to another. This implies that

$$|\text{in}\rangle = S|\text{out}\rangle \quad \text{and} \quad |\text{out}\rangle = S^\dagger|\text{in}\rangle, \tag{10.26}$$

where the S operator contains the entire information (including information about interactions) of the evolution from the far past to the far future. In addition, for field operators, we find that

$$\phi_{\text{out}} = S^\dagger \phi_{\text{in}} S. \tag{10.27}$$

Hence, the 'in' and 'out' fields are constructed by the appropriate field operators that provide the correct creation and annihilation operators to give 'in' and 'out' states. Note that the 'in' and 'out' states are states in the Heisenberg picture (see Section 11.1), which mean that the states are time-independent and describe the system of particles in its entire evolution.

10.2 The LSZ formalism for real scalar fields

In general, the motivation for the LSZ formalism can be stated as: to write the S matrix elements as vacuum expectation values (VEVs) for T-ordered products of field operators. In particular, the LSZ formalism can be used as a simple derivation of Feynman rules (see Section 11.7). However, this derivation is less intuitive than the original approach based on propagator theory.

Let us continue to consider a real scalar field ϕ. In addition, consider a general state $|\alpha, \mathbf{p}_n; \text{in}\rangle$, where $\alpha = \{\mathbf{p}_1, \mathbf{p}_2, \ldots, \mathbf{p}_{n-1}\}$, and another state $|\beta; \text{out}\rangle$, where $\beta = \{\mathbf{q}_1, \mathbf{q}_2, \ldots, \mathbf{q}_m\}$, i.e. the 'in' state contains at least a particle with 3-momentum $\mathbf{p}_n$ and possibly other particles with 3-momenta summarized by the symbol α, and the 'out' state can contain anything from the vacuum to a set of particles with 3-momenta summarized by the symbol β. Now, we want to use the LSZ formalism for the amplitude $\langle\beta; \text{out}|\alpha, \mathbf{p}_n; \text{in}\rangle$. Extracting the particle with 3-momentum $\mathbf{p}_n$ from the 'in' state using a creation operator, we obtain by using Eqs. (10.23) and (10.24)

$$\langle \beta; \text{out}|\alpha, \mathbf{p}_n; \text{in}\rangle = \left\langle \beta; \text{out} \left| a_{\text{in}}^\dagger(\mathbf{p}_n) \right| \alpha; \text{in} \right\rangle$$

$$= \left\langle \beta; \text{out} \left| a_{\text{out}}^\dagger(\mathbf{p}_n) \right| \alpha; \text{in} \right\rangle + \left\langle \beta; \text{out} \left| a_{\text{in}}^\dagger(\mathbf{p}_n) - a_{\text{out}}^\dagger(\mathbf{p}_n) \right| \alpha; \text{in} \right\rangle$$

$$= \langle \beta, \hat{\mathbf{p}}_n; \text{out}|\alpha; \text{in}\rangle$$

$$- \mathrm{i} \left\langle \beta; \text{out} \left| \int f_{\mathbf{p}_n}(x) \overleftrightarrow{\partial}_0 [\phi_{\text{in}}(x) - \phi_{\text{out}}(x)] \, \mathrm{d}^3 x \right| \alpha; \text{in} \right\rangle,$$

$$(10.28)$$

where the elastic term $\langle \beta, \hat{\mathbf{p}}_n; \text{out}|\alpha; \text{in}\rangle$ gives forward scattering. Note that the equality sign between the second and third lines in Eq. (10.28) holds due to the conjugate versions of Eqs. (10.23) and (10.24) as well as the fact that the field ϕ is real. In addition, note that the operator $a_{\text{out}}^\dagger(\mathbf{p}_n)$ acting on the 'out' state, i.e. to the left, behaves as an annihilation operator, and not as a creation operator. Now, what is the meaning of the symbol $\beta, \hat{\mathbf{p}}_n$? Actually, there are two possible cases.

- If $\mathbf{p}_n \in \beta$, then it means remove the particle with 3-momentum $\mathbf{p}_n$ from the set of particles described by the symbol β.
- If $\mathbf{p}_n \notin \beta$, then it means that the state described by the symbol $\beta, \hat{\mathbf{p}}_n$ is annihilated, i.e. there is no forward scattering. Thus, we have $\langle \beta, \hat{\mathbf{p}}_n; \text{out}|\alpha; \text{in}\rangle = 0$.

Then, replace the asymptotic fields ϕ_{in} and ϕ_{out} by limits [cf. Eqs. (10.16) and (10.19)], i.e.

$$\langle \beta; \text{out}|\alpha, \mathbf{p}_n; \text{in}\rangle = \langle \beta, \hat{\mathbf{p}}_n; \text{out}|\alpha; \text{in}\rangle + \frac{\mathrm{i}}{\sqrt{Z}}$$

$$\times \left(\lim_{x^0 \to \infty} - \lim_{x^0 \to -\infty} \right) \int f_{\mathbf{p}_n}(x) \overleftrightarrow{\partial}_0 \langle \beta; \text{out}|\phi(x)|\alpha; \text{in}\rangle \, \mathrm{d}^3 x.$$

$$(10.29)$$

Next, we investigate the integral, which can, in general, be written as

$$\left(\lim_{x^0 \to \infty} - \lim_{x^0 \to -\infty} \right) \int f(x) \overleftrightarrow{\partial}_0 g(x) \, \mathrm{d}^3 x = \int_{-\infty}^{\infty} \frac{\partial}{\partial x^0} \left[f(x) \overleftrightarrow{\partial}_0 g(x) \right] \, \mathrm{d}^4 x$$

$$= \int_{-\infty}^{\infty} \left\{ f(x) \partial_0^2 g(x) - [\partial_0^2 f(x)] g(x) \right\} \, \mathrm{d}^4 x,$$

$$(10.30)$$

where the function g is, for example, given by $g(x) = \langle \beta; \text{out}|\phi(x)|\alpha; \text{in}\rangle$. Since the function f is a normalizable solution to the Klein–Gordon equation $(\Box + m^2) f(x) = 0$, we can write $\partial_0^2 f(x) = (\nabla^2 - m^2) f(x)$. In addition, suppose that $\mathbf{p}_n \notin \beta$, i.e. there is no elastic term, then we have

$$\langle \beta; \text{out}|\alpha, \mathbf{p}_n; \text{in}\rangle = \frac{\mathrm{i}}{\sqrt{Z}} \int_{-\infty}^{\infty} \left[f_{\mathbf{p}_n}(x) \left(\partial_0^2 + m^2 \right) - \nabla^2 f_{\mathbf{p}_n}(x) \right]$$

$$\times \langle \beta; \text{out}|\phi(x)|\alpha; \text{in}\rangle \, \mathrm{d}^4 x.$$

$$(10.31)$$

Finally, integrating two times by parts, we obtain

$$\langle \beta; \text{out}|\alpha, \mathbf{p}_n; \text{in}\rangle = \frac{i}{\sqrt{Z}} \int_{-\infty}^{\infty} f_{p_n}(x) \left(\Box_x + m^2\right) \langle \beta; \text{out}|\phi(x)|\alpha; \text{in}\rangle \, d^4x.$$

$$(10.32)$$

Next, rewriting the old symbol β as the new symbol γ, $\mathbf{q}_m$, we study the matrix element in the integrand of Eq. (10.32) and call this matrix element M

$$M = \langle \gamma, \mathbf{q}_m; \text{out}|\phi(x)|\alpha; \text{in}\rangle$$

$$= \langle \gamma; \text{out}|\phi(x)|\alpha, \hat{\mathbf{q}}_m; \text{in}\rangle + \langle \gamma; \text{out}|a_{\text{out}}(\mathbf{q}_m)\phi(x) - \phi(x)a_{\text{in}}(\mathbf{q}_m)|\alpha; \text{in}\rangle$$

$$= \langle \gamma; \text{out}|\phi(x)|\alpha, \hat{\mathbf{q}}_m; \text{in}\rangle - i \int \langle \gamma; \text{out}|\phi_{\text{out}}(y)\phi(x) - \phi(x)\phi_{\text{in}}(y)|\alpha; \text{in}\rangle$$

$$\times \overset{\leftrightarrow}{\partial_0} f_{q_m}^*(y) \, d^3y. \quad (10.33)$$

Now, assuming no forward scattering, i.e. $\mathbf{q}_m \notin \alpha$, the matrix element M can be written as

$$M = \frac{-i}{\sqrt{Z}} \left(\lim_{y^0 \to \infty} - \lim_{y^0 \to -\infty} \right) \int \langle \gamma; \text{out}|T\phi(y)\phi(x)|\alpha; \text{in}\rangle \overset{\leftrightarrow}{\partial_0} f_{q_m}^*(y) \, d^3y, \quad (10.34)$$

where again the T-ordered product is defined as [cf. Eq. (5.4)]

$$T\phi(y)\phi(x) = \phi(y)\phi(x)\theta(y^0 - x^0) + \phi(x)\phi(y)\theta(x^0 - y^0). \quad (10.35)$$

Thus, performing similar steps as in Eqs. (10.29)–(10.32), we have

$$M = \frac{i}{\sqrt{Z}} \int_{-\infty}^{\infty} \left[\left(\Box_y + m^2\right) \langle \gamma; \text{out}|T\phi(y)\phi(x)|\alpha; \text{in}\rangle \right] f_{q_m}^*(y) \, d^4y. \quad (10.36)$$

Now, we repeat the procedure $n - 1$ and $m - 1$ times, and thus, we finally obtain the famous *LSZ reduction formula* for (real) Klein–Gordon fields

$$\langle \mathbf{q}_1, \mathbf{q}_2, \ldots, \mathbf{q}_m; \text{out}|\mathbf{p}_1, \mathbf{p}_2, \ldots, \mathbf{p}_n; \text{in}\rangle = \left(\frac{i}{\sqrt{Z}} \right)^{n+m} \prod_{i=1}^{n} \prod_{j=1}^{m} \int f_{p_i}(x_i) f_{q_j}^*(y_j)$$

$$\times \left(\Box_{x_i} + m^2\right) \left(\Box_{y_j} + m^2\right) \langle \Omega|T\phi(y_1)\phi(y_2)\ldots\phi(y_m)\phi(x_1)\phi(x_2)\ldots\phi(x_n)|\Omega\rangle$$

$$\times d^4x_i d^4y_j. \quad (10.37)$$

Similarly, the LSZ reduction formula can be derived for Dirac fields. Equation (10.37) is the main result of this chapter, which immediately relates S matrix elements with the corresponding VEVs for T-ordered products of field operators, and in the next chapter, we are going to derive and discuss the VEVs for T-ordered products of field operators.

Exercise 10.1 Derive the LSZ reduction formula for Dirac fields.

10.3 Proton–meson scattering

As an example of how to use the LSZ reduction formula, let us consider proton–meson scattering. Assume that the incoming proton with mass m (which is described by a Dirac field) and meson with mass μ (which is described by a real Klein–Gordon field) have 3-momenta $\mathbf{p}$ and $\mathbf{q}$, respectively, whereas the outgoing proton and meson have 3-momenta $\mathbf{p}'$ and $\mathbf{q}'$, respectively. See Fig. 10.1. In addition, the incoming proton has spin s, while the outgoing proton has spin s'. Thus, we want to compute the S matrix element

$$S_{fi} = \langle \mathbf{q}', (\mathbf{p}', s'); \text{out} | \mathbf{q}, (\mathbf{p}, s); \text{in} \rangle, \tag{10.38}$$

where i and f stand for initial and final, respectively. Using the LSZ reduction formula, this S matrix element can be written as

$$S_{fi} = \delta_{fi} + \frac{1}{ZZ_2} \int \left(\Box_{x'} + \mu^2 \right) \left(\Box_x + \mu^2 \right) f_{q'}^*(x') f_q(x) \bar{U}_{p',s'}(y')$$

$$\times \left[\left(i \overset{\rightarrow}{\partial}_{y'} - m \mathbb{1}_4 \right) \langle \Omega | T \psi(y') \bar{\psi}(y) \phi(x) \phi(x') | \Omega \rangle \right] \left(-i \overset{\leftarrow}{\partial}_y - m \mathbb{1}_4 \right)$$

$$\times U_{p,s}(y) \, \mathrm{d}^4 x \, \mathrm{d}^4 x' \, \mathrm{d}^4 y \, \mathrm{d}^4 y', \tag{10.39}$$

where Z is the wave function renormalization constant for the meson (Klein–Gordon) field and Z_2 is the wave function renormalization constant for the proton (Dirac) field. Note that $\delta_{fi} \neq 0$ only for forward scattering. In this case, we can choose

$$f_q(x) = \frac{1}{\sqrt{2(2\pi)^3}} e^{-iq \cdot x}, \tag{10.40}$$

$$U_{p,s}(x) = \frac{1}{\sqrt{(2\pi)^3}} \sqrt{\frac{m}{E(\mathbf{p})}} e^{-ip \cdot x} \tag{10.41}$$

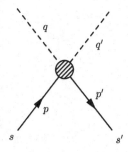

Figure 10.1 Proton–meson scattering

for Klein–Gordon and Dirac fields, respectively. Thus, we obtain

$$\delta_{fi} = q_0\delta(\mathbf{q} - \mathbf{q}')\delta(\mathbf{p} - \mathbf{p}')\delta_{ss'}. \tag{10.42}$$

Guide to additional recommended reading

The following books (see the indicated pages) and their authors have similar treatments of the content in the present chapter.

- A. Z. Capri, *Relativistic Quantum Mechanics and Introduction to Quantum Field Theory*, World Scientific (2002), pp. 143–151.
- M. Kaku, *Quantum Field Theory – A Modern Introduction*, Oxford (1993), pp. 141–147.
- M. E. Peskin and D. V. Schroeder, *An Introduction to Quantum Field Theory*, Addison-Wesley (1995), pp. 222–230.
- S. S. Schweber, *An Introduction to Relativistic Quantum Field Theory*, Dover (2005), pp. 742–764.

11

Perturbation theory

In general, perturbation theory is a mathematical method that is used to find approximative solutions to problems that cannot be solved exactly. Therefore, the starting point of perturbation theory is the exact solution to a related problem. Perturbation theory can be applied if the original problem can be reformulated by adding a 'small' term to the exactly solvable problem. Thus, perturbation theory gives rise to an expression for the original solution in terms of a power series in some 'small' parameter, which measures the deviation from the exactly solvable problem. The leading term in this power series is the solution to the exactly solvable problem, whereas the higher-order terms may be found by some iterative procedure. In order for the perturbation theory to work properly, the higher-order terms in the 'small' parameter need to become successively smaller, i.e. the power series should converge. However, normally in quantum field theory, the terms will not anymore become successively smaller at some specific order, since the number of possible Feynman diagrams will grow so fast that the terms will instead become successively larger. Thus, perturbation theory will only be successful up to this order, and then, it will break down.

This chapter is devoted to the study of perturbation theory, which is, nevertheless, a very powerful tool in quantum field theory. Note that the discussion in Sections 11.1–11.6 will mainly be performed for a real scalar field, i.e. a neutral Klein–Gordon field, but it can, of course, be naturally extended to other types of fields, such as a Dirac field, if the appropriate changes are made. For example, in Section 11.7, we will present a discussion on Feynman rules and diagrams for four different models, which contain other fields than neutral Klein–Gordon fields. The last two sections, i.e. Sections 11.8 and 11.9, are rather independent of the type of fields that we are dealing with.

11.1 Three different pictures

Normally, in quantum field theory, one considers three different pictures for the time evolution of the states, the observables (i.e. the operators or the dynamical variables of the system), and the Hamiltonian of the system. These pictures are the *Heisenberg picture*, the *Schrödinger picture*, and the *interaction picture*.

In the Heisenberg picture, the operators are time-dependent and the time evolution of the dynamical variables of the system are governed by the Hamiltonian H according to

$$\phi(x) = e^{iHt}\phi(0, \mathbf{x})e^{-iHt}, \tag{11.1}$$

$$\pi(x) = e^{iHt}\pi(0, \mathbf{x})e^{-iHt}, \tag{11.2}$$

where $\phi(x)$ and $\pi(x)$ are the field operators for a neutral Klein–Gordon field. In this case, the states $|\psi\rangle_H$ and the Hamiltonian H are time-independent. In addition, the observables evolve in time according to the *Heisenberg equations of motion*, i.e.

$$\frac{d}{dt}A(t) = i[H, A(t)], \tag{11.3}$$

where the observable $A(t)$ can, for example, be the field $\phi(x)$ or $\pi(x)$. Note that Eq. (11.3) follows from either Eq. (11.1) or Eq. (11.2).

In the Schrödinger picture, the operators are time-independent, whereas the states evolve in time according to the Schrödinger equation

$$i\frac{\partial}{\partial t}|\psi(t)\rangle_S = H|\psi(t)\rangle_S, \tag{11.4}$$

where again the Hamiltonian H is time-independent. The different quantities in the Schrödinger picture are related to those in the Heisenberg picture by the following relations

$$\phi_S(\mathbf{x}) \equiv e^{-iHt}\phi(x)e^{iHt} = \phi(0, \mathbf{x}), \tag{11.5}$$

$$\pi_S(\mathbf{x}) \equiv e^{-iHt}\pi(x)e^{iHt} = \pi(0, \mathbf{x}), \tag{11.6}$$

$$|\psi(t)\rangle_S = e^{-iHt}|\psi(0)\rangle_S = e^{-iHt}|\psi\rangle_H. \tag{11.7}$$

In general, we identify the Hamiltonian H with the generator P^0 of the Poincaré group $\mathcal{P}_+^\uparrow$. Thus, if we consider the one-parameter subgroup $U(t) = e^{-iHt}$ of $U(\mathcal{P}_+^\uparrow)$, then we can write Eq. (11.3) in integrated form as

$$A(t) = U^{-1}(t)AU(t) = e^{iHt}Ae^{-iHt}, \tag{11.8}$$

cf. Eqs. (11.1) and (11.2). Note that $A \equiv A(0)$.

However, in perturbation theory, we introduce yet another picture, which will be our final picture – the *interaction picture*[1] – with operators and states defined as

---

[1] The interaction picture is sometimes called the *Dirac picture* or, in quantum field theory, the *Dyson picture*.

$$\phi_{in}(x) \equiv e^{iH_{in}t} \phi_S(\mathbf{x}) e^{-iH_{in}t} = e^{iH_{in}t} e^{-iHt} \phi(x) e^{iHt} e^{-iH_{in}t} = U(t)\phi(x)U^{-1}(t),$$
(11.9)

$$\pi_{in}(x) \equiv e^{iH_{in}t} \pi_S(\mathbf{x}) e^{-iH_{in}t} = e^{iH_{in}t} e^{-iHt} \pi(x) e^{iHt} e^{-iH_{in}t} = U(t)\pi(x)U^{-1}(t),$$
(11.10)

$$|\psi\rangle_{in} \equiv e^{iH_{in}t} |\psi(t)\rangle_S - U(t)|\psi\rangle_H,$$
(11.11)

where $U(t) \equiv e^{iH_{in}t} e^{-iHt}$ is the *unitary time-evolution operator*. Here, H is total Hamiltonian and H_{in} is the time-independent free Hamiltonian.

11.2 The unitary time-evolution operator

The field ϕ and the asymptotic free field ϕ_{in} satisfy the same commutation relations and both ϕ and π as well as ϕ_{in} and π_{in} are complete sets of fields. Therefore, assuming Eqs. (11.9) and (11.10), there exists an operator U, which is different from the S operator, such that

$$\phi(x) = U^{-1}(t)\phi_{in}(x)U(t),$$
(11.12)

$$\pi(x) = U^{-1}(t)\pi_{in}(x)U(t),$$
(11.13)

i.e. the operator U relates the ordinary fields (in the Heisenberg picture) and the asymptotic free fields (in the interaction picture), whereas the S operator relates the 'in' and 'out' fields ϕ_{in} and ϕ_{out}. Thus, we need to derive an equation for the operator U, which can be performed, since we have the Heisenberg equations of motion for the fields ϕ_{in} and π_{in} [cf. Eq. (11.3)]

$$\frac{\partial \phi_{in}}{\partial t} = i\left[H_{in}(\phi_{in}, \pi_{in}), \phi_{in}\right],$$
(11.14)

$$\frac{\partial \pi_{in}}{\partial t} = i\left[H_{in}(\phi_{in}, \pi_{in}), \pi_{in}\right],$$
(11.15)

where H_{in} is the Hamiltonian for the asymptotic free in field. In order to find the equation for the operator U, we express the time-derivative of the asymptotic field ϕ_{in} on the left-hand side of Eq. (11.14) using Eq. (11.12)

$$
\begin{aligned}
\dot{\phi}_{in} &= \frac{\partial \phi_{in}}{\partial t} = \frac{\partial}{\partial t}\left[U(t)\phi(x)U^{-1}(t)\right] = \dot{U}\phi U^{-1} + U\dot{\phi}U^{-1} + U\phi\dot{U}^{-1} \\
&= \{UU^{-1} = 1 \quad \Rightarrow \quad \dot{U}U^{-1} + U\dot{U}^{-1} = 0 \quad \Rightarrow \quad \dot{U}^{-1} = -U^{-1}\dot{U}U^{-1}\} \\
&= \dot{U}\phi U^{-1} + U\dot{\phi}U^{-1} - U\phi U^{-1}\dot{U}U^{-1} = \dot{U}U^{-1}\phi_{in} + iU[H, \phi]U^{-1} - \phi_{in}\dot{U}U^{-1} \\
&= \left[\dot{U}U^{-1}, \phi_{in}\right] + i\left[H(\phi_{in}, \pi_{in}), \phi_{in}\right],
\end{aligned}
$$
(11.16)

where H is the total Hamiltonian given by

$$H(\phi_{in}, \pi_{in}) = H_{in}(\phi_{in}, \pi_{in}) + H_I(\phi_{in}, \pi_{in}).$$
(11.17)

Then, the interaction Hamiltonian can be expressed as

$$H_I(t) \equiv H_I(\phi_{\text{in}}, \pi_{\text{in}}) = H(\phi_{\text{in}}, \pi_{\text{in}}) - H_{\text{in}}(\phi_{\text{in}}, \pi_{\text{in}}), \tag{11.18}$$

which means that we have

$$\dot{\phi}_{\text{in}} = \dot{\phi}_{\text{in}} + \left[\dot{U}U^{-1} + iH_I, \phi_{\text{in}}\right]. \tag{11.19}$$

Thus, we obtain the equation

$$\left[\dot{U}U^{-1} + iH_I, \phi_{\text{in}}\right] = 0. \tag{11.20}$$

Now, since the commutator is equal to zero, we can write

$$\dot{U}U^{-1} + iH_I = -iE_0(t)\mathbb{1}, \tag{11.21}$$

where $-iE_0(t)$ is a complex number. Thus, we obtain the equation for the operator U as

$$i\dot{U} = (H_I + E_0\mathbb{1})\,U, \tag{11.22}$$

where the interaction Hamiltonian H_I contains **all** interactions.

Next, we introduce the shifted interaction Hamiltonian

$$H_I' = H_I + E_0\mathbb{1}, \tag{11.23}$$

which means that Eq. (11.22) simplifies to

$$i\dot{U} = H_I'U. \tag{11.24}$$

In order to solve for U in Eq. (11.24), we need initial conditions. Therefore, consider the operator

$$U(t, t') = U(t)U^{-1}(t'), \tag{11.25}$$

which fulfils the relations $U(t, t') = U(t, t'')U(t'', t')$ and $U^{-1}(t, t') = U(t', t)$ and is a solution to the Schrödinger equation for the time-evolution operator $U(t, t')$

$$i\frac{\partial}{\partial t}U(t, t') = H_I'(t)U(t, t'), \tag{11.26}$$

where the operator $U(t, t')$ satisfies the initial condition $U(t, t) = \mathbb{1}$. Note that one can show that the operator $U(t, t')$ can be written as

$$U(t, t') = e^{iH_{\text{in}}(t-t_0)}e^{-iH(t-t')}e^{-iH_{\text{in}}(t'-t_0)}, \tag{11.27}$$

which shows that this operator is unitary. Converting Eq. (11.26) to an integral equation, we have

$$U(t, t') = \mathbb{1} - i\int_{t'}^{t} H_I'(t_1)U(t_1, t')\,\mathrm{d}t_1. \tag{11.28}$$

This equation can be solved with the *Neumann–Liouville method of iterations.* Iterating this equation gives

$$U(t, t') = \mathbb{1} - i \int_{t'}^{t} H_I'(t_1)\, dt_1 + (-i)^2 \int_{t'}^{t} H_I'(t_1)\, dt_1 \int_{t'}^{t_1} H_I'(t_2)\, dt_2 + \cdots$$

$$+ (-i)^n \int_{t'}^{t} H_I'(t_1)\, dt_1 \int_{t'}^{t_1} H_I'(t_2)\, dt_2 \ldots \int_{t'}^{t_{n-1}} H_I'(t_n)\, dt_n + \cdots . \quad (11.29)$$

Note that the times are ordered as follows $t_1 \geq t_2 \geq \cdots \geq t_n \geq \cdots$. Thus, we can express the product of the shifted interaction Hamiltonian at different times as $T(H_I'(t_1) H_I'(t_2) \ldots H_I'(t_n))$ for both bosons and fermions. In the case of fermions, the shifted interaction Hamiltonian H_I' always includes a product of an even number of fermions. Thus, for the $H_I'^2$ term, we can symmetrize the integrals in the following way

$$\int_{t'}^{t} \int_{t'}^{t_1} T\left(H_I'(t_1) H_I'(t_2)\right) dt_1 dt_2 = \int_{t'}^{t} \int_{t'}^{t_2} T\left(H_I'(t_1) H_I'(t_2)\right) dt_2 dt_1$$

$$= \frac{1}{2} \int_{t'}^{t} \int_{t'}^{t} T\left(H_I'(t_1) H_I'(t_2)\right) dt_1 dt_2. \quad (11.30)$$

Similarly, for the $H_I'^n$ term, we have

$$\int_{t'}^{t} \int_{t'}^{t_1} \cdots \int_{t'}^{t_{n-1}} T\left(H_I'(t_1) H_I'(t_2) \ldots H_I'(t_n)\right) dt_1 dt_2 \ldots dt_n$$

$$= \frac{1}{n!} \int_{t'}^{t} \int_{t'}^{t} \cdots \int_{t'}^{t} T\left(H_I'(t_1) H_I'(t_2) \ldots H_I'(t_n)\right) dt_1 dt_2 \ldots dt_n. \quad (11.31)$$

Finally, we arrive at the time-evolution operator

$$U(t, t') = \mathbb{1} + \sum_{n=1}^{\infty} \frac{(-i)^n}{n!} \int_{t'}^{t} \int_{t'}^{t} \cdots \int_{t'}^{t} T\left(H_I'(t_1) H_I'(t_2) \ldots H_I'(t_n)\right) dt_1 dt_2 \ldots dt_n$$

$$= T \exp\left(-i \int_{t'}^{t} H_I'(\tau)\, d\tau\right) = T \exp\left(-i \int \mathcal{H}_I'(\phi_{\text{in}}(x), \pi_{\text{in}}(x))\, d^4 x\right),$$

$$(11.32)$$

where $\mathcal{H}_I'$ is the Hamiltonian density corresponding to the shifted interaction Hamiltonian H_I'. Note that one always makes computations with the first line in Eq. (11.32). The T-ordered exponential form is only a compact way of writing the expression for the time-evolution operator $U(t, t')$.

11.3 Perturbation of VEVs for *T*-ordered products

Consider the VEV for the *T*-ordered product of n fields (or the n-point correlation function or the n-point Green's function)

$$\tau(x_1, x_2, \ldots, x_n) = \langle \Omega | T \phi(x_1)\phi(x_2) \ldots \phi(x_n) | \Omega \rangle, \qquad (11.33)$$

where $|\Omega\rangle$ is again the physical vacuum state and the field $\phi(x_i)$ can be expressed in terms of the operator $U(t_i)$ and the asymptotic field $\phi_{in}(x_i)$ according to Eq. (11.12), i.e.

$$\phi(x_i) = U^{-1}(t_i)\phi_{in}(x_i)U(t_i), \quad i = 1, 2, \ldots, n. \qquad (11.34)$$

Note also that, for example, one can write $U(t_1)U^{-1}(t_2) = U(t_1, t_2)$. Thus, we have

$$\tau(x_1, x_2, \ldots, x_n) = \langle \Omega | T U^{-1}(t_1)\phi_{in}(x_1)U(t_1, t_2)\phi_{in}(x_2)U(t_2, t_3) \ldots$$
$$U(t_{n-1}, t_n)\phi_{in}(x_n)U(t_n)|\Omega\rangle. \qquad (11.35)$$

Now, introduce t such that $-t < t_i$ and $t > t_i$ for all i, then we find that

$$\tau(x_1, x_2, \ldots, x_n) = \langle \Omega | T U^{-1}(t)U(t, t_1)\phi_{in}(x_1)U(t_1, t_2) \ldots$$
$$U(t_{n-1}, t_n)\phi_{in}(x_n)U(t_n, -t)U(-t)|\Omega\rangle. \qquad (11.36)$$

Applying the limit $t \to \infty$, we obtain

$$\tau(x_1, x_2, \ldots, x_n)$$
$$= \lim_{t \to \infty} \langle \Omega | U^{-1}(t) T \left[\phi_{in}(x_1)\phi_{in}(x_2) \ldots \phi_{in}(x_n)U(t, -t) \right] U(-t)|\Omega\rangle$$
$$= \lim_{t \to \infty} \langle \Omega | U^{-1}(t) T \left[\phi_{in}(x_1)\phi_{in}(x_2) \ldots \phi_{in}(x_n)e^{-i \int_{-t}^{t} H_I'(t')\,dt'} \right] U(-t)|\Omega\rangle, \qquad (11.37)$$

where the operators $U^{-1}(t)$ and $U(-t)$ have been extracted from the *T*-ordered product.

Next, we want to show that the physical vacuum state $|\Omega\rangle$ is an eigenstate of the operators $U(-t) = U^{-1}(t)$ and $U(t)$ when $t \to \infty$. In addition, we have to assume that the shifted interaction Hamiltonian does not contain any derivative coupling, i.e. $H_I'(\phi_{in}, \pi_{in}) = H_I'(\phi_{in})$. Consider an arbitrary state $|\alpha, \mathbf{p}\rangle \neq |\Omega\rangle$. Then, using Eq. (10.23), we can expand the following matrix element as

$$\langle \alpha, \mathbf{p}; \text{in}|U(-t)|\Omega\rangle$$

$$= \langle \alpha; \text{in}|a_{\text{in}}(\mathbf{p})U(-t)|\Omega\rangle$$

$$= -i \int f_p^*(-t', \mathbf{x}) \frac{\overleftrightarrow{\partial}}{\partial t'} \langle \alpha; \text{in}|\phi_{\text{in}}(-t', \mathbf{x})U(-t)|\Omega\rangle \, d^3x$$

$$= -i \int f_p^*(-t', \mathbf{x}) \frac{\overleftrightarrow{\partial}}{\partial t'} \langle \alpha; \text{in}|U(-t')\phi(-t', \mathbf{x})U^{-1}(-t')U(-t)|\Omega\rangle \, d^3x$$

$$= -i \int f_p^*(-t', \mathbf{x}) \frac{\partial}{\partial t'} \langle \alpha; \text{in}|U(-t')\phi(-t', \mathbf{x})U^{-1}(-t')U(-t)|\Omega\rangle \, d^3x$$

$$+ i \int \left\{ \frac{\partial}{\partial t'} \left[f_p^*(-t', \mathbf{x}) \right] \right\} \langle \alpha; \text{in}|U(-t')\phi(-t', \mathbf{x})U^{-1}(-t')U(-t)|\Omega\rangle \, d^3x$$

$$= i \int f_p^*(-t', \mathbf{x}) \langle \alpha; \text{in}|\dot{U}(-t')\phi(-t', \mathbf{x})U^{-1}(-t')U(-t)|\Omega\rangle \, d^3x$$

$$+ i \int f_p^*(-t', \mathbf{x}) \langle \alpha; \text{in}|U(-t')\dot{\phi}(-t', \mathbf{x})U^{-1}(-t')U(-t)|\Omega\rangle \, d^3x$$

$$+ i \int f_p^*(-t', \mathbf{x}) \langle \alpha; \text{in}|U(-t')\phi(-t', \mathbf{x})\dot{U}^{-1}(-t')U(-t)|\Omega\rangle \, d^3x$$

$$- i \int \frac{\partial f_p^*}{\partial t'}(-t', \mathbf{x}) \langle \alpha; \text{in}|U(-t')\phi(-t', \mathbf{x})U^{-1}(-t')U(-t)|\Omega\rangle \, d^3x. \quad (11.38)$$

Now, it holds that $U^{-1}(-t')U(-t) = \mathbb{1}$ when $t = t' \to \infty$ and we also have

$$\lim_{t \to \infty} \phi(-t, \mathbf{x}) = \lim_{t \to \infty} \phi_{\text{in}}(-t, \mathbf{x}), \quad (11.39)$$

i.e. the field ϕ and the asymptotic field ϕ_{in} are equal to each other when $t \to \infty$. Thus, for $t = t' \to \infty$, Eq. (11.38) simplifies drastically and we obtain

$$\lim_{t \to \infty} \langle \alpha, \mathbf{p}; \text{in}|U(-t)|\Omega\rangle = \lim_{t \to \infty} \langle \alpha; \text{in}|U(-t)a(\mathbf{p})|\Omega\rangle$$

$$+ i \lim_{t \to \infty} \int f_p^*(-t, \mathbf{x}) \langle \alpha; \text{in}|\dot{U}(-t)\phi(-t, \mathbf{x})$$

$$+ U(-t)\phi(-t, \mathbf{x})\dot{U}^{-1}(-t)U(-t)|\Omega\rangle \, d^3x, \quad (11.40)$$

where we have again used Eq. (10.23). Furthermore, we have the following relations

$$a(\mathbf{p})|\Omega\rangle = 0, \quad (11.41)$$

$$\dot{U}\phi + U\phi\dot{U}^{-1}U = \dot{U}U^{-1}\phi_{\text{in}}U + \phi_{\text{in}}U\dot{U}^{-1}U = \left(\dot{U}U^{-1}\phi_{\text{in}} + \phi_{\text{in}}U\dot{U}^{-1} \right) U$$

$$= \left(\dot{U}U^{-1}\phi_{\text{in}} - \phi_{\text{in}}\dot{U}U^{-1} \right) U = \left[\dot{U}U^{-1}, \phi_{\text{in}} \right] U. \quad (11.42)$$

However, inserting Eq. (11.23) into Eq. (11.21), it holds that $\dot{U}U^{-1} = -iH_I'(\phi_{\text{in}})$, which means that Eq. (11.42) can be rewritten as

$$\dot{U}\phi + U\phi\dot{U}^{-1}U = -i\left[H_I'(\phi_{\text{in}}), \phi_{\text{in}}\right]U = 0, \tag{11.43}$$

since $H_I'(\phi_{\text{in}})$ does not contain π_{in}. Thus, inserting Eqs. (11.41) and (11.43) into Eq. (11.40), we finally find that

$$\lim_{t\to\infty} \langle \alpha, \mathbf{p}; \text{in}|U(-t)|\Omega\rangle = 0, \tag{11.44}$$

which implies that $\lim_{t\to\infty} U(-t)|\Omega\rangle$ is orthogonal to any state that contains one or more particles.

Now, we can write

$$\lim_{t\to\infty} U(-t)|\Omega\rangle = \lambda_-|\Omega\rangle, \quad |\lambda_-| = 1, \tag{11.45}$$

and similarly, we can also write

$$\lim_{t\to\infty} U(t)|\Omega\rangle = \lambda_+|\Omega\rangle, \quad |\lambda_+| = 1, \tag{11.46}$$

which implies that

$$\lim_{t\to\infty} \langle\Omega|U^{-1}(t) = \lambda_+^*\langle\Omega|. \tag{11.47}$$

Finally, using Eqs. (11.45) and (11.47), Eq. (11.37) can be rewritten as

$$\tau(x_1, x_2, \ldots, x_n) = \lim_{t\to\infty} \langle\Omega|U^{-1}(t)|\Omega\rangle\langle\Omega|T[\ldots]|\Omega\rangle\langle\Omega|U(-t)|\Omega\rangle$$
$$= \lambda_+^*\lambda_- \lim_{t\to\infty} \langle\Omega|T[\ldots]|\Omega\rangle. \tag{11.48}$$

What is the quantity $\lambda_+^*\lambda_-$? In fact, it is given by

$$\lambda_+^*\lambda_- = \lim_{t\to\infty} \langle\Omega|U^{-1}(t)|\Omega\rangle\langle\Omega|U(-t)|\Omega\rangle = \lim_{t\to\infty} \langle\Omega|U^{-1}(t)U(-t)|\Omega\rangle$$

$$= \lim_{t\to\infty} \langle\Omega|U(-t, t)|\Omega\rangle = \lim_{t\to\infty} \langle\Omega|T\exp\left(i\int_{-t}^{t} H_I'(t')\,dt'\right)|\Omega\rangle$$

$$= \lim_{t\to\infty} \frac{1}{\langle\Omega|T\exp\left(-i\int_{-t}^{t} H_I'(t')\,dt'\right)|\Omega\rangle}, \tag{11.49}$$

since $\lambda_+^*\lambda_- = 1/(\lambda_+\lambda_-^*)$. Inserting the expression for $\lambda_+^*\lambda_-$ into Eq. (11.48), we obtain

$$\tau(x_1, x_2, \ldots, x_n) = \lim_{t\to\infty} \frac{\langle\Omega|T\phi_{\text{in}}(x_1)\phi_{\text{in}}(x_2)\ldots\phi_{\text{in}}(x_n)e^{-i\int_{-t}^{t} H_I'(t')\,dt'}|\Omega\rangle}{\langle\Omega|Te^{-i\int_{-t}^{t} H_I'(t')\,dt'}|\Omega\rangle}. \tag{11.50}$$

Thus, we have found the relation between VEVs for T-ordered products of operator fields and the corresponding VEVs for T-ordered products of asymptotic operator fields and exponentials of the interaction Hamiltonian. Finally, replacing the shifted interaction Hamiltonian H_I' with the original interaction Hamiltonian H_I

and cancelling the complex number $\exp\left(-i\int_{-t}^{t} E_0(t')\,dt'\right)$, we have the VEV for the T-ordered product of n fields as

$$
\tau(x_1, x_2, \ldots, x_n)
$$
$$
= \frac{\sum_{r=0}^{\infty} \frac{(-i)^r}{r!} \int \langle\Omega|T\phi_{\text{in}}(x_1)\phi_{\text{in}}(x_2)\ldots\phi_{\text{in}}(x_n)\mathcal{H}'_I(y_1)\mathcal{H}'_I(y_2)\ldots\mathcal{H}'_I(y_r)|\Omega\rangle d^4 y_1 d^4 y_2 \ldots d^4 y_r}{\sum_{r=0}^{\infty} \frac{(-i)^r}{r!} \int \langle\Omega|T\mathcal{H}'_I(y_1)\mathcal{H}'_I(y_2)\ldots\mathcal{H}'_I(y_r)|\Omega\rangle d^4 y_1 d^4 y_2 \ldots d^4 y_r}.
$$

$$(11.51)$$

In the next section, we will investigate the relation between the physical vacuum and the free theory ground state. Then, in the two following sections, we will study specific correlation functions and how to compute VEVs for T-ordered products of asymptotic operator fields in general.

11.4 The relation between the physical vacuum $|\Omega\rangle$ and the free theory ground state $|0\rangle$

The physical vacuum $|\Omega\rangle$ is the ground state of the total Hamiltonian H. Starting with the ground state of the free theory $|0\rangle$, which is the ground state of the free Hamiltonian H_{in}, and evolving it with H, we find that

$$
e^{-iHt}|0\rangle = \sum_n e^{-iE_n t}|n\rangle\langle n|0\rangle, \tag{11.52}
$$

where the $|n\rangle$s are the eigenstates of H and the E_ns are the corresponding eigenvalues. Now, $|\Omega\rangle$ must have some overlap with $|0\rangle$, i.e. $\langle\Omega|0\rangle \neq 0$, otherwise the interaction Hamiltonian $H_I = H - H_{\text{in}}$ would not be a small perturbation. Thus, the series in Eq. (11.52) contains $|\Omega\rangle$ and we have

$$
e^{-iHt}|0\rangle = e^{-iE_0 t}|\Omega\rangle\langle\Omega|0\rangle + \sum_{n\neq 0} e^{-iE_n t}|n\rangle\langle n|0\rangle, \tag{11.53}
$$

where the energy $E_0 = \langle\Omega|H|\Omega\rangle$ is the VEV (or ground state energy) of the total Hamiltonian and the zero of energy is defined as $H_{\text{in}}|0\rangle = 0$. Now, since $E_n > E_0$ for all $n \neq 0$, we can cancel the contributions from the terms with $n \neq 0$ by applying the limit $t \rightarrow (1 - i\epsilon)\infty$, where ϵ is positive and small. In this limit, the exponential factor $e^{-iE_n t}$ goes to zero slowest for $n = 0$. Thus, we obtain

$$
|\Omega\rangle = \lim_{t \rightarrow (1-i\epsilon)\infty} \left(e^{-iE_0 t}\langle\Omega|0\rangle\right)^{-1} e^{-iHt}|0\rangle. \tag{11.54}
$$

In the limit when t is very large, we can add a small arbitrary quantity t_0 to t without changing the final result, which means that Eq. (11.54) can be written as

$$|\Omega\rangle = \lim_{t \to (1-i\epsilon)\infty} \left(e^{-iE_0(t+t_0)}\langle\Omega|0\rangle\right)^{-1} e^{-iH(t+t_0)}|0\rangle = \{H_{in}|0\rangle = 0\}$$

$$= \lim_{t \to (1-i\epsilon)\infty} \left(e^{-iE_0[t_0-(-t)]}\langle\Omega|0\rangle\right)^{-1} e^{-iH[t_0-(-t)]}e^{-iH_{in}(-t-t_0)}|0\rangle$$

$$= \lim_{t \to (1-i\epsilon)\infty} \left(e^{-iE_0[t_0-(-t)]}\langle\Omega|0\rangle\right)^{-1} U(t_0, -t)|0\rangle. \tag{11.55}$$

Thus, this implies that one can simply obtain the physical vacuum $|\Omega\rangle$ by time-evolving the ground state of the free theory $|0\rangle$ from time $-\infty$ to time t_0 with the operator $U(t_0, -\infty)$, i.e. $|\Omega\rangle \sim U(t_0, -\infty)|0\rangle$. Similarly, the state $\langle\Omega|$ can be written as

$$\langle\Omega| = \lim_{t \to (1-i\epsilon)\infty} \langle 0|U(t, t_0) \left(e^{-iE_0(t-t_0)}\langle 0|\Omega\rangle\right)^{-1}. \tag{11.56}$$

Let us now again compute the n-point correlation function $\tau(x_1, x_2, \ldots, x_n)$. In order to do so, we assume that $t > x_1^0 > x_2^0 > \cdots > x_n^0 > t_0$. Then, inserting Eqs. (11.55) and (11.56) into the first line of Eq. (11.37) as well as neglecting the small ϵ, we obtain

$$\tau(x_1, x_2, \ldots, x_n) = \lim_{t \to \infty} \left(|\langle 0|\Omega\rangle|^2 e^{-2iE_0t}\right)^{-1}$$

$$\times \langle 0|U(t, t_0)U^{-1}(t)T \left[\phi_{in}(x_1)\phi_{in}(x_2)\ldots\phi_{in}(x_n)U(t, -t)\right]U(-t)U(t_0, -t)|0\rangle. \tag{11.57}$$

Next, choosing the arbitrary time $t_0 = 0$, we find that

$$\tau(x_1, x_2, \ldots, x_n)$$

$$= \lim_{t \to \infty} \left(|\langle 0|\Omega\rangle|^2 e^{-2iE_0t}\right)^{-1}$$

$$\times \langle 0|U(t)U^{-1}(t)T \left[\phi_{in}(x_1)\phi_{in}(x_2)\ldots\phi_{in}(x_n)U(t, -t)\right]U(-t)U^{-1}(-t)|0\rangle$$

$$= \lim_{t \to \infty} \left(|\langle 0|\Omega\rangle|^2 e^{-2iE_0t}\right)^{-1} \langle 0|T\phi_{in}(x_1)\phi_{in}(x_2)\ldots\phi_{in}(x_n)U(t, -t)|0\rangle. \tag{11.58}$$

Finally, in order to normalize the above equation, we divide it by 1 in the form

$$1 = \langle\Omega|\Omega\rangle = \lim_{t \to \infty} \left(|\langle 0|\Omega\rangle|^2 e^{-2iE_0t}\right)^{-1} \langle 0|U(t, -t)|0\rangle, \tag{11.59}$$

and thus, we obtain

$$\tau(x_1, x_2, \ldots, x_n) = \lim_{t \to \infty} \frac{\langle 0|T\phi_{in}(x_1)\phi_{in}(x_2)\ldots\phi_{in}(x_n)U(t, -t)|0\rangle}{\langle 0|U(t, -t)|0\rangle}$$

$$= \lim_{t \to \infty} \frac{\langle 0|T\phi_{in}(x_1)\phi_{in}(x_2)\ldots\phi_{in}(x_n)e^{-i\int_{-t}^{t} H_I'(t')\,dt'}|0\rangle}{\langle 0|Te^{-i\int_{-t}^{t} H_I'(t')\,dt'}|0\rangle}, \tag{11.60}$$

where the effect of the normalization constant $\langle 0|Te^{-i\int_{-t}^{t} H_I'(t')\,dt'}|0\rangle$ is to take out the so-called *disconnected part* of the VEV. Note that this formula for the n-point correlation function is exact. However, it is well-suited for performing perturbation theory. One can simply keep as many terms as are required in the series expansions of the exponential functions. In addition, comparing Eqs. (11.50) and (11.60), we observe that the formulas are the same except that Eq. (11.60) is simpler, since it contains the vacuum of the free theory instead of the physical vacuum.

11.5 Specific correlation functions

The **complete** n-point correlation function (or n-point Green's function) is given by

$$\tau(x_1, x_2, \ldots, x_n) = \langle \Omega|T\phi(x_1)\phi(x_2)\ldots\phi(x_n)|\Omega\rangle. \tag{11.61}$$

This expression represents the sum of all Feynman diagrams (or graphs) with n particles created or destroyed (annihilated) at spacetime points $x_1, x_2, \ldots, x_n$, see Fig. 11.1. Generally, a Feynman diagram is an intuitive graphical or pictorial representation of a perturbative contribution to the correlation function or the total amplitude for a physical process, which can occur in many different ways, from an initial state to a final state. Note that only the sum of all possible Feynman diagrams for a given physical process (i.e. particle interaction) represents such a process, since particles do not choose a specific Feynman diagram when they interact. Especially, the two-point correlation function $\tau(x, y)$ represents the physical amplitude for propagation of a particle between the spacetime points y and x, whereas the four-point correlation function $\tau(x_1, x_2, x_3, x_4)$ represents the physical amplitude for interaction of four particles at the spacetime points x_1, x_2, x_3, and x_4, which can, for example, be interpreted as particles are created at two spacetime points, each propagates to one of the other spacetime points, and then they

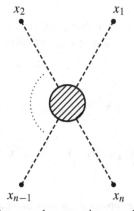

Figure 11.1 The complete n-point correlation function.

are annihilated. Actually, disconnected Feynman diagrams can contain *bubble* (or *vacuum*) *diagrams* that have **no** external points.[2] Note that any subdiagram, which is not connected to an external point, is called a disconnected part of a diagram. Thus, a diagram with **no** disconnected parts is called a *connected Feynman diagram*. Note that any Feynman diagram can be separated into a *connected part* and a disconnected part. This holds also for the contributions to the n-point correlation function. Thus, we can write

$$\tau(x_1, x_2, \ldots, x_n) = \frac{G(x_1, x_2, \ldots, x_n)}{N}, \tag{11.62}$$

where

$$G(x_1, x_2, \ldots, x_n) = \langle 0|T\phi_{\text{in}}(x_1)\phi_{\text{in}}(x_2)\ldots\phi_{\text{in}}(x_n)e^{-i\int_{-\infty}^{\infty} H_I'(t')\,dt'}|0\rangle, \tag{11.63}$$

$$N = \langle 0|Te^{-i\int_{-\infty}^{\infty} H_I'(t')\,dt'}|0\rangle \tag{11.64}$$

represent all Feynman diagrams and the disconnected part, respectively. Using Eq. (11.60) together with the power series in Eq. (11.32), the sum of all Feynman diagrams to all orders in perturbation theory can be written as

$$G(x_1, x_2, \ldots, x_n) = \sum_{r=0}^{\infty} G_r(x_1, x_2, \ldots, x_n), \tag{11.65}$$

where the contribution to rth order is given by

$$G_r(x_1, x_2, \ldots, x_n) = \frac{(-i)^r}{r!} \int \langle 0|T\phi_{\text{in}}(x_1)\phi_{\text{in}}(x_2)\ldots\phi_{\text{in}}(x_n)$$
$$\times \mathcal{H}_I'(y_1)\mathcal{H}_I'(y_2)\ldots\mathcal{H}_I'(y_r)|0\rangle \, d^4 y_1 d^4 y_2 \ldots d^4 y_r, \tag{11.66}$$

and we have the disconnected part

$$N = \sum_{r=0}^{\infty} \frac{(-i)^r}{r!} \int \langle 0|T\mathcal{H}_I'(y_1)\mathcal{H}_I'(y_2)\ldots\mathcal{H}_I'(y_r)|0\rangle \, d^4 y_1 d^4 y_2 \ldots d^4 y_r. \tag{11.67}$$

Now, in order to identify the connected part, the number of ways of selecting p terms from a set of r terms implies that Eq. (11.66) can be rewritten as

$$G_{c,p}(x_1, x_2, \ldots, x_n)$$
$$= \sum_{r=p}^{\infty} \frac{(-i)^r}{r!} \int \langle 0|T\phi_{\text{in}}(x_1)\phi_{\text{in}}(x_2)\ldots\phi_{\text{in}}(x_n)\mathcal{H}_I'(y_1)\mathcal{H}_I'(y_2)\ldots\mathcal{H}_I'(y_p)|0\rangle_c$$
$$\times \frac{r!}{p!(r-p)!} \langle 0|T\mathcal{H}_I'(y_{p+1})\mathcal{H}_I'(y_{p+2})\ldots\mathcal{H}_I'(y_r)|0\rangle \, d^4 y_1 d^4 y_2 \ldots d^4 y_r. \tag{11.68}$$

---

[2] Bubble diagrams do not cause any transitions. One can show that such diagrams may be omitted altogether, at any rate in elementary applications.

where the subscript c denotes the connected part, which is of order p. Next, setting $k = r - p$, where p is fixed, we obtain

$$G_{c,p}(x_1, x_2, \ldots, x_n)$$
$$= \frac{(-i)^p}{p!} \int \langle 0|T\phi_{in}(x_1)\phi_{in}(x_2)\ldots\phi_{in}(x_n)\mathcal{H}'_I(y_1)\mathcal{H}'_I(y_2)\ldots\mathcal{H}'_I(y_p)|0\rangle_c$$
$$\times d^4y_1 d^4y_2 \ldots d^4y_p$$
$$\times \sum_{k=0}^{\infty} \frac{(-i)^k}{k!} \int \langle 0|T\mathcal{H}'_I(z_1)\mathcal{H}'_I(z_2)\ldots\mathcal{H}'_I(z_k)|0\rangle \, d^4z_1 d^4z_2 \ldots d^4z_k, \quad (11.69)$$

where the last line is exactly the same as the result in Eq. (11.67), which means that using Eqs. (11.62), (11.65), and (11.66), we find that

$$\tau(x_1, x_2, \ldots, x_n) = \frac{\sum_{r=0}^{\infty} G_r(x_1, x_2, \ldots, x_n)}{N} = \frac{\sum_{p=0}^{\infty} G_{c,p}(x_1, x_2, \ldots, x_n)}{N}$$
$$\equiv \frac{N\sum_{p=0}^{\infty} \tau_p(x_1, x_2, \ldots, x_n)}{N} = \sum_{p=0}^{\infty} \tau_p(x_1, x_2, \ldots, x_n),$$
$$(11.70)$$

since both $\sum_{r=0}^{\infty} G_r(x_1, x_2, \ldots, x_n)$ and $\sum_{p=0}^{\infty} G_{c,p}(x_1, x_2, \ldots, x_n)$ contain all Feynman diagrams. Thus, the n-point correlation function takes on the form

$$\tau(x_1, x_2, \ldots, x_n)$$
$$= \sum_{p=0}^{\infty} \tau_p(x_1, x_2, \ldots, x_n) = \sum_{p=0}^{\infty} \frac{(-i)^p}{p!} \int \langle 0|T\phi_{in}(x_1)\phi_{in}(x_2)\ldots\phi_{in}(x_n)$$
$$\times \mathcal{H}'_I(\phi_{in}(y_1))\mathcal{H}'_I(\phi_{in}(y_2))\ldots\mathcal{H}'_I(\phi_{in}(y_p))|0\rangle_c \, d^4y_1 d^4y_2 \ldots d^4y_p, \quad (11.71)$$

which is the sum of the contributions of all connected Feynman diagrams.

First, we will investigate the two-point correlation function $\tau(x, y)$. It is important to distinguish between the interacting theory two-point correlation function and the free theory two-point correlation function. To all orders in perturbation theory, using Eq. (11.60), the complete two-point ($n = 2$) correlation function (for neutral Klein–Gordon fields) is given by

$$\tau(x, y) = \langle \Omega|T\phi(x)\phi(y)|\Omega\rangle = \frac{\langle 0|T\phi_{in}(x)\phi_{in}(y)e^{-i\int_{-\infty}^{\infty} H'_I(t')\,dt'}|0\rangle}{\langle 0|Te^{-i\int_{-\infty}^{\infty} H'_I(t')\,dt'}|0\rangle}, \quad (11.72)$$

whereas to zeroth ($p = 0$) order in perturbation theory, we have the two-point ($n = 2$) correlation function

$$\tau_0(x, y) = \langle 0|T\phi_{\text{in}}(x)\phi_{\text{in}}(y)|0\rangle_c = i\Delta_F(x - y) = \int \frac{d^4 p}{(2\pi)^4} e^{-ip\cdot(x-y)} \frac{i}{p^2 - m^2 + i\epsilon}, \tag{11.73}$$

which is exactly the *Feynman propagator* for a neutral Klein–Gordon field, i.e. the VEV for a T-ordered product of two asymptotically free fields. Similarly, using as interaction theory the so-called $\lambda\phi^4$-theory, i.e.

$$\mathcal{H}'_I(\phi) = \frac{\lambda}{4!}\phi^4, \tag{11.74}$$

we obtain, to first ($p = 1$) order in perturbation theory, the two-point ($n = 2$) correlation function

$$\tau_1(x, y) = -\frac{i\lambda}{4!} \int \langle 0|T\phi_{\text{in}}(x)\phi_{\text{in}}(y)\phi_{\text{in}}(z)^4|0\rangle_c \, d^4z, \tag{11.75}$$

while, to second ($p = 2$) order in perturbation theory, we have

$$\tau_2(x, y) = -\frac{\lambda^2}{2!(4!)^2} \int \langle 0|T\phi_{\text{in}}(x)\phi_{\text{in}}(y)\phi_{\text{in}}(z)^4\phi_{\text{in}}(z')^4|0\rangle_c \, d^4z d^4z'. \tag{11.76}$$

Second, we will study another important correlation function, which is the four-point correlation function. In this case, we have, to zeroth ($p = 0$) order in perturbation theory, the four-point ($n = 4$) correlation function

$$\tau_0(x_1, x_2, x_3, x_4) = \langle 0|T\phi_{\text{in}}(x_1)\phi_{\text{in}}(x_2)\phi_{\text{in}}(x_3)\phi_{\text{in}}(x_4)|0\rangle_c, \tag{11.77}$$

whereas, to first ($p = 1$) order in perturbation theory, we obtain

$$\tau_1(x_1, x_2, x_3, x_4) = -\frac{i\lambda}{4!} \int \langle 0|T\phi_{\text{in}}(x_1)\phi_{\text{in}}(x_2)\phi_{\text{in}}(x_3)\phi_{\text{in}}(x_4)\phi_{\text{in}}(y)^4|0\rangle_c \, d^4y, \tag{11.78}$$

and, to second ($p = 2$) order in perturbation theory, we find

$$\tau_2(x_1, x_2, x_3, x_4)$$
$$= -\frac{\lambda^2}{2!(4!)^2} \int \langle 0|T\phi_{\text{in}}(x_1)\phi_{\text{in}}(x_2)\phi_{\text{in}}(x_3)\phi_{\text{in}}(x_4)\phi_{\text{in}}(y_1)^4\phi_{\text{in}}(y_2)^4|0\rangle_c \, d^4y_1 d^4y_2. \tag{11.79}$$

Finally, we mention that it is sometimes convenient to work with correlation functions in momentum space

$$(2\pi)^4\delta(p_1 + p_2 + \cdots + p_n)\tau(p_1, p_2, \ldots, p_n)$$
$$= \int \prod_{i=1}^{n} e^{-ip_i \cdot x_i} \tau(x_1, x_2, \ldots, x_n) \, d^4x_i \tag{11.80}$$

as well as the *amputated correlation functions* that are related to the correlation functions in momentum space $\tau(p_1, p_2, \ldots, p_n)$ by removing the propagators on external lines

$$\tau_{\text{amp}}(p_1, p_2, \ldots, p_n) = \left(\prod_{i=1}^{n} \frac{1}{i\Delta_F(p_i)}\right) \tau(p_1, p_2, \ldots, p_n), \qquad (11.81)$$

where $p_1 + p_2 + \cdots + p_n = 0$. Actually, for Klein–Gordon (spin-0) particles, the amputated correlation functions are the transition amplitudes (or the T matrix elements) from which cross-sections can be calculated. This can be shown using Eq. (10.37).

Next, we will present the so-called *Wick's theorem* that can be used to decompose the connected parts of the VEV for T-ordered products of asymptotically free fields into a product of two-point correlation functions, i.e. Feynman propagators.

11.6 Wick's theorem

Wick's theorem[3] can be used to simplify and compute expressions of the form

$$\langle 0|T\phi_{\text{in}}(x_1)\phi_{\text{in}}(x_2) \ldots \phi_{\text{in}}(x_n)|0\rangle_c. \qquad (11.82)$$

In what follows, we will drop the subscript c. As we have discussed, $n = 2$ leads to the Feynman propagator (or the two-point correlation function). Now, expanding the asymptotically free fields in positive- and negative-frequency parts gives

$$\phi_{\text{in}}(x) = \phi_{\text{in},+}(x) + \phi_{\text{in},-}(x), \qquad (11.83)$$

where $\phi_{\text{in},+}(x) \sim a(\mathbf{p})$ and $\phi_{\text{in},-}(x) \sim a^\dagger(\mathbf{p})$. See Eq. (6.35). Thus, in the case that $x^0 > y^0$, we can write the T-ordered product of two asymptotically free fields as

$$T\phi_{\text{in}}(x)\phi_{\text{in}}(y)$$
$$= \phi_{\text{in},+}(x)\phi_{\text{in},+}(y) + \phi_{\text{in},+}(x)\phi_{\text{in},-}(y) + \phi_{\text{in},-}(x)\phi_{\text{in},+}(y) + \phi_{\text{in},-}(x)\phi_{\text{in},-}(y).$$
$$(11.84)$$

Next, we want to write all the annihilation operators to the right of all the creation operators, i.e. we want to normal-order the above expression. This can be performed using a commutator of the asymptotically free fields, and we find that

$$T\phi_{\text{in}}(x)\phi_{\text{in}}(y)$$
$$= \phi_{\text{in},+}(x)\phi_{\text{in},+}(y) + \phi_{\text{in},-}(y)\phi_{\text{in},+}(x) + \phi_{\text{in},-}(x)\phi_{\text{in},+}(y) + \phi_{\text{in},-}(x)\phi_{\text{in},-}(y)$$
$$+ \left[\phi_{\text{in},+}(x), \phi_{\text{in},-}(y)\right]$$
$$=: \phi_{\text{in}}(x)\phi_{\text{in}}(y) : + \left[\phi_{\text{in},+}(x), \phi_{\text{in},-}(y)\right]. \qquad (11.85)$$

---

[3] For the interested reader, please see: G. C. Wick, The evaluation of the collision matrix, *Phys. Rev.* **80**, 268–272 (1950).

Here, it is in place to mention some of the general properties of normal-ordering, which was first introduced in Eq. (4.54) [or Eq. (4.40)]. For bosonic fields (e.g. Klein–Gordon fields and electromagnetic fields), it holds that

$$: a_1 a_2 : \ = a_1 a_2, \tag{11.86}$$

$$: a_1^\dagger a_2^\dagger : \ = a_1^\dagger a_2^\dagger, \tag{11.87}$$

$$: a_1^\dagger a_2 : \ = a_1^\dagger a_2, \tag{11.88}$$

$$: a_1 a_2^\dagger : \ = a_2^\dagger a_1, \tag{11.89}$$

where a_1 and a_2 are annihilation operators and $a_1^\dagger$ and $a_2^\dagger$ are creation operators. Note that Eqs. (11.86)–(11.88) also hold for fermionic fields, but with the as changed to bs (or ds). However, for fermionic fields (e.g. Dirac fields), Eq. (11.89) has to be replaced by

$$: b_1 b_2^\dagger : \ = -b_2^\dagger b_1, \tag{11.90}$$

where b_1 is an annihilation operator and $b_2^\dagger$ is a creation operator (cf. Sections 7.2 and 7.3), and similarly, for the ds. In general, if f, g, and h are polynomials of some annihilation and creation operators, then we have the following relations

$$: \alpha f + \beta g : \ = \alpha : f : + \beta : g :, \tag{11.91}$$

$$: f(g+h) : \ = \ : fg : + : fh :, \tag{11.92}$$

where α and β are some constants. Then, let us define the quantity

$$\overbrace{\phi_{\mathrm{in}}(x)\phi_{\mathrm{in}}(y)} \equiv \begin{cases} \left[\phi_{\mathrm{in},+}(x), \phi_{\mathrm{in},-}(y)\right], & x^0 > y^0 \\ \left[\phi_{\mathrm{in},+}(y), \phi_{\mathrm{in},-}(x)\right], & y^0 > x^0 \end{cases}, \tag{11.93}$$

which is the *contraction* of two fields. This quantity is exactly the Feynman propagator (for a neutral Klein–Gordon field), i.e.

$$\overbrace{\phi_{\mathrm{in}}(x)\phi_{\mathrm{in}}(y)} = \mathrm{i}\Delta_F(x-y). \tag{11.94}$$

In addition, we find the following contractions

$$\overbrace{\phi_{\mathrm{in}}(x)\phi_{\mathrm{in}}^\dagger(y)} = \overbrace{\phi_{\mathrm{in}}^\dagger(y)\phi_{\mathrm{in}}(x)} = \mathrm{i}\Delta_F(x-y), \tag{11.95}$$

$$\overbrace{\psi_{\mathrm{in},\alpha}(x)\bar{\psi}_{\mathrm{in},\beta}(y)} = -\overbrace{\bar{\psi}_{\mathrm{in},\beta}(y)\psi_{\mathrm{in},\alpha}(x)} = \mathrm{i}S_{F\alpha\beta}(x-y), \tag{11.96}$$

$$\overbrace{A_\mu(x)A_\nu(y)} = \mathrm{i}D_{F\mu\nu}(x-y) \tag{11.97}$$

for a complex Klein–Gordon field, a Dirac field, and a radiation field, respectively. Thus, using Eqs. (11.85) and (11.93), we obtain the relation between time-ordering and normal-ordering for two fields as

$$T\phi_{\text{in}}(x)\phi_{\text{in}}(y)$$

$$=\,:\phi_{\text{in}}(x)\phi_{\text{in}}(y)\,+\,\overline{\phi_{\text{in}}(x)\phi_{\text{in}}(y)}\,:\; =\; :\phi_{\text{in}}(x)\phi_{\text{in}}(y)\,:\,+\,\overline{\phi_{\text{in}}(x)\phi_{\text{in}}(y)}\,. \quad (11.98)$$

In general, for n fields, we have *Wick's theorem*

$$T\phi_{\text{in}}(x_1)\phi_{\text{in}}(x_2)\ldots\phi_{\text{in}}(x_n)$$

$$=\,:\phi_{\text{in}}(x_1)\phi_{\text{in}}(x_2)\ldots\phi_{\text{in}}(x_n)\,+\,\text{all possible contractions}:. \quad (11.99)$$

For example, for $n = 4$ and using $\phi_i \equiv \phi_{\text{in}}(x_i)$, we find that

$$T\phi_1\phi_2\phi_3\phi_4$$

$$=\,:\phi_1\phi_2\phi_3\phi_4 + \overline{\phi_1\phi_2}\phi_3\phi_4 + \phi_1\phi_2\overline{\phi_3\phi_4} + \phi_1\overline{\phi_2\phi_3}\phi_4 + \overline{\phi_1\phi_2\phi_3}\phi_4$$

$$+\,\phi_1\phi_2\overline{\phi_3\phi_4} + \overline{\phi_1\phi_2}\,\overline{\phi_3\phi_4} + \phi_1\overline{\phi_2\phi_3}\phi_4 + \overline{\phi_1\phi_2\phi_3\phi_4} + \overline{\phi_1\phi_2\phi_3\phi_4}\,:. \quad (11.100)$$

Now, using the fact that $\langle 0|:A:|0\rangle = 0$ for any operator A [cf. Eq. (4.55)],[4] i.e. only terms which contain contractions of all survive, we obtain the VEV for a T-ordered product of four asymptotically free fields as

$$\tau_0(x_1,x_2,x_3,x_4) = \langle 0|T\phi_1\phi_2\phi_3\phi_4|0\rangle = \overline{\phi_1\phi_2}\,\overline{\phi_3\phi_4} + \overline{\phi_1\phi_2\phi_3\phi_4} + \overline{\phi_1\phi_2\phi_3\phi_4}$$

$$= i\Delta_F(x_1 - x_2)i\Delta_F(x_3 - x_4) + i\Delta_F(x_1 - x_3)i\Delta_F(x_2 - x_4)$$

$$+ i\Delta_F(x_1 - x_4)i\Delta_F(x_2 - x_3), \quad (11.101)$$

which can be graphically represented by Feynman diagrams

$$\tau_0(x_1,x_2,x_3,x_4) = \qquad\qquad (11.102)$$

Independently of the interaction theory, we note that

$$\langle 0|T\phi_{\text{in}}(x_1)\phi_{\text{in}}(x_2)\ldots\phi_{\text{in}}(x_n)|0\rangle = 0 \quad\text{if } n \text{ is odd,} \qquad (11.103)$$

---

[4] Except for an operator that is proportional to the unit operator.

since using Wick's theorem, all resulting terms will at least contain one field that is not contracted, whereas

$$\langle 0|T\phi_{in}(x_1)\phi_{in}(x_2)\ldots\phi_{in}(x_n)|0\rangle = \sum_{perm.} \prod \langle 0|T\phi_{in}(x_i)\phi_{in}(x_j)|0\rangle \quad \text{if } n \text{ is even,}$$

(11.104)

i.e. the VEV of Eq. (11.99) is given by the sum of terms containing contractions of all the fields. For example, for $n = 4$, we obtain only the three last terms in Eq. (11.100) as the result for Eq. (11.104), which can also be observed in Eq. (11.101). Especially, for $\lambda\phi^4$-theory, it holds that $\tau(x_1, x_2, \ldots, x_n) = 0$ if n is odd. In addition, since $H_I'(\phi_{in})$ contains a normal-ordered product of asymptotically free fields at the same spacetime point, the propagators that come from contractions of these fields with other fields always have one spacetime point in common, which is called a *vertex*.

Finally, using Wick's theorem for the first ($p = 1$) order two- and four-point correlation functions in Eqs. (11.75) and (11.78), we obtain for the two-point function

$$\tau_1(x, y) = -\frac{i\lambda}{2!} \int i\Delta_F(x - z)i\Delta_F(z - z)i\Delta_F(y - z) \, d^4z \qquad (11.105)$$

and for the four-point function

$$\tau_1(x_1, x_2, x_3, x_4) = -i\lambda \int i\Delta_F(x_1 - y)i\Delta_F(x_2 - y)i\Delta_F(x_3 - y)i\Delta_F(x_4 - y) \, d^4y,$$

(11.106)

where the factor 4! in Eq. (11.78) is cancelled, since there are a corresponding number of ways to contract the fields $\phi_{in}(x_i)$ ($i = 1, 2, 3, 4$), i.e. the external legs at spacetime points x_i, with each of the four fields in $\phi_{in}(y)^4$. The result of Eq. (11.106) illustrates how a vertex is associated with an interaction term.

Finally, when converting from position space to momentum space, we perform the x-integrations and obtain a δ-function at each vertex, which represents conservation of 4-momentum. When all the x-integrations are performed, there is one δ-function left representing the overall 4-momentum conservation, cf. Eq. (11.80). In addition, there is a 4-momentum integration for each internal loop. Thus, all Feynman rules can be represented in momentum space.

To this end, using Wick's theorem, we are now able to compute VEVs for T-ordered products of asymptotically free fields, which can be used to derive n-point correlation functions [using Eq. (11.71)] to all orders in perturbation theory. In turn, the results can be employed to find S matrix elements through the LSZ reduction formula (10.37).

11.7 Feynman rules and diagrams

In this section, we will present and discuss the Feynman rules and diagrams for four different models of the interaction theory. In general, the zeroth-order two-point correlation function $\tau_0(x, y)$ for asymptotically free fields will lead to the Feynman propagator of the free theory, and the first-order three-point and/or four-point correlation functions, i.e. $\tau_1(x_1, x_2, x_3)$ and/or $\tau_1(x_1, x_2, x_3, x_4)$, which are normally the first non-trivial contributions including the interaction theory, will give the vertex Feynman rules for the interaction theory. The four models for the interaction theory are $\lambda\phi^4$-theory, Yukawa theory,[5] QED, and finally, Yang–Mills theory. In the case of $\lambda\phi^4$-theory, as we have already observed, the first-order four-point correlation function gives the vertex Feynman rule, whereas in the cases of Yukawa theory and QED, the first-order three-point correlation functions give the vertex Feynman rules, and finally, in the case of Yang–Mills theory, both the first-order three- and four-point correlation functions give the vertex Feynman rules. Note that it can be shown that all four models are so-called *renormalizable*,[6] which is a good property for a model.

Finally, before we present the Feynman rules of the four different models for the interaction theory, we introduce the following nomenclature:

- **internal line:** an internal line for a particle is a factor of the corresponding virtual particle's propagator, which connects two vertices;

- **vertex:** a vertex is a factor, which is derived from an interaction term in the Lagrangian and connects three or more internal and/or external lines;

- **external line:** an external line has definite 4-momentum and spin. An *incoming* external line extends from 'the past', i.e. $t \to -\infty$, to a vertex and represents an initial state, whereas an *outgoing* external line extends from a vertex to 'the future', i.e. $t \to \infty$, and represents a final state.

In addition, we have the following definitions for ordinary Feynman diagrams:

- **forest diagram:** a forest diagram is a Feynman diagram in which all internal lines have 4-momenta that are totally determined by external lines and 4-momentum conservation is imposed at each vertex;

- **tree diagram:** a tree diagram is a *connected* forest diagram and **all** tree diagrams are forest diagrams, but the opposite is **not** true;

---

[5] Yukawa theory is named after H. Yukawa, who was awarded the Nobel Prize in physics in 1949 'for his prediction of the existence of mesons on the basis of theoretical work on nuclear forces'.

[6] A model is renormalizable if its physical quantities given by parameters such as masses and coupling constants are finite when cut-offs are removed (cf. the discussion in Section 8.3, and in addition, see the discussion in Chapter 13).

- **loop diagram:** a Feynman diagram, which is **not** a forest diagram, is a loop diagram, since the number of 4-momentum integrals that are undetermined by 4-momentum conservation is equal to the number of closed loops in the diagram.

11.7.1 $\lambda\phi^4$-theory

The Lagrangian for a neutral Klein–Gordon field with $\lambda\phi^4$-theory as interaction theory is given by

$$\mathcal{L} = \frac{1}{2}\partial_\mu\phi\partial^\mu\phi - \frac{1}{2}m^2\phi^2 - \frac{\lambda}{4!}\phi^4, \qquad (11.107)$$

which leads to the interaction Hamiltonian

$$\mathcal{H}'_I(\phi) = \frac{\lambda}{4!}\phi^4. \qquad (11.108)$$

Therefore, to all orders in perturbation theory, the sum of all possible Feynman diagrams with two external points is given by

$$\langle 0|T\phi_{\text{in}}(x)\phi_{\text{in}}(y)e^{-i\int_{-\infty}^{\infty} H'_I(t')\,dt'}|0\rangle. \qquad (11.109)$$

Thus, in position space, the Feynman rules are:

- propagator:

$$y \bullet\!-\!-\!-\!-\!-\!-\!-\!-\!\bullet x = i\Delta_F(x - y), \qquad (11.110)$$

- vertex:

$$= -i\lambda \int d^4z, \qquad (11.111)$$

- external point:

$$x \bullet\!-\!-\!-\!-\!-\!-\!-\!- = 1, \qquad (11.112)$$

- divide by a potential symmetry factor (due to equivalent Feynman diagrams), whereas, in momentum space, they are:

- propagator:

$$----\underset{p}{----} = \frac{i}{p^2 - m^2 + i\epsilon}, \tag{11.113}$$

- vertex:

$$= -i\lambda, \tag{11.114}$$

- external point:

$$x \,\bullet\!----\underset{p}{----} = e^{-ip \cdot x}, \tag{11.115}$$

- impose 4-momentum conservation at each vertex,
- integrate over each undetermined loop 4-momentum:

$$\int \frac{d^4 p}{(2\pi)^4}, \tag{11.116}$$

- divide by a potential symmetry factor (due to equivalent Feynman diagrams).

Now, using the Feynman rules for $\lambda\phi^4$-theory, we have

$$\langle 0|T\phi_{\text{in}}(x)\phi_{\text{in}}(y)e^{-i\int_{-\infty}^{\infty} H'_I(t')\,dt'}|0\rangle$$

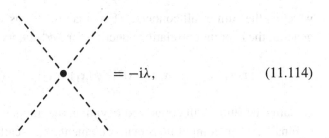

$$= \left(y \,\bullet\!-------\bullet\, x \;+\; y \,\bullet\!---\bullet\!---\bullet\, x \;+\; y \,\bullet\!--\bullet\!-\bullet\!--\bullet\, x \;+ \dots \right)$$

$$\times \exp\left(\text{sum of bubble diagrams}\right), \tag{11.117}$$

where the exponential contains the sum of all disconnected Feynman diagrams. In addition, we find that

$$\left\langle 0 \left| T e^{-i\int_{-\infty}^{\infty} H'_I(t')\,dt'} \right| 0 \right\rangle = \exp\left(\text{sum of bubble diagrams}\right), \tag{11.118}$$

which is exactly the same exponential. Thus, we obtain the two-point correlation function for $\lambda\phi^4$-theory to all orders in perturbation theory

$$\tau(x, y) = \langle\Omega|T\phi(x)\phi(y)|\Omega\rangle$$

$$= \quad y \bullet\text{-------}\bullet x \; + \; y \bullet\text{----}\bullet\text{---}\bullet x \; + \; y \bullet\text{--}\langle\text{-}\rangle\text{--}\bullet x \; + \cdots,$$

$$(11.119)$$

which is the sum of all connected Feynman diagrams with two external points. In general, the n-point correlation function for $\lambda\phi^4$-theory

$$\tau(x_1, x_2, \ldots, x_n) = \langle\Omega|T\phi(x_1)\phi(x_2)\ldots\phi(x_n)|\Omega\rangle \qquad (11.120)$$

contains the sum of all connected Feynman diagrams with n external points.

Finally, when computing S matrix elements (i.e. scattering amplitudes) in different models, it is more convenient to have Feynman rules for *external lines* instead of Feynman rules for *external points* (which are useful for correlation functions). In $\lambda\phi^4$-theory, the Feynman rules for an external line in position and momentum space, respectively, are:

- external line in position space:

$$\overbrace{\phi(x)|\mathbf{p}\rangle} = \text{----}\underset{p}{\overset{x}{\text{---}}}\langle\text{-} = e^{-ip\cdot x}, \qquad (11.121)$$

- external line in momentum space:

$$\overbrace{\phi|\mathbf{p}\rangle} = \text{-----}\langle\text{-} = 1. \qquad (11.122)$$

11.7.2 Yukawa theory

Yukawa theory is a simplified model of QED. The Lagrangian for Yukawa theory is given by

$$\mathcal{L}_Y = \mathcal{L}_D + \mathcal{L}_{KG} - g\bar{\psi}\psi\phi, \qquad (11.123)$$

where $\mathcal{L}_D = \bar{\psi}(i\partial\!\!\!/ - m\mathbb{1}_4)\psi$ is the free Dirac Lagrangian [cf. Eq. (7.10)], $\mathcal{L}_{KG} = \frac{1}{2}\left(\partial_\mu\phi\partial^\mu\phi - m_\phi^2\phi^2\right)$ is the free neutral Klein–Gordon Lagrangian [cf. Eq. (6.1)], and g is the coupling constant of the Yukawa interaction. Thus, for Yukawa theory in momentum space, the Feynman rules are given by:

- Klein–Gordon and Dirac propagators:

$$\overset{\sqcap}{\phi(x)\phi(y)} = i\Delta_F(x-y): \quad \overset{\text{------}}{\underset{q}{}} \quad = \frac{i}{q^2 - m_\phi^2 + i\epsilon} \qquad (11.124)$$

$$\overset{\sqcap}{\psi(x)\bar\psi(y)} = iS_F(x-y): \quad \overset{\longrightarrow}{\underset{p}{}} \quad = \frac{i\,(\slashed{p}+m\mathbb{1}_4)}{p^2 - m^2 + i\epsilon}, \qquad (11.125)$$

- vertex:

$$= -ig, \qquad (11.126)$$

- external lines (leg contractions):

$$\overset{\sqcap}{\phi|\mathbf{q}\rangle} = \overset{\text{--}}{\underset{q}{}}\!\!<\; = 1 \quad \overset{\sqcap}{\langle\mathbf{q}|\phi} = \;>\!\!\overset{\text{--}}{\underset{q}{}} = 1 \qquad (11.127)$$

$$\overset{\sqcap}{\psi|\mathbf{p},s\rangle} = \overset{\longrightarrow}{\underset{p}{}}\!\!<\; = u(\mathbf{p},s) \quad \overset{\sqcap}{\langle\mathbf{p},s|\bar\psi} = \;>\!\!\overset{\longrightarrow}{\underset{p}{}} = \bar{u}(\mathbf{p},s) \qquad (11.128)$$

$$\overset{\sqcap}{\bar\psi|\mathbf{p},s\rangle} = \overset{p}{\underset{\longleftarrow}{}}\!\!<\; = \bar{v}(\mathbf{p},s) \quad \overset{\sqcap}{\langle\mathbf{p},s|\psi} = \;>\!\!\overset{p}{\underset{\longleftarrow}{}} = v(\mathbf{p},s), \qquad (11.129)$$

- impose 4-momentum conservation at each vertex,
- integrate over each undetermined loop 4-momentum,
- figure out the overall sign of the diagram,
- fermion loops receive an additional factor of -1.

Note that Feynman diagrams in Yukawa theory never have any symmetry factors, since the three fields in the Hamiltonian cannot substitute for one another in contractions.

11.7.3 Quantum electrodynamics

The Lagrangian of QED is given by

$$\mathcal{L}_{\text{QED}} = \bar\psi\left(i\slashed{D} - m\mathbb{1}_4\right)\psi - \frac{1}{4}F^{\mu\nu}F_{\mu\nu} = \bar\psi\left(i\slashed{\partial} - m\mathbb{1}_4\right)\psi - \frac{1}{4}F^{\mu\nu}F_{\mu\nu} - e\bar\psi\gamma^\mu\psi A_\mu, \qquad (11.130)$$

which leads to the interaction Hamiltonian

$$H_I' = e\int \bar\psi\gamma^\mu\psi A_\mu\,\mathrm{d}^3x, \qquad (11.131)$$

i.e. we have replaced the neutral Klein–Gordon field ϕ in Yukawa theory with the electromagnetic vector field A_μ. Thus, in the case of QED, we obtain the following new Feynman rules in momentum space:

- Dirac propagator (internal fermion line):

$$\psi(x)\bar{\psi}(y) = iS_F(x-y): \qquad \xrightarrow{\quad p \quad} \quad = \frac{i\,(\not{p}+m\mathbb{1}_4)}{p^2-m^2+i\epsilon}, \qquad (11.132)$$

- photon propagator (internal photon line):

$$A_\mu(x)A_\nu(y) = iD_{F\mu\nu}(x-y): \qquad {}^\nu\!\!\!\sim\!\!\!\sim\!\!\!\sim\!\!\!\sim\!\!\!\sim\!\!\!\!{}^\mu \quad = \frac{-ig_{\mu\nu}}{q^2+i\epsilon}, \qquad (11.133)$$

- QED vertex:

$$\sim\!\!\!\sim\!\!\!\sim \mu = -ie\gamma^\mu, \qquad (11.134)$$

- external photon lines:

$$A^\mu|\mathbf{k}\rangle = {}^\mu\!\!\!\sim\!\!\!\sim\!\!\!\sim = \varepsilon^\mu(\mathbf{k}) \qquad (11.135)$$

$$\langle\mathbf{k}|A^\mu = \sim\!\!\!\sim\!\!\!\sim{}^\mu = \varepsilon^{*\mu}(\mathbf{k}), \qquad (11.136)$$

where ε^μ is the polarization vector of the initial- or final-state photon,

- impose 4-momentum conservation at each vertex,
- integrate over each undetermined loop 4-momentum, i.e. free internal 4-momentum,
- closed fermion loops receive an additional factor of -1 and a trace should be performed over the Dirac spinor indices,
- figure out the overall sign of the diagram,
- divide by a potential symmetry factor (due to equivalent Feynman diagrams).

Note that, in QED, it can be shown that the order of a loop diagram ℓ, i.e. the number of closed loops, is given by

$$\ell = 1 + f + b - n, \qquad (11.137)$$

where f is the number of internal fermion lines, b is the number of internal photon lines, and n is the number of vertices.

11.7.4 Yang–Mills theory

The Lagrangian for Yang–Mills theory (or non-Abelian gauge theory) is given by

$$
\mathcal{L}_{\text{YM}} = \bar{\psi} \left(i\partial\!\!\!/ - m\mathbb{1}_4 \right) \psi - \frac{1}{4} \left(\partial_\mu A_\nu^a - \partial_\nu A_\mu^a \right)^2 + g A_\mu^a \bar{\psi} \gamma^\mu t^a \psi
$$

$$
- g f^{abc} \left(\partial_\mu A_\nu^a \right) A^{b\,\mu} A^{c\,\nu} - \frac{1}{4} g^2 f^{eab} f^{ecd} A_\mu^a A_\nu^b A^{c\,\mu} A^{d\,\nu}, \qquad (11.138)
$$

where g is the coupling constant as well as the t^as and the f^{abc}s are the generators and the structure constants of the underlying Lie algebra corresponding to the Lie group of the non-Abelian gauge theory, respectively (see Appendix A). Note that the first three terms in Eq. (11.138) are similar to the terms in Eq. (11.130), but the last two terms are new and due to the non-Abelian property of Yang–Mills theory (cf. the discussion on the Yang–Mills Lagrangian in Section 9.4). The Dirac fermion propagator and the gauge boson (in QED: photon) propagator are the same as for QED multiplied with an identity matrix in the gauge group space. Similarly, for the polarization of external particles, each external particle has an orientation in the gauge group space. Thus, for Yang–Mills theory in momentum space, the other Feynman rules are given by:

- fermion–boson vertex:

$$
\mu, a = i g \gamma^\mu t^a, \qquad (11.139)
$$

- 3-boson vertex:

$$
\mu, a = g f^{abc} \left[g^{\mu\nu} (k - p)^\rho + g^{\nu\rho} (p - q)^\mu + g^{\rho\mu} (q - k)^\nu \right], \qquad (11.140)
$$

- 4-boson vertex:

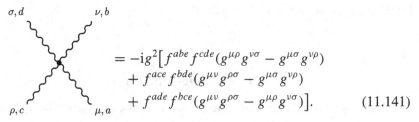

$$= -ig^2\big[f^{abe}f^{cde}(g^{\mu\rho}g^{\nu\sigma} - g^{\mu\sigma}g^{\nu\rho})$$
$$+ f^{ace}f^{bde}(g^{\mu\nu}g^{\rho\sigma} - g^{\mu\sigma}g^{\nu\rho})$$
$$+ f^{ade}f^{bce}(g^{\mu\nu}g^{\rho\sigma} - g^{\mu\rho}g^{\nu\sigma})\big]. \tag{11.141}$$

Note that in the Feynman rule for the 3-boson vertex, i.e. Eq. (11.140), all three 4-momenta k, p, and q flow into the vertex.

11.8 Kinematics for binary reactions

Consider the kinematics for binary reactions with two particles a and b in the initial state and two particles c and d in the final state, see Fig. 11.2.[7] In this case, we have $4 \cdot 4 = 16$ free parameters, which come from the four components of each 4-momentum. However, we have some conditions that restrict the number of free parameters that are due to the kinematics of binary reactions. According to the special theory of relativity, we have the conditions $p_i^2 = m_i^2$, where $i = a, b, c, d$, which leads to four conditions. In addition, we have Poincaré invariance, which consists of four conditions from translation invariance (i.e. conservation of 4-momentum $p_a^\mu + p_b^\mu = p_c^\mu + p_d^\mu$) and six conditions from Lorentz invariance [i.e. three conditions from rotations (or conservation of angular momentum) and three conditions from boosts]. In total, we have 14 conditions restricting the 16 free parameters, and hence, only two independent variables. Let us define the Mandelstam variables as

$$s = (p_a + p_b)^2, \tag{11.142}$$
$$t = (p_a - p_c)^2, \tag{11.143}$$
$$u = (p_a - p_d)^2, \tag{11.144}$$

which fulfil the relation

$$s + t + u = m_a^2 + m_b^2 + m_c^2 + m_d^2. \tag{11.145}$$

Therefore, only two out of the three Mandelstam variables are linearly independent from each other. Thus, we can choose to work with s and t, s and u, or t and u. The variable s is usually called the *centre-of-mass energy*, whereas the variable t is the *momentum transfer*. No special name exists for the variable u.

---

[7] In this book, the left-hand side of a Feynman diagram represents the 'past', whereas the right-hand side of a Feynman diagram represents the 'future'.

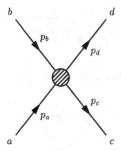

Figure 11.2 Kinematics for the binary reaction $a + b \rightarrow c + d$.

Next, we will discuss two simple inertial frames for binary reactions, which are the *laboratory (lab) system* and the *centre-of-mass (CM) system*. The lab system is defined as particle b being at rest, which implies that $p_b = (m_b, \mathbf{0})$. Inserting this into the definition of the Mandelstam variable s, we obtain

$$s = m_a^2 + m_b^2 + 2p_a \cdot p_b = m_a^2 + m_b^2 + 2E_a^{\text{lab}} m_b, \tag{11.146}$$

where E_a^{lab} is the energy of particle a. Solving for this energy, we find that

$$E_a^{\text{lab}} = \frac{s - m_a^2 - m_b^2}{2m_b}. \tag{11.147}$$

The CM system is defined as the centre-of-mass being at rest before and after the reaction, i.e. $\mathbf{p}_a^{\text{CM}} + \mathbf{p}_b^{\text{CM}} = \mathbf{p}_c^{\text{CM}} + \mathbf{p}_d^{\text{CM}} = \mathbf{0}$, which means that $\mathbf{p}^{\text{CM}} \equiv \mathbf{p}_a^{\text{CM}} = -\mathbf{p}_b^{\text{CM}}$. Again, inserting this into the definition of the Mandelstam variable s, we obtain

$$s = (p_a + p_b)^2 = \left(E_a^{\text{CM}} + E_b^{\text{CM}} \right)^2 - \left(\mathbf{p}_a^{\text{CM}} + \mathbf{p}_b^{\text{CM}} \right)^2 = \left(E_a^{\text{CM}} + E_b^{\text{CM}} \right)^2$$

$$= \left(\sqrt{m_a^2 + \mathbf{p}^{\text{CM}2}} + \sqrt{m_b^2 + \mathbf{p}^{\text{CM}2}} \right)^2. \tag{11.148}$$

Note that since the Mandelstam variable s is defined as an inner product, i.e. the inner product of the 4-momentum $p_a + p_b$ with itself, it is a *Lorentz invariant*, which means that it has the same value in **all** inertial frames. Especially, the values of $s \equiv W^2$ given in Eqs. (11.146) and (11.148) are the same. In addition, one can show that

$$|\mathbf{p}^{\text{CM}}| = |\mathbf{p}^{\text{lab}}| \frac{m_b}{\sqrt{s}}, \tag{11.149}$$

since $m_a^2 + m_b^2 + 2E_a^{\text{lab}} m_b = m_a^2 + \mathbf{p}^{\text{CM}2} + m_b^2 + \mathbf{p}^{\text{CM}2} + 2\sqrt{m_a^2 + \mathbf{p}^{\text{CM}2}} \sqrt{m_b^2 + \mathbf{p}^{\text{CM}2}}$. Equation (11.149) can also be written as

$$|\mathbf{p}^{\text{CM}}| = \sqrt{E_a^{\text{lab}^2} - m_a^2} \sqrt{\frac{m_b^2}{s}} = \frac{1}{2\sqrt{s}} \sqrt{s^2 + m_a^4 + m_b^4 - 2m_a^2 s - 2m_b^2 s - 2m_a^2 m_b^2}.$$

$$(11.150)$$

Furthermore, introducing the function

$$\lambda(x, y, z) \equiv x^2 + y^2 + z^2 - 2xy - 2xz - 2yz, \qquad (11.151)$$

which is called *Källén's quadratic form*, we obtain

$$|\mathbf{p}^{\text{CM}}| = \frac{1}{2\sqrt{s}} \sqrt{\lambda\left(s, m_a^2, m_b^2\right)}. \qquad (11.152)$$

In the case that a Feynman diagram of a binary reaction only contains one virtual particle,[8] it is conventional to describe that particle as being in a specific '*channel*'. There are three different channels named after the three Mandelstam variables s, t, and u. The different channels and their corresponding Feynman diagrams as well as their characteristic dependences on the Mandelstam variables of the transition amplitude $\mathcal{M}$ are the following:

s-channel: $\mathcal{M} \propto \dfrac{1}{s - m_\phi^2}, \quad s = (p + p')^2 = (k + k')^2,$

$$(11.153)$$

t-channel: $\mathcal{M} \propto \dfrac{1}{t - m_\phi^2}, \quad t = (p - k)^2 = (p' - k')^2,$

$$(11.154)$$

u-channel: $\mathcal{M} \propto \dfrac{1}{u - m_\phi^2}, \quad u = (p - k')^2 = (p' - k)^2,$

$$(11.155)$$

---

[8] A *virtual particle* is described by an internal line in a Feynman diagram (i.e. not an external line). This means that virtual particles as opposed to real particles are not directly detectable.

where ϕ is the virtual particle and m_ϕ is its mass. In general, a single reaction will obtain contributions from more than one channel and these contributions must be added coherently, i.e. on amplitude level.

11.9 The S matrix, the T matrix, cross-sections, and decay rates

In this section, we will discuss scattering in terms of transition probabilities. Consider the reaction $\alpha \rightarrow \beta$, where α is the initial state and β is the final state, respectively. According to Eqs. (5.48) and (5.49), the transition amplitude for this reaction is given by

$$W_{\alpha\beta} = \langle \beta|S|\alpha \rangle, \tag{11.156}$$

where the S *matrix* is constructed as the matrix elements of the S operator and is defined as[9]

$$S_{\alpha\beta} \equiv \langle \beta; \text{out}|\alpha; \text{in} \rangle = \langle \beta; \text{in}|S|\alpha; \text{in} \rangle = \langle \beta; \text{out}|S|\alpha; \text{out} \rangle \equiv \langle \beta|S|\alpha \rangle. \tag{11.157}$$

Thus, using translational invariance as described in Eq. (5.61), we have

$$\langle \beta|S|\alpha \rangle (p_\alpha - p_\beta) = 0, \tag{11.158}$$

where p_α and p_β are the 4-momenta of the initial and final states, respectively. Equation (11.158) has the solution

$$\langle \beta|S|\alpha \rangle = (2\pi)^4 \delta(p_\alpha - p_\beta)\langle \beta|\hat{S}|\alpha \rangle, \tag{11.159}$$

where the factor $(2\pi)^4$ is a usual convention. Next, in the case that there are no interactions, the S operator is equal to the unit operator, i.e. $S = \mathbb{1}$, which stands for 'nothing happens', whereas if interactions are present, then we can write the S operator as

$$S = \mathbb{1} + iT, \tag{11.160}$$

where T is the T *operator*, which describes the 'reaction'. This T operator should not be confused with time-ordering. In addition, we define the T *matrix* as

$$\langle \beta|T|\alpha \rangle = (2\pi)^4 \delta(p_\alpha - p_\beta)\langle \beta|\hat{T}|\alpha \rangle, \tag{11.161}$$

where $\langle \beta|\hat{T}|\alpha \rangle$ is the matrix element of the $\hat{T}$ operator, or the *scattering amplitude*, which describes the interesting physics. Then, the *transition probability* for the reaction $\alpha \rightarrow \beta$ to occur is given by the modulus squared of the transition amplitude

$$P(\alpha \rightarrow \beta) = |W_{\alpha\beta}|^2 = |\langle \beta|S|\alpha \rangle|^2. \tag{11.162}$$

[9] Note that Eqs. (11.156) and (11.157) describe the same quantity, and therefore, the notation $W_{\alpha\beta}$ is obsolete. We will use the notation $S_{\alpha\beta}$ throughout the rest of this book.

Actually, considering the interesting physics only, the transition probability is given by

$$P(\alpha \to \beta) = |\langle \beta | T | \alpha \rangle|^2 = |(2\pi)^4 \delta(p_\alpha - p_\beta)|^2 |\langle \beta | \hat{T} | \alpha \rangle|^2. \qquad (11.163)$$

Hence, we have to find a way to calculate the 'square' of the δ-function. Let $(2\pi)^4 \delta(p) = \int e^{ip \cdot x} d^4x$, where p is an arbitrary 4-momentum vector. Then, we can write

$$|(2\pi)^4 \delta(p)|^2 = (2\pi)^4 \delta(p) \int e^{ip \cdot x} d^4x. \qquad (11.164)$$

Next, using $\delta(x) f(x) = \delta(x) f(0)$, where $f(x)$ is an arbitrary function, we have

$$|(2\pi)^4 \delta(p)|^2 = (2\pi)^4 \delta(p) \int d^4x = \lim_{V,T \to \infty} (2\pi)^4 \delta(p) VT, \qquad (11.165)$$

where VT is the spacetime volume. Note that this way of 'squaring' the δ-function is only a formal calculation and not mathematically rigorous. Anyway, we thus obtain

$$P(\alpha \to \beta) = \lim_{V,T \to \infty} VT (2\pi)^4 \delta(p_\alpha - p_\beta) |\langle \beta | \hat{T} | \alpha \rangle|^2. \qquad (11.166)$$

The *cross-section* for a reaction between colliding beams (or bundles) of particles of type a and type b is defined as

$$\sigma \equiv \frac{\text{number of scattering events}}{\rho_a \ell_a \rho_b \ell_b \Phi}, \qquad (11.167)$$

where ρ_a and ρ_b are the number densities of particles of type a and b, respectively, ℓ_a and ℓ_b are the lengths of the bundles of particles of type a and b, respectively, and finally, Φ is the cross-sectional area common to the two bundles of particles. The dimension of a cross-section is $[\sigma] = L^2$, where L is the length dimension, and the cross-section σ is symmetric with respect to the particle bundles a and b. Conventionally, cross-sections are measured in the unit *barn* (b), where $1 \text{ b} = 10^{-28} \text{ m}^2$. Now, the number of scattering events N is given by

$$N = \sigma \ell_a \ell_b \int \rho_a(x) \rho_b(x) d^2x. \qquad (11.168)$$

In the case that the densities of particles a and b are constant, i.e. $\rho_a = \text{const.}$ and $\rho_b = \text{const.}$, we obtain

$$N = \frac{\sigma N_a N_b}{\Phi}. \qquad (11.169)$$

Actually, using a formal argumentation, the cross-section σ is the effective area under which particle bundle a with energy E_a sees particle bundle b with energy E_b at scattering. If V is the reaction volume during the time T, then $2E_b V$ particles of type b have the effective cross-section $2E_b V\sigma$. In addition, if the relative velocity between particle bundle a and particle bundle b is v_{ab}, i.e., $v_{ab} = |v_a - v_b|$, then all particles in a cylinder with volume $2E_b\sigma V T v_{ab}$ will strike the particles b. The number of such particles a is the number of collisions, which is $2E_b\sigma V T v_{ab} 2E_a$. However, this number is also given by Eq. (11.166), which leads to the equation

$$VT(4E_a E_b v_{ab}\sigma)$$

$$= VT\left[\prod_{f=c,d,\dots}\int\frac{d^3 p_f}{(2\pi)^3}\frac{1}{2E_f}\right](2\pi)^4\delta\left(p_a + p_b - \sum_{f=c,d,\dots}p_f\right)\overline{|\langle\beta|\hat{T}|\alpha\rangle|^2},$$

$$(11.170)$$

where the 'averaged' modulus squared of the scattering amplitude is given by

$$\overline{|\langle\beta|\hat{T}|\alpha\rangle|^2} = \frac{1}{2s_a + 1}\frac{1}{2s_b + 1}\sum_{m_{s_a}}\sum_{m_{s_b}}\sum_{m_{s_c}, m_{s_d}, \dots}$$

$$\times\left|\langle c, s_c, m_{s_c}; d, s_d, m_{s_d}; \dots|\hat{T}|a, s_a, m_{s_a}; b, s_b, m_{s_b}\rangle\right|^2, \quad (11.171)$$

i.e. one has averaged over the initial states and summed over the final states. Thus, we obtain the cross-section

$$\sigma_{a+b\to c+d+\cdots} = \frac{1}{4E_a E_b v_{ab}}\left[\prod_{f=c,d,\dots}\int\frac{d^3 p_f}{(2\pi)^3}\frac{1}{2E_f}\right]$$

$$\times(2\pi)^4\delta\left(p_a + p_b - \sum_{f=c,d,\dots}p_f\right)\overline{|\langle\beta|\hat{T}|\alpha\rangle|^2}. \quad (11.172)$$

Finally, the *total cross-section* $\sigma_{a+b}(s)$ evaluated at the energy $\sqrt{s} = W$ of the incoming particles is given by

$$\sigma_{a+b}(s) = \sum_{n=2}^{\infty}\sigma_{a+b\to(n)}(s), \quad (11.173)$$

where it is assumed that n particles are produced in the final state β. Note that the sum in Eq. (11.173) is performed for all $n \geq 2$ and all types of particles that are possible, i.e. all open channels.

Similarly, the concept of *decay rate* can be defined. It is given by

$$\Gamma \equiv \frac{\text{number of decays per unit time}}{\text{number of } a \text{ particles}}. \quad (11.174)$$

In order to calculate the decay rate of particles Γ, which is the transition probability per unit time per particle, let the initial state α be a particle state a, i.e. $|\alpha\rangle = |a\rangle$, which is assumed to be in its rest frame, and the final state β be a many-particle state $|\beta\rangle = |b, c, \ldots\rangle$. Now, the phase-space element in the volume element d^3x is $\mathrm{d}^3x\mathrm{d}^3p$. Each $\mathrm{d}x\mathrm{d}p$ has the volume h in phase-space, according to Heisenberg's uncertainty principle. The number of particles in d^3x is $2E\mathrm{d}^3x$. Thus, the density of states is

$$\frac{\mathrm{d}^3x\mathrm{d}^3p}{h^3 2E\mathrm{d}^3x} = \frac{\mathrm{d}^3p}{(2\pi)^3}\frac{1}{2E}\frac{1}{\hbar^3} = \{\hbar = 1\} = \frac{\mathrm{d}^3p}{2E(2\pi)^3}. \tag{11.175}$$

Therefore, we obtain the decay rate

$$\Gamma_{a\to b+c+\ldots} = \lim_{V,T\to\infty} \frac{VT}{2m_a VT} \left[\prod_{f=b,c,\ldots} \int \frac{\mathrm{d}^3p_f}{(2\pi)^3}\frac{1}{2E_f} \right]$$

$$\times (2\pi)^4\delta\left(p_a - \sum_{f=b,c,\ldots} p_f\right) |\langle b, c, \ldots |\hat{T}|a\rangle|^2$$

$$= \frac{1}{2m_a} \left[\prod_{f=b,c,\ldots} \int \frac{\mathrm{d}^3p_f}{(2\pi)^3}\frac{1}{2E_f} \right] (2\pi)^4\delta\left(p_a - \sum_{f=b,c,\ldots} p_f\right)$$

$$\times |\langle b, c, \ldots |\hat{T}|a\rangle|^2, \tag{11.176}$$

where the number of particles a in the spacetime volume VT is $2m_a VT$. Note that if the particles have spin $s_a, s_b, s_c, \ldots$, whose polarizations are not observed, then one has to sum over final states and average of the initial state, i.e. one has to make the replacement

$$|\langle b, c, \ldots |\hat{T}|a\rangle|^2 \to \overline{|\langle b, c, \ldots |\hat{T}|a\rangle|^2}$$

$$= \frac{1}{2s_a + 1} \sum_{m_{s_a}} \sum_{m_{s_b}, m_{s_c}, \ldots} |\langle b, s_b, m_{s_b}; c, s_c, m_{s_c}; \ldots |\hat{T}|a, s_a, m_{s_a}\rangle|^2. \tag{11.177}$$

Thus, we have the decay rate

$$\Gamma_{a\to b+c+\ldots} = \frac{1}{2m_a} \left[\prod_{f=b,c,\ldots} \int \frac{\mathrm{d}^3p_f}{(2\pi)^3}\frac{1}{2E_f} \right] (2\pi)^4\delta\left(p_a - \sum_{f=b,c,\ldots} p_f\right)$$

$$\times \overline{|\langle b, c, \ldots |\hat{T}|a\rangle|^2}. \tag{11.178}$$

In addition, we define the *lifetime* as

$$\tau = \frac{1}{\Gamma}. \tag{11.179}$$

In the case of *binary reactions*, which are reactions of the form $a + b \rightarrow c + d + \cdots$, i.e. two states as in-states and an arbitrary number of states as out-states, the probability amplitude is

$$\langle p_c, p_d, \ldots; \text{out}| p_a, p_b; \text{in} \rangle = \langle p_c, p_d, \ldots |S| p_a, p_b \rangle. \tag{11.180}$$

In addition, we define the *invariant matrix element* $\mathcal{M}$ (which is, in principle, the T matrix) for binary reactions as

$$\langle p_c, p_d, \ldots |iT| p_a, p_b \rangle = (2\pi)^4 \delta \left(p_a + p_b - \sum_{f=c,d,\ldots} p_f \right) i \mathcal{M}(a+b \rightarrow c+d+\cdots). \tag{11.181}$$

Thus, one obtains the cross-section for binary reactions

$$d\sigma = \frac{1}{2E_a 2E_b v_{ab}} \left[\prod_{f=c,d,\ldots} \frac{d^3 p_f}{(2\pi)^3} \frac{1}{2E_f} \right] |\mathcal{M}(a + b \rightarrow c + d + \cdots)|^2$$

$$\times (2\pi)^4 \delta \left(p_a + p_b - \sum_{f=c,d,\ldots} p_f \right). \tag{11.182}$$

Similarly, the formula for the decay rate for $a \rightarrow b + c + \cdots$ is given by

$$d\Gamma = \frac{1}{2m_a} \left[\prod_{f=b,c,\ldots} \frac{d^3 p_f}{(2\pi)^3} \frac{1}{2E_f} \right] |\mathcal{M}(a \rightarrow b+c+\cdots)|^2$$

$$\times (2\pi)^4 \delta \left(p_a - \sum_{f=b,c,\ldots} p_f \right). \tag{11.183}$$

Now, in the special case that we have only two particles in the final state, i.e. for $a + b \rightarrow c + d$, we can write the total cross-section in terms of the *differential cross-section* as

$$\sigma_{a+b \rightarrow c+d}(s) = \int d\sigma(s) = \int \frac{d\sigma}{d\Omega} d\Omega, \tag{11.184}$$

where $d\Omega$ is the solid angle element. For example, in the CM system, we have

$$d\Omega_{\text{CM}} = -d\phi \, d\cos\theta, \tag{11.185}$$

since $d\Omega_{\text{CM}} = -d\phi \, (-\sin\theta) \, d\theta = \sin\theta \, d\phi \, d\theta$ and $\int_\Omega d\Omega_{\text{CM}} = \int_0^{2\pi} d\phi \int_0^\pi \sin\theta \, d\theta = 2\pi \cdot 2 = 4\pi$. Actually, in the CM system, the Mandelstam variable t can be expressed in the scattering angle θ, since

$$t = (p_a - p_c)^2 = m_a^2 + m_c^2 - 2p_a \cdot p_c = m_a^2 + m_b^2$$
$$- 2\left(E_a^{CM} E_c^{CM} - \mathbf{p}_a^{CM} \cdot \mathbf{p}_c^{CM}\right). \tag{11.186}$$

Inserting Eq. (11.152) as well as $\left|\mathbf{p}_c^{CM}\right| = \sqrt{\lambda\left(s, m_c^2, m_d^2\right)}/\left(2\sqrt{s}\right)$, since $s = (p_c + p_d)^2$, into the scalar product $\mathbf{p}_a^{CM} \cdot \mathbf{p}_c^{CM} = \left|\mathbf{p}_a^{CM}\right|\left|\mathbf{p}_c^{CM}\right|\cos\theta$, we can solve for θ and we obtain

$$\cos\theta = \frac{2s\left(t - m_a^2 - m_c^2 + 2E_a^{CM} E_c^{CM}\right)}{\sqrt{\lambda\left(s, m_a^2, m_b^2\right)\lambda\left(s, m_c^2, m_d^2\right)}}. \tag{11.187}$$

Thus, we have

$$t = \frac{\sqrt{\lambda\left(s, m_a^2, m_b^2\right)\lambda\left(s, m_c^2, m_d^2\right)}}{2s}\cos\theta + m_a^2 + m_c^2 - 2E_a^{CM} E_c^{CM}. \tag{11.188}$$

Next, using Eq. (11.188) and $d\Omega_{CM} = -2\pi\,d\cos\theta$ [since Eq. (11.188) is independent of the azimuthal angle ϕ], we find that

$$\left(\frac{d\sigma}{d\Omega}\right)_{CM} = \frac{d\sigma}{dt}\left(\frac{dt}{d\Omega}\right)_{CM} = -\frac{1}{4\pi s}\sqrt{\lambda\left(s, m_a^2, m_b^2\right)\lambda\left(s, m_c^2, m_d^2\right)}\frac{d\sigma}{dt}. \tag{11.189}$$

In order to calculate $d\sigma/dt$, we use *Källén's trick*. The phase-space integration has six variables, but four δ-functions, which means that effectively only two integrations are left. One of these integrations is the scattering angle θ or the Mandelstam variable t, whereas the other one is the azimuthal angle ϕ, which has been integrated out, because of cylindrical symmetry. Hence, if one inserts one extra δ-function such that $\delta(t - (p_a - p_c)^2)$ in the integrand, then all integrations can be performed. Thus, we obtain

$$\frac{d\sigma}{dt} = \frac{1}{2\sqrt{\lambda\left(s, m_a^2, m_b^2\right)}}\overline{|\langle p_c, p_d|\hat{T}|p_a, p_b\rangle|^2}J\left(s, m_a^2, m_b^2\right), \tag{11.190}$$

where $\overline{|\langle p_c, p_d|\hat{T}|p_a, p_b\rangle|^2}$ does not depend on the angle ϕ and

$$J = J\left(s, m_a^2, m_b^2\right)$$
$$= \int \frac{d^3 p_c}{2E_c(2\pi)^3}\frac{d^3 p_d}{2E_d(2\pi)^3}\delta(t - (p_a - p_c)^2)(2\pi)^4\delta(p_a + p_b - p_c - p_d). \tag{11.191}$$

Then, calculate the integral J in the CM system, which means that $\mathbf{p}_a + \mathbf{p}_b = 0$, and define $W = E_c + E_d$ and $p = \left|\mathbf{p}_c^{CM}\right| = \left|\mathbf{p}_d^{CM}\right|$. This implies that

$$\delta(p_a + p_b - p_c - p_d) \equiv \delta(E_a + E_b - W)\delta(\mathbf{p}_a + \mathbf{p}_b - \mathbf{p}_c - \mathbf{p}_d)$$
$$= \delta(W - E_a - E_b)\delta(\mathbf{p}_c + \mathbf{p}_d). \tag{11.192}$$

Thus, we obtain

$$
\begin{aligned}
J &= \frac{1}{(2\pi)^2} \int \frac{d^3 p_c}{2E_c(\mathbf{p}_c)} \frac{d^3 p_d}{2E_d(\mathbf{p}_d)} \delta\left(W(\mathbf{p}_c, \mathbf{p}_d) - \sqrt{s}\right) \delta(\mathbf{p}_c + \mathbf{p}_d) \delta(t - (p_a - p_c)^2) \\
&= \frac{1}{(2\pi)^2} \int \frac{d^3 p_c}{4 F_c F_d} \delta\left(W(\mathbf{p}_c) - \sqrt{s}\right) \delta(t - (p_a - p_c)^2).
\end{aligned}
\tag{11.193}
$$

However, $d^3 p_c = -p^2\, dp d\phi d\cos\theta$ and $(p_a - p_c)^2 = m_a^2 + m_c^2 - 2E_a E_c + 2|\mathbf{p}_a||\mathbf{p}_c|\cos\theta$ as well as using $\int \delta(a - bx)\, dx = 1/|b|$, implies that

$$
J = -\frac{1}{2\pi} \int \frac{p\, dp}{4 E_c E_d} \delta(W(p) - \sqrt{s}) \frac{1}{2|\mathbf{p}_a|}.
\tag{11.194}
$$

In addition, using Eq. (11.152), we have that $2|\mathbf{p}_a| = \sqrt{\lambda\left(s, m_a^2, m_b^2\right)}/\sqrt{s}$, and performing the variable substitution

$$
dW = dp \frac{dW}{dp} = dp \left(\frac{p}{E_c} + \frac{p}{E_d}\right) = p\, dp \frac{W}{E_c E_d},
\tag{11.195}
$$

we finally find that

$$
J = -\frac{1}{2\pi} \int \frac{dW}{4W} \delta\left(W - \sqrt{s}\right) \frac{\sqrt{s}}{\lambda\left(s, m_a^2, m_b^2\right)} = -\frac{1}{8\pi} \frac{1}{\sqrt{\lambda\left(s, m_a^2, m_b^2\right)}}.
\tag{11.196}
$$

Thus, inserting the integral J into Eq. (11.190) and the result of this into Eq. (11.189), we obtain

$$
\frac{d\sigma}{dt} = -\frac{1}{16\pi \lambda\left(s, m_a^2, m_b^2\right)} |\langle p_c, p_d | \hat{T} | p_a, p_b \rangle|^2,
\tag{11.197}
$$

$$
\left(\frac{d\sigma}{d\Omega}\right)_{\text{CM}} = \frac{1}{64\pi^2 s} \sqrt{\frac{\lambda\left(s, m_c^2, m_d^2\right)}{\lambda\left(s, m_a^2, m_b^2\right)}} |\langle p_c, p_d | \hat{T} | p_a, p_b \rangle|^2.
\tag{11.198}
$$

For elastic scattering, i.e. $a = c$ and $b = d$, we have the differential cross-section in the CM system

$$
\left(\frac{d\sigma}{d\Omega}\right)_{\text{CM}} = \frac{1}{64\pi^2 s} \frac{1}{(2s_a + 1)(2s_b + 1)} \sum_{\text{spin}} |\langle p_c, p_d | \hat{T} | p_a, p_b \rangle|^2
\tag{11.199}
$$

or expressed in the invariant matrix element for the unpolarized case

$$
\left(\frac{d\sigma}{d\Omega}\right)_{\text{CM}} = \frac{1}{2E_a 2E_b v_{ab}} \frac{|\mathbf{p}_c|}{(2\pi)^2 4 E_{\text{CM}}} |\mathcal{M}(a + b \rightarrow c + d)|^2,
\tag{11.200}
$$

where $E_{CM} = E_a + E_b$ is the CM energy. In addition, if the masses of the different particles are equal, i.e. $m_a = m_b = m_c = m_d$, then we find that

$$\left(\frac{d\sigma}{d\Omega}\right)_{CM} = \frac{|\mathcal{M}(a+b \to c+d)|^2}{64\pi^2 E_{CM}^2}. \tag{11.201}$$

For example, in $\lambda\phi^4$-theory, we have using the Feynman rules

$$= (4!)\left(-\frac{i\lambda}{4!}\right)\int e^{-i(p_a+p_b-p_c-p_d)\cdot x}\, d^4x$$

$$= -i\lambda(2\pi)^4\delta(p_a + p_b - p_c - p_d)$$

$$= i\mathcal{M}(a+b \to c+d)(2\pi)^4\delta(p_a + p_b - p_c - p_d), \tag{11.202}$$

which means that the invariant matrix element is

$$\mathcal{M}(a+b \to c+d) = -\lambda. \tag{11.203}$$

Inserting the result of Eq. (11.203) into Eq. (11.201), we obtain the differential cross-section

$$\left(\frac{d\sigma}{d\Omega}\right)_{CM} = \frac{\lambda^2}{64\pi^2 E_{CM}^2}. \tag{11.204}$$

Finally, integrating Eq. (11.204), implies that the total cross-section for the reaction $a+b \to c+d$ in $\lambda\phi^4$-theory is

$$\sigma = \frac{1}{2}\int\left(\frac{d\sigma}{d\Omega}\right)_{CM} d\Omega = \frac{1}{2}\cdot 4\pi \cdot \frac{\lambda^2}{64\pi^2 E_{CM}^2} = \frac{\lambda^2}{32\pi E_{CM}^2}, \tag{11.205}$$

where the factor of $1/2$ is due to two identical particles in the final state and compensates for the double counting in the angular integral.

Problems

(1) Using Wick's theorem, derive the Feynman rules for $\lambda\phi^3$-theory. In addition, discuss the similarities and differences compared with $\lambda\phi^4$-theory.
(2) For a free Dirac field ψ, compute the following correlation functions
   (a) $\langle 0|T[\psi(x)\psi(y)\psi(z)]|0\rangle$,
   (b) $\langle 0|T[\psi(x)\psi(y)\psi(z)\psi(w)]|0\rangle$,
   where $|0\rangle$ is the ground state (vacuum) of the free Dirac theory.
(3) $\lambda\phi^4$-*theory*. For a Hermitian scalar field with Lagrangian density

$$\mathcal{L} = :\frac{1}{2}\partial_\mu\phi\partial^\mu\phi - \frac{1}{2}m^2\phi^2 - \frac{1}{4!}\lambda\phi^4:,$$

compute the S matrix element for the transition $|\mathbf{p}\rangle \to |\mathbf{q}, \mathbf{r}, \mathbf{s}\rangle$ to first order in the coupling constant λ.

(4) *Decay of a scalar particle.* Consider the following Lagrangian, involving two real scalar fields Φ and ϕ:

$$\mathcal{L} = \frac{1}{2}\partial_\mu \Phi \partial^\mu \Phi - \frac{1}{2}M^2\Phi^2 + \frac{1}{2}\partial_\mu \phi \partial^\mu \phi - \frac{1}{2}m^2\phi^2 - \mu\Phi\phi^2.$$

The last term is an interaction that allows a Φ particle to decay into two ϕ particles, provided that $M > 2m$. Assuming that this condition is fulfilled, calculate the lifetime of the Φ particle to lowest order in μ.

(5) *Pseudoscalar Yukawa theory.* The neutral pion, π^0, usually decays into two photons, but it can also decay into an electron–positron pair. The effective Hamiltonian density for the latter decay is

$$\mathcal{H}_I(x) = ig\bar{\psi}(x)\gamma^5\psi(x)\phi(x),$$

where ϕ is the (real) π^0 field and ψ is the electron field.

(a) Show that $\mathcal{H}_I$ is Hermitian.
(b) Calculate the decay rate for the decay $\pi^0 \to e^- + e^+$ and estimate the coupling constant g using the latest experimental data (see e.g. pdg.lbl.gov).

(6) *Rutherford scattering.* The cross-section for scattering of an electron by the Coulomb field of a nucleus can be computed, to lowest order, without quantizing the electromagnetic field. Instead, treat the field as a given, classical potential $A^\mu(x)$. The interaction Hamiltonian is

$$H_I(t) = \int d^3x \, e\bar{\psi}(x)\gamma^\mu\psi(x)A_\mu(x),$$

where $\psi(x)$ is the usual quantized Dirac field.

(a) Show that the T matrix element for electron scattering off a localized classical potential is, to lowest order,

$$\langle p'|iT|p\rangle = -ie\bar{u}(p')\gamma^\mu u(p)\tilde{A}_\mu(p' - p),$$

where $\tilde{A}^\mu(q)$ is the four-dimensional Fourier transform of $A^\mu(x)$.
(b) If $A^\mu(x)$ is time-independent, then its Fourier transform contains a δ-function of energy. It is then natural to define

$$\langle p'|iT|p\rangle \equiv i\mathcal{M}(2\pi)\delta(E_f - E_i),$$

where E_i and E_f are the initial and final energies of the particle, respectively, and to adopt a new Feynman rule for computing $\mathcal{M}$:

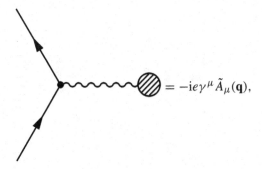 $= -ie\gamma^\mu\tilde{A}_\mu(\mathbf{q}),$

where $\tilde{A}^\mu(\mathbf{q})$ is the three-dimensional Fourier transform of $A^\mu(x)$. Given this definition of $\mathcal{M}$, show that the cross-section for scattering off a time-independent, localized potential is

$$d\sigma = \frac{1}{v_i} \frac{1}{2E_i} \frac{d^3 p_f}{(2\pi)^3} \frac{1}{2E_f} |\mathcal{M}(i \to f)|^2 2\pi \delta(E_f - E_i),$$

where v_i is the particle's initial velocity. Integrate over $|p_f|$ to find a simple expression for $d\sigma/d\Omega$.

(c) Specialize to the case of electron scattering from a Coulomb potential $[A^0 = Ze/(4\pi r)]$. In the non-relativistic limit, derive the Rutherford formula,

$$\frac{d\sigma}{d\Omega} = \frac{\alpha^2 Z^2}{4m^2 v^4 \sin^4(\theta/2)}.$$

Guide to additional recommended reading

The following books (see the indicated pages) and their authors have similar treatments of the content in the present chapter.

- A. Z. Capri, *Relativistic Quantum Mechanics and Introduction to Quantum Field Theory*, World Scientific (2002), pp. 153–176.
- T.-P. Cheng and L.-F. Li, *Gauge Theory of Elementary Particle Physics*, Oxford (1996), pp. 3–11.
- F. Gross, *Relativistic Quantum Mechanics and Field Theory*, Wiley (1993), pp. 57–87, 238–286.
- M. Kaku, *Quantum Field Theory – A Modern Introduction*, Oxford (1993), pp. 147–159.
- M. E. Peskin and D. V. Schroeder, *An Introduction to Quantum Field Theory*, Addison-Wesley (1995), pp. 77–130.
- F. Schwabl, *Advanced Quantum Mechanics*, Springer (1999), pp. 321–323, 323–339, 339–346.
- H. Snellman, *Elementary Particle Physics*, KTH (2004), pp. 34–38, 75–88.
- F. J. Ynduráin, *Relativistic Quantum Mechanics and Introduction to Field Theory*, Springer (1996), pp. 222–223.
- For the interested reader: G. C. Wick, The evaluation of the collision matrix, *Phys. Rev.* **80**, 268–272 (1950).

12

Elementary processes of quantum electrodynamics

In this chapter, we will investigate six of the most important elementary processes in QED. The reactions are the following.

(1) $e^+ + e^- \rightarrow \mu^+ + \mu^-$
(2) $e^- + \mu^- \rightarrow e^- + \mu^-$ (electron–muon scattering, related by crossing symmetry to reaction (1))
(3) $e^+ + e^- \rightarrow e^+ + e^-$ (Bhabha scattering)
(4) $e^- + e^- \rightarrow e^- + e^-$ (Møller scattering, related by crossing symmetry to reaction (3))
(5) $e^- + \gamma \rightarrow e^- + \gamma$ (Compton scattering)
(6) $e^+ + e^- \rightarrow 2\gamma$ (related by crossing symmetry to reaction (5))

In general, the scheme for the calculation of (unpolarized) cross-sections for QED processes is the following.

- Draw all possible Feynman diagrams for the given process to a specific order in perturbation theory.
- Use the Feynman rules of QED to write down the contributions to the invariant matrix element $\mathcal{M}$.
- Modulus square the invariant matrix element $\mathcal{M}$ and average or sum over spins, using completeness relations, to obtain the (unpolarized) amplitude $\overline{|\mathcal{M}|^2}$.
- Evaluate traces using trace formulas, collect terms, and simplify the expression for $\overline{|\mathcal{M}|^2}$ as much as possible.
- Choose a particular inertial frame (e.g. the CM system or the lab system) and draw a picture of the kinematics for the process in that system.
- Express all 4-momenta in terms of a suitably chosen set of variables.
- Insert the expression for $\overline{|\mathcal{M}|^2}$ into the cross-section formula and integrate over phase-space (i.e. 4-momentum) variables that are undetermined in order to obtain the differential cross-section.

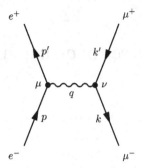

Figure 12.1 Feynman diagram for the reaction $e^+ + e^- \to \mu^+ + \mu^-$.

- Finally, integrate the differential cross-section to obtain the total (unpolarized) cross-section for the process.

Note that processes involving loop diagrams may require additional steps, but such processes will not be considered here.

12.1 $e^+ + e^- \to \mu^+ + \mu^-$

Let us start the investigation with the reaction $e^+ + e^- \to \mu^+ + \mu^-$, which is the simplest of all QED processes, but also one of the most important processes in high-energy physics.[1] In fact, this process is used to calibrate e^+e^--colliders and it is important for the study of all reactions produced in such colliders. Using the Feynman rules of QED, we can draw the Feynman diagram for the reaction $e^+ + e^- \to \mu^+ + \mu^-$ at tree level (an s-channel diagram), i.e. to leading order in perturbation theory, which is presented in Fig. 12.1. Furthermore, we can immediately write down the invariant matrix element for this reaction, which is given by

$$i\mathcal{M}(s, s' \to r, r') = \bar{v}^{s'}(p')\,(-ie\gamma^\mu)\,u^s(p)\left(\frac{-ig_{\mu\nu}}{q^2}\right)\bar{u}^r(k)\,(-ie\gamma^\nu)\,v^{r'}(k')$$

$$= \frac{ie^2}{q^2}\bar{v}^{s'}(p')\gamma^\mu u^s(p)\bar{u}^r(k)\gamma_\mu v^{r'}(k'), \qquad (12.1)$$

where p (s) and p' (s') are the 4-momenta (spins) of the incoming electron and positron, respectively, whereas k (r) and k' (r') are the 4-momenta (spins) of the outgoing muon and antimuon, respectively. In addition, the 4-momentum transfer (i.e. the 4-momentum of the photon) is $q = p + p' = k + k'$. 'Squaring' this matrix element gives

---

[1] Note that μ^- is the muon, whereas μ^+ is its antiparticle, i.e. the antimuon.

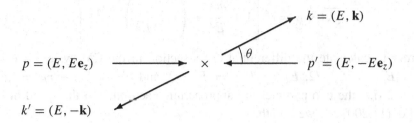

Figure 12.2 The kinematics in the CM system for the reaction $e^+ + e^- \rightarrow \mu^+ + \mu^-$.

$$|\mathcal{M}(s, s' \rightarrow r, r')|^2$$
$$= \frac{e^4}{q^4} \left[\bar{v}^{s'}(p')\gamma^\mu u(p)^s \bar{u}^s(p)\gamma^\nu v^{s'}(p') \right] \left[\bar{u}^r(k)\gamma_\mu v^{r'}(k')\bar{v}^{r'}(k')\gamma_\nu u^r(k) \right].$$
(12.2)

In order to obtain the unpolarized amplitude, we throw away spin information, which leads to a simpler expression. This is performed by the average

$$\overline{|\mathcal{M}|^2} = \frac{1}{4}\sum_{\text{spin}} |\mathcal{M}|^2 = \frac{1}{2}\sum_s \frac{1}{2}\sum_{s'}\sum_r\sum_{r'} |\mathcal{M}(s, s' \rightarrow r, r')|^2.$$
(12.3)

In addition, using completeness relations, we find that

$$\overline{|\mathcal{M}|^2} = \frac{e^4}{4q^4}\mathrm{tr}\left[(p\!\!\!/' - m_e \mathbb{1}_4)\gamma^\mu (p\!\!\!/ + m_e \mathbb{1}_4)\gamma^\nu \right] \mathrm{tr}\left[(k\!\!\!/ + m_\mu \mathbb{1}_4)\gamma_\mu (k\!\!\!/' - m_\mu \mathbb{1}_4)\gamma_\nu \right].$$
(12.4)

Next, using trace technology (cf. Section 3.3), we have

$$\mathrm{tr}\left[(p\!\!\!/' - m_e \mathbb{1}_4)\gamma^\mu (p\!\!\!/ + m_e \mathbb{1}_4)\gamma^\nu \right] = 4\left[p'^\mu p^\nu + p'^\nu p^\mu - g^{\mu\nu}(p \cdot p' + m_e^2) \right]$$
(12.5)

and similarly for the other trace. Thus, we obtain the unpolarized amplitude

$$\overline{|\mathcal{M}|^2} = \frac{8e^4}{q^4}\left[(p \cdot k)(p' \cdot k') + (p \cdot k')(p' \cdot k) + m_\mu^2 (p \cdot p') \right],$$
(12.6)

where we have neglected the electron mass, i.e. $m_e = 0$, since $m_e \ll m_\mu$. In the CM system (see Fig. 12.2), we have the 4-momenta $p = (E, E\mathbf{e}_z)$, $p' = (E, -E\mathbf{e}_z)$, $k = (E, \mathbf{k})$, and $k' = (E, -\mathbf{k})$, where the angle between the 3-vectors $\mathbf{k}$ and $\mathbf{e}_z$ is denoted θ and given by $\mathbf{k} \cdot \mathbf{e}_z = |\mathbf{k}| \cos\theta$ as well as $|\mathbf{k}| = \sqrt{E^2 - m_\mu^2}$. Inserting these quantities expressed in the CM system into Eq. (12.6), we obtain

$$\overline{|\mathcal{M}|^2}_{\mathrm{CM}} = e^4 \left[1 + \frac{m_\mu^2}{E^2} + \left(1 - \frac{m_\mu^2}{E^2} \right) \cos^2 \theta \right]. \tag{12.7}$$

In order to calculate the differential cross-section in the CM system, we insert $E_A = E_{e^+} = E_{\mathrm{CM}}/2$, $E_B = E_{e^-} = E_{\mathrm{CM}}/2$, and $v_{ab} = |v_a - v_b| = |v_{e^+} - v_{e^-}| = 2$ (i.e. the two particles are approaching each other at the speed of light) into Eq. (11.200) and we find that

$$\begin{aligned}
\left(\frac{d\sigma}{d\Omega} \right)_{\mathrm{CM}} &= \frac{1}{2E_{\mathrm{CM}}^2} \frac{|\mathbf{k}|}{16\pi^2 E_{\mathrm{CM}}} \overline{|\mathcal{M}|^2}_{\mathrm{CM}} \\
&= \frac{\alpha^2}{4E_{\mathrm{CM}}^2} \sqrt{1 - \frac{m_\mu^2}{E^2}} \left[1 + \frac{m_\mu^2}{E^2} + \left(1 - \frac{m_\mu^2}{E^2} \right) \cos^2 \theta \right],
\end{aligned} \tag{12.8}$$

where again $\alpha = e^2/(4\pi)$ is Sommerfeld's fine-structure constant. Now, the total cross-section is given by integrating the differential cross-section. The result is

$$\sigma = \int_\Omega \left(\frac{d\sigma}{d\Omega} \right)_{\mathrm{CM}} d\Omega_{\mathrm{CM}} = \frac{4\pi\alpha^2}{3E_{\mathrm{CM}}^2} \sqrt{1 - \frac{m_\mu^2}{E^2}} \left(1 + \frac{1}{2} \frac{m_\mu^2}{E^2} \right). \tag{12.9}$$

Finally, in the relativistic limit, i.e. $E \gg m_\mu$, we find that

$$\left(\frac{d\sigma}{d\Omega} \right)_{\mathrm{CM}} \to \frac{\alpha^2}{4E_{\mathrm{CM}}^2} \left(1 + \cos^2 \theta \right) = \frac{\alpha^2}{E_{\mathrm{CM}}^2} \left[\frac{1}{2} - \frac{1}{4}\theta^2 + \mathcal{O}(\theta^4) \right], \tag{12.10}$$

$$\sigma \to \frac{4\pi\alpha^2}{3E_{\mathrm{CM}}^2}. \tag{12.11}$$

In Fig. 12.3, we show results from the DELCO Collaboration in 1978 for production cross-section ratios, which, in fact, give the cross-section ratio $\sigma(e^+ + e^- \to \tau^+ + \tau^-)/\sigma(e^+ + e^- \to \mu^+ + \mu^-)$ close to the threshold of $\tau^+\tau^-$-production.[2] Using the data of the DELCO Collaboration, it is possible to obtain an experimental value for the mass of the tau. The result is $m_\tau = \left(1782^{+2}_{-7} \right)$ MeV. Today, other experimental collaborations such as the BES and KEDR Collaborations have measured the cross-section $\sigma(e^+ + e^- \to \tau^+ + \tau^-)$ to a much better precision. For example, in Fig. 12.4, we present the results from the KEDR Collaboration in 2007, which yield $m_\tau = \left(1776.80^{+0.25}_{-0.23} \pm 0.15 \right)$ MeV. In summary, the Particle Data Group (PDG) average value of 2010 for the mass of the tau is given by $m_\tau = (1776.82 \pm 0.16)$ MeV,[3] and we observe that the value of the KEDR

---

[2] Note that τ^- is the tau, whereas τ^+ is its antiparticle, i.e. the antitau. The electron (first generation), the muon (second generation), and the tau (third generation) are the three different and known generations of so-called *charged leptons*.

[3] Particle Data Group, K. Nakamura *et al.*, The 2010 Review of Particle Physics, *J. Phys. G* **37**, 075021 (2010), pdg.lbl.gov.

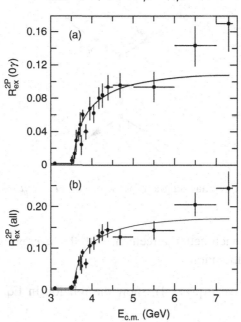

Figure 12.3 The production cross-section ratios, $R_{eX}^{2P} = \sigma(e^+ + e^- \rightarrow e + X)/$ $\sigma(e^+ + e^- \rightarrow \mu^+ + \mu^-)$, as functions of the centre-of-mass energy E_{CM} for $e + X$ events with no detected photons are shown in (a) and for all $e + X$ events in (b). Reprinted figure with permission from the DELCO Collaboration, W. Bacino *et al.*, *Phys. Rev. Lett.* **41**, 13–15 (1978). © 1978 by the American Physical Society.

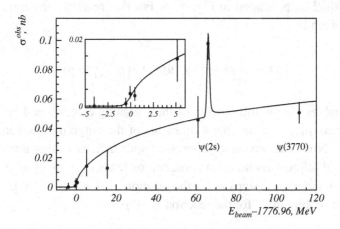

Figure 12.4 The observed cross-section $\sigma^{obs} = \sigma(e^+ + e^- \rightarrow \tau^+ + \tau^-)$ as a function of the beam energy E_{beam} by the KEDR Collaboration, V. V. Anashin *et al.*, *Nucl. Phys. B (Proc. Suppl.)* **169**, 125–131 (2007), hep-ex/0611046. © 2007 by Elsevier Science B.V.

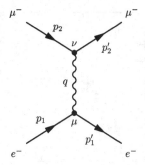

Figure 12.5 Feynman diagram for the reaction $e^- + \mu^- \to e^- + \mu^-$.

Collaboration is in much better agreement with this average value than the value of the DELCO Collaboration.

Exercise 12.1 Derive Eq. (12.4), show the relation in Eq. (12.5), and finally, derive Eq. (12.6).

12.2 $e^- + \mu^- \to e^- + \mu^-$

Next, we study the reaction $e^- + \mu^- \to e^- + \mu^-$, which is called electron–muon scattering. Note that this reaction is different from the reaction $e^+ + e^- \to \mu^+ + \mu^-$, but closely related to it. Again, using the Feynman rules of QED, we can draw the Feynman diagram for the reaction $e^- + \mu^- \to e^- + \mu^-$ at tree level (a t-channel diagram), which is presented in Fig. 12.5. For this reaction, the invariant matrix element is given by

$$i\mathcal{M} = \frac{ie^2}{q^2}\bar{u}\left(p_1'\right)\gamma^\mu u(p_1)\bar{u}\left(p_2'\right)\gamma_\mu u(p_2), \qquad (12.12)$$

where p_1 and p_2 are the 4-momenta of the incoming electron and muon, respectively, whereas p_1' and p_2' are the 4-momenta of the outgoing electron and muon, respectively. Note that we have suppressed spin indices in this invariant matrix element. In addition, 4-momentum conservation leads to $p_1 + p_2 = p_1' + p_2'$. Furthermore, using the same steps as for the reaction $e^+ + e^- \to \mu^+ + \mu^-$, we obtain the unpolarized amplitude for the reaction $e^- + \mu^- \to e^- + \mu^-$

$$\overline{|\mathcal{M}|^2} = \frac{e^4}{4q^4}\mathrm{tr}\left[\left(\not{p}_1' + m_e\mathbb{1}_4\right)\gamma^\mu(\not{p}_1 + m_e\mathbb{1}_4)\gamma^\nu\right]$$
$$\times \mathrm{tr}\left[\left(\not{p}_2' + m_\mu\mathbb{1}_4\right)\gamma_\mu(\not{p}_2 + m_\mu\mathbb{1}_4)\gamma_\nu\right]. \qquad (12.13)$$

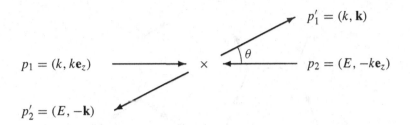

Figure 12.6 The kinematics in the CM system for the reaction $e^- + \mu^- \to e^- + \mu^-$.

What is the relation of this amplitude to the amplitude for the reaction $e^+ + e^- \to \mu^+ + \mu^-$? If we make the replacements $p \to p_1$, $p' \to -p_1'$, $k \to p_2'$, and $k' \to -p_2$ in Eq. (12.13), then we have the expression in Eq. (12.4). Thus, in the limit $m_e \to 0$, we find the amplitude

$$\overline{|\mathcal{M}|^2} = \frac{8e^4}{q^4} \left[(p_1 \cdot p_2')(p_1' \cdot p_2) + (p_1 \cdot p_2)(p_1' \cdot p_2') - m_\mu^2 (p_1 \cdot p_1') \right]. \quad (12.14)$$

Finally, in the CM system (see Fig. 12.6) with the 4-momenta $p_1 = (k, k\mathbf{e}_z)$, $p_2 = (E, -k\mathbf{e}_z)$, $p_1' = (k, \mathbf{k})$, and $p_2' = (E, -\mathbf{k})$, which means that $E_{CM} = E + k$, we obtain

$$\overline{|\mathcal{M}|^2} = \frac{2e^4}{k^2(1 - \cos\theta)^2} \left[(E + k)^2 + (E + k\cos\theta)^2 - m_\mu^2(1 - \cos\theta) \right]. \quad (12.15)$$

Note that the centre-of-mass energy $E_{CM} = E + k$ is different for this reaction compared with $E_{CM} = 2E_{e^+} = 2E_{e^-}$ for $e^+ + e^- \to \mu^+ + \mu^-$ due to the fact that the initial particles have the same mass in the first case, but not in the second case. In the limit $m_\mu \to 0$, the differential cross-section becomes

$$\left(\frac{d\sigma}{d\Omega} \right)_{CM} = \frac{\alpha^2}{2E_{CM}^2(1 - \cos\theta)^2} \left[4 + (1 + \cos\theta)^2 \right]. \quad (12.16)$$

Note that

$$\left(\frac{d\sigma}{d\Omega} \right)_{CM} \to \frac{\alpha^2}{E_{CM}^2} \left[\frac{16}{\theta^4} - \frac{4}{3\theta^2} + \frac{37}{90} - \frac{1}{3780}\theta^2 + \mathcal{O}(\theta^4) \right] \quad \text{when } \theta \to 0,$$
$$(12.17)$$

which means that the differential cross-section diverges in the limit $\theta \to 0$. The reason is that the dominating contribution to the reaction $e^- + \mu^- \to e^- + \mu^-$ stems from the virtual photon (see Fig. 12.5), which contribution is divergent, since $q^2 = (p_1 - p_1')^2 \approx 0$. Thus, it is the different kinematics for the reactions $e^+ + e^- \to \mu^+ + \mu^-$ and $e^- + \mu^- \to e^- + \mu^-$, which make the cross-sections different for the two reactions.

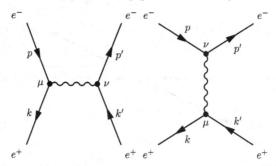

Figure 12.7 Feynman diagrams for Bhabha scattering. *Left diagram:* annihilation (*s*-channel). *Right diagram:* scattering (*t*-channel).

Introducing the Mandelstam variables $s = (p + p')^2$, $t = (k - p)^2$, and $u = (k' - p)^2$, we can express the averaged amplitudes for the two reactions as

$$e^+ + e^- \rightarrow \mu^+ + \mu^- : \quad \overline{|\mathcal{M}|^2} = \frac{8e^4}{s^2}\left[\left(\frac{t}{2}\right)^2 + \left(\frac{u}{2}\right)^2\right] = 2e^4\frac{t^2 + u^2}{s^2},$$

$$\tag{12.18}$$

$$e^- + \mu^- \rightarrow e^- + \mu^- : \quad \overline{|\mathcal{M}|^2} = \frac{8e^4}{t^2}\left[\left(\frac{s}{2}\right)^2 + \left(\frac{u}{2}\right)^2\right] = 2e^4\frac{s^2 + u^2}{t^2}, \tag{12.19}$$

which means that we can make the replacement $s \leftrightarrow t$ in one of the two amplitudes in order to obtain the other. Therefore, one says that the reaction $e^- + \mu^- \rightarrow e^- + \mu^-$ is related by *crossing symmetry* to the reaction $e^+ + e^- \rightarrow \mu^+ + \mu^-$. In addition, note that the reaction $e^+ + e^- \rightarrow \mu^+ + \mu^-$ (the *s*-channel) is $t \leftrightarrow u$ symmetric, whereas the reaction $e^- + \mu^- \rightarrow e^- + \mu^-$ (the *t*-channel) is $s \leftrightarrow u$ symmetric.

12.3 $e^+ + e^- \rightarrow e^+ + e^-$

The reaction $e^+ + e^- \rightarrow e^+ + e^-$ is called *Bhabha scattering*,[4] which is an electron–positron scattering process. There are two leading-order (or tree level) Feynman diagrams contributing to the invariant matrix element of this reaction, one annihilation diagram (an *s*-channel diagram) and one scattering diagram (a *t*-channel diagram). The two Feynman diagrams are presented in Fig. 12.7.[5] Normally, Bhabha scattering is used as a luminosity monitor in electron–positron colliders. Accurate measurement of luminosity is a necessity for accurate measurements of cross-sections.

---

[4] Bhabha scattering is named after the Indian physicist H. J. Bhabha.
[5] Note that Bhabha scattering has two allowed Feynman diagrams, whereas the reaction
$e^+ + e^- \rightarrow \mu^+ + \mu^-$ has only one allowed Feynman diagram, since the initial and final particles are **not** the same for this reaction, i.e. electrons and muons are not the same fermions.

For Bhabha scattering, the invariant matrix element is given by

$$i\mathcal{M} = \bar{v}(k)(-ie\gamma^\mu)v(k')\left[\frac{-ig_{\mu\nu}}{(k-k')^2}\right]\bar{u}(p')(-ie\gamma^\nu)u(p)$$

$$+ (-1)\bar{v}(k)(-ie\gamma^\mu)u(p)\left[\frac{-ig_{\mu\nu}}{(k+p)^2}\right]\bar{u}(p')(-ie\gamma^\nu)v(k')$$

$$= ie^2\left[\bar{v}(k)\gamma^\mu v(k')\frac{1}{(k-k')^2}\bar{u}(p')\gamma_\mu u(p) - \bar{v}(k)\gamma^\mu u(p)\right.$$

$$\left.\times \frac{1}{(k+p)^2}\bar{u}(p')\gamma_\mu v(k')\right], \tag{12.20}$$

where k and k' are the 4-momenta of the incoming and outgoing positron, respectively, whereas p and p' are the 4-momenta of the incoming and outgoing electron, respectively. Note that there is a relative sign factor -1 between the two contributions that is implicit in the normal products of the contributions which becomes explicit if the creation and annihilation operators are brought into the same normal-ordering in both cases. In addition, 4-momentum conservation leads to $k + p = k' + p'$. Calculating the modulus squared of the matrix element in Eq. (12.20) gives

$$|\mathcal{M}|^2 = e^4\left|\frac{[\bar{v}(k)\gamma^\mu v(k')][\bar{u}(p')\gamma_\mu u(p)]}{(k-k')^2}\right|^2$$

$$- e^4\left\{\frac{[\bar{v}(k)\gamma^\mu v(k')][\bar{u}(p')\gamma_\mu u(p)]}{(k-k')^2}\right\}^*\left\{\frac{[\bar{v}(k)\gamma^\nu u(p)][\bar{u}(p')\gamma_\nu v(k')]}{(k+p)^2}\right\}$$

$$- e^4\left\{\frac{[\bar{v}(k)\gamma^\mu v(k')][\bar{u}(p')\gamma_\mu u(p)]}{(k-k')^2}\right\}\left\{\frac{[\bar{v}(k)\gamma^\nu u(p)][\bar{u}(p')\gamma_\nu v(k')]}{(k+p)^2}\right\}^*$$

$$+ e^4\left|\frac{[\bar{v}(k)\gamma^\nu u(p)][\bar{u}(p')\gamma_\nu v(k')]}{(k+p)^2}\right|^2, \tag{12.21}$$

where the first term describes scattering (by photon exchange), the last term describes annihilation, and the two middle terms describe interference between scattering and annihilation.

Again, in order to calculate the unpolarized cross-section, we average over the spins of the incoming particles and sum over the spins of the outgoing particles, i.e.

$$\overline{|\mathcal{M}|^2} = \frac{1}{(2s_{e^+} + 1)(2s_{e^-} + 1)}\sum_{\text{spin}}|\mathcal{M}|^2. \tag{12.22}$$

We use Eq. (12.22) for each term in Eq. (12.21) separately. For the scattering contribution, we have

$$|\mathcal{M}|^2_{\text{scattering}} = \frac{e^4}{(k-k')^4}[\bar{v}(k')\gamma^\mu v(k)][\bar{v}(k)\gamma^\nu v(k')][\bar{u}(p)\gamma_\mu u(p')][\bar{u}(p')\gamma_\nu u(p)].$$
(12.23)

Now, we assume that s and s' are the spins of the incoming and outgoing electron, respectively, and r and r' are the spins of the incoming and outgoing positron, respectively. Then, we perform Eq. (12.22) for Eq. (12.23) and find that

$$\overline{|\mathcal{M}|^2}_{\text{scattering}} = \frac{e^4}{4(k-k')^4}\left\{\sum_{r'}\bar{v}^{r'}(k')\gamma^\mu\left[\sum_r v^r(k)\bar{v}^r(k)\right]\gamma^\nu v^{r'}(k')\right\}$$

$$\times\left\{\sum_s \bar{u}^s(p)\gamma_\mu\left[\sum_{s'}u^{s'}(p')\bar{u}^{s'}(p')\right]\gamma_\nu u^s(p)\right\}.$$
(12.24)

Next, using completeness relations, we obtain

$$\overline{|\mathcal{M}|^2}_{\text{scattering}} = \frac{e^4}{4(k-k')^4}\text{tr}[(\slashed{k}' - m_e\mathbb{1}_4)\gamma^\mu(\slashed{k} - m_e\mathbb{1}_4)\gamma^\nu]$$

$$\times \text{tr}[(\slashed{p} + m_e\mathbb{1}_4)\gamma_\mu(\slashed{p}' + m_e\mathbb{1}_4)\gamma_\nu],$$
(12.25)

which, using trace technology, can be simplified to

$$\overline{|\mathcal{M}|^2}_{\text{scattering}} = \frac{8e^4}{(k-k')^4}\left[(k'\cdot p')(k\cdot p) + (k'\cdot p)(k\cdot p')\right.$$

$$\left. - m_e^2 p'\cdot p - m_e^2 k'\cdot k + 2m_e^4\right].$$
(12.26)

Neglecting the electron mass (i.e. $m_e \simeq 0$) yields

$$\overline{|\mathcal{M}|^2}_{\text{scattering}} \simeq \frac{8e^4}{(k-k')^4}\left[(k'\cdot p')(k\cdot p) + (k'\cdot p)(k\cdot p')\right].$$
(12.27)

Finally, using the Mandelstam variables $s = (k+p)^2$, $t = (k-k')^2$, and $u = (k-p')^2$, which become $s \simeq 2k\cdot p = 2k'\cdot p'$, $t \simeq -2k\cdot k' = -2p\cdot p'$, and $u \simeq -2k\cdot p' = -2k'\cdot p$ in the relativistic limit (i.e. $E \gg m_e$), we find that

$$\overline{|\mathcal{M}|^2}_{\text{scattering}} = \frac{8e^4}{t^2}\left(\frac{s}{2}\frac{s}{2} + \frac{u}{2}\frac{u}{2}\right) = 2e^4\frac{s^2 + u^2}{t^2}.$$
(12.28)

Similarly, for the interference and annihilation terms, we obtain

$$\overline{|\mathcal{M}|^2}_{\text{interference}} = 4e^4\frac{u^2}{st} \quad \text{and} \quad \overline{|\mathcal{M}|^2}_{\text{annihilation}} = 2e^4\frac{t^2 + u^2}{s^2}.$$
(12.29)

Note that the scattering and annihilation diagrams are related by crossing symmetry, since the initial and final state particles are the same. Hence, it is sufficient to permute the 4-momenta or to make the replacement $s \leftrightarrow t$. Thus, for Bhabha scattering in the relativistic limit, we find that

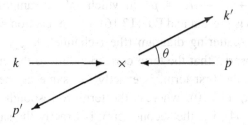

Figure 12.8 The kinematics in the CM system for the reaction $e^+ + e^- \rightarrow e^+ + e^-$

$$\overline{|\mathcal{M}|^2} = \overline{|\mathcal{M}|^2}_{\text{scattering}} + \overline{|\mathcal{M}|^2}_{\text{interference}} + \overline{|\mathcal{M}|^2}_{\text{annihilation}}$$
$$= 2e^4 \left(\frac{s^2 + u^2}{t^2} + \frac{2u^2}{st} + \frac{t^2 + u^2}{s^2} \right). \tag{12.30}$$

Note that the averaged amplitude for Bhabha scattering is symmetric with respect to the replacement $s \leftrightarrow t$.

Finally, in the relativistic limit, using Eqs. (11.201) and (12.30) as well as $d\Omega_{\text{CM}} = -2\pi\, d\cos\theta$, $s = E^2_{\text{CM}}$, and $e^2 = 4\pi\alpha$, the differential cross-section is given by

$$\frac{d\sigma}{d\cos\theta} = -\frac{\pi\alpha^2}{s} \left[u^2 \left(\frac{1}{s} + \frac{1}{t} \right)^2 + \left(\frac{t}{s} \right)^2 + \left(\frac{s}{t} \right)^2 \right]. \tag{12.31}$$

Note that, since $s + t + u \simeq 0$ in the relativistic limit, we have $u^2 = (s + t)^2$, and thus, we can write Eq. (12.31) as

$$\frac{d\sigma}{d\cos\theta} = -\frac{\pi\alpha^2}{s} \frac{s^4 + t^4 + u^4}{s^2 t^2}. \tag{12.32}$$

In fact, in the relativistic limit and in the CM system (see Fig. 12.8), using the relation

$$\left(\frac{d\sigma}{d\Omega} \right)_{\text{CM}} = -\frac{1}{2\pi} \frac{d\sigma}{d\cos\theta} \tag{12.33}$$

as well as $s = 4E^2 \simeq 4p^2$, $t = -4p^2 \sin^2 \frac{\theta}{2}$, and $u = -4p^2 \cos^2 \frac{\theta}{2}$, we find that the differential cross-section in Eq. (12.31) is transformed to

$$\left(\frac{d\sigma}{d\Omega} \right)_{\text{CM}} = \frac{\alpha^2}{8E^2} \left(\frac{1 + \cos^2\theta}{2} + \frac{1 + \cos^4 \frac{\theta}{2}}{\sin^4 \frac{\theta}{2}} - \frac{2\cos^4 \frac{\theta}{2}}{\sin^2 \frac{\theta}{2}} \right), \tag{12.34}$$

where again the three terms in this formula correspond to the annihilation diagram, the scattering diagram, and the interference between the two diagrams, respectively. This result should be compared with the corresponding results in Eq. (12.10) for the reaction $e^+ + e^- \rightarrow \mu^+ + \mu^-$ in which only the annihilation diagram (the s-channel diagram) is present and Eq. (12.16) for the reaction $e^- + \mu^- \rightarrow e^- + \mu^-$ in which only the scattering diagram (the t-channel diagram) is present. Since $E_{CM} = 2E$, we observe that the term corresponding to the annihilation diagram in Eq. (12.34), i.e. the first term, is exactly the same as the differential cross-section given in Eq. (12.10), whereas the term corresponding to the scattering diagram in Eq. (12.34), i.e. the second term, is exactly the same as the differential cross-section given in Eq. (12.16). Actually, using the trigonometric identities $\sin^2 \frac{\theta}{2} = (1 - \cos \theta)/2$ and $\cos^2 \frac{\theta}{2} = (1 + \cos \theta)/2$, the differential cross-section in Eq. (12.34) can be written more compactly as

$$\left(\frac{d\sigma}{d\Omega} \right)_{CM} = \frac{\alpha^2}{16E^2} \frac{(3 + \cos^2 \theta)^2}{(1 - \cos \theta)^2}. \tag{12.35}$$

Note that, for small θ, we obtain

$$\left(\frac{d\sigma}{d\Omega} \right)_{CM} = \frac{\alpha^2}{E^2} \left[\frac{4}{\theta^4} - \frac{4}{3\theta^2} + \frac{29}{45} - \frac{977}{7560}\theta^2 + \mathcal{O}(\theta^4) \right]. \tag{12.36}$$

In addition, in the limits $\theta \rightarrow \pi$, $\theta \rightarrow \pi/2$, and $\theta \rightarrow \pi/4$, we have $(d\sigma/d\Omega)_{CM} = \alpha^2/(4E^2)$, $(d\sigma/d\Omega)_{CM} = 9\alpha^2/(16E^2)$, and $(d\sigma/d\Omega)_{CM} \approx 8.92\alpha^2/E^2$, respectively.

In experiments, the theoretically predicted behaviour has been confirmed using a large range of energies and angles, and the interaction has been tested down to very short distances (or equivalently, very high energies), which can be compared with those probed in the $e^+ + e^- \rightarrow \mu^+ + \mu^-$ (or $e^+ + e^- \rightarrow \tau^+ + \tau^-$) experiments. In Fig. 12.9, we show results from the CELLO and PLUTO Collaborations of the differential cross-section for the reaction $e^+ + e^- \rightarrow e^+ + e^-$.

Exercise 12.2 Compute the interference and annihilation terms of the averaged amplitude given in Eq. (12.29).

Exercise 12.3 Derive the differential cross-section in Eq. (12.31) [or equivalently, Eq. (12.35)].

12.4 $e^- + e^- \rightarrow e^- + e^-$

The reaction $e^- + e^- \rightarrow e^- + e^-$ is called *Møller scattering* (or electron–electron scattering) and describes basically all electron–electron interactions. Nowadays, Møller scattering is less used in experimental particle physics, having been replaced by electron–positron scattering. In QED, at tree level, there are two Feynman diagrams (see Fig. 12.10) describing this process: a t-channel diagram, where the

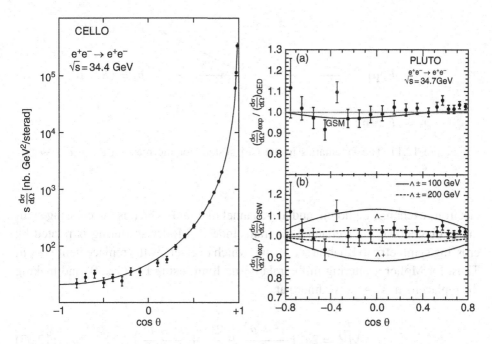

Figure 12.9 *Left figure:* the differential cross-section as a function of $\cos\theta$ at $\sqrt{s} = 34.4$ GeV for the reaction $e^+ + e^- \rightarrow e^+ + e^-$ in the central detector and endcap calorimeter range by the CELLO Collaboration, H.-J. Behrend *et al.*, *Phys. Lett.* **103B**, 148–152 (1981). Only statistical errors are plotted. The QED prediction (solid curve) is normalized to the endcap points. © 1981 by Elsevier Science B.V. *Right figure:* the differential cross-section as a function of $\cos\theta$ at $\sqrt{s} = 34.7$ GeV for the reaction $e^+ + e^- \rightarrow e^+ + e^-$, the curve in (a) shows the Standard Model prediction normalized to lowest order QED expectation and the curves in (b) show the expected distributions for several values of the QED-cut-off parameters normalized to the Standard Model with $M_Z = 93$ GeV and $\sin^2\vartheta_W = 0.217$, by the PLUTO Collaboration, C. Berger *et al.*, *Z. Phys. C* **27**, 341–349 (1985). © 1985 by Springer-Verlag.

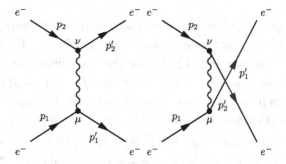

Figure 12.10 Feynman diagrams for Møller scattering. *Left diagram*: t-channel. *Right diagram*: u-channel.

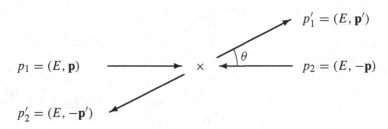

Figure 12.11 The kinematics in the CM system for the reaction $e^- + e^- \rightarrow e^- + e^-$.

electrons exchange a photon, and a u-channel diagram, which is the crossing symmetry diagram of the t-channel diagram. Indeed, Møller scattering is related by crossing symmetry to Bhabha scattering, which is given by the replacement $s \leftrightarrow u$. Thus, for Møller scattering in the relativistic limit, using Eq. (12.30) and making the replacement $s \leftrightarrow u$, we find that

$$\overline{|\mathcal{M}|^2} = 2e^4 \left(\frac{s^2 + u^2}{t^2} + \frac{2s^2}{tu} + \frac{s^2 + t^2}{u^2} \right). \tag{12.37}$$

In general, without performing the approximations used in Eq. (12.37), one can show that the averaged amplitude for Møller scattering is (after lengthy computations) given by

$$\overline{|\mathcal{M}|^2} = 2e^4 \left[\frac{s^2 + u^2 - 8m_e^2(s + u) + 24m_e^4}{t^2} + \frac{2\left(s^2 - 8m_e^2 s + 12m_e^4\right)}{tu} \right.$$
$$\left. + \frac{s^2 + t^2 - 8m_e^2(s + t) + 24m_e^4}{u^2} \right]. \tag{12.38}$$

In fact, the averaged amplitude for Møller scattering is symmetric with respect to the replacement $t \leftrightarrow u$. This holds for both Eqs. (12.37) and (12.38).

In the CM system (see Fig. 12.11), for the kinematics, we introduce the 4-momenta for the four electrons as $p_1 = (E, \mathbf{p})$ and $p_2 = (E, -\mathbf{p})$ for the incoming electrons as well as $p_1' = (E, \mathbf{p}')$ and $p_2' = (E, -\mathbf{p}')$ for the outgoing electrons, and using 4-momentum conservation, we find that $p_1 + p_2 = p_1' + p_2'$. Thus, we have the Mandelstam variables $s = (p_1 + p_2)^2 = 4E^2$, $t = \left(p_1 - p_1'\right)^2 = -(\mathbf{p} - \mathbf{p}')^2 = -4p^2 \sin^2 \frac{\theta}{2}$, and $u = \left(p_1 - p_2'\right)^2 = -(\mathbf{p} + \mathbf{p}')^2 = -4p^2 \cos^2 \frac{\theta}{2}$. Note that in this case $|\mathbf{p}| = |\mathbf{p}'| = p$ and $\mathbf{p} \cdot \mathbf{p}' = p^2 \cos \theta$, where $p = \sqrt{E^2 - m_e^2}$. Inserting the Mandelstam variables into Eq. (12.38), we obtain the differential cross-section

$$\left(\frac{d\sigma}{d\Omega}\right)_{CM} = \frac{\alpha^2}{4E^2}\left[\frac{(2E^2 - m_e^2)^2}{(E^2 - m_e^2)^2}\left(\frac{4}{\sin^4\theta} - \frac{3}{\sin^2\theta}\right) + 1 + \frac{4}{\sin^2\theta}\right], \quad (12.39)$$

which is called the *Møller formula*.

In the non-relativistic limit, i.e. $E \simeq m_e$, where $s = 4E^2 \simeq 4m_e^2$ and $v^2 = p^2/E^2 = (E^2 - m_e^2)/E^2$, using the Møller formula (12.39), we find that

$$\left(\frac{d\sigma}{d\Omega}\right)_{CM, nr} = \frac{\alpha^2}{4m_e^2 v^4}\left(\frac{4}{\sin^4\theta} - \frac{3}{\sin^2\theta}\right)$$

$$= \frac{\alpha^2}{16m_e^2 v^4}\left(\frac{1}{\sin^4\frac{\theta}{2}} + \frac{1}{\cos^4\frac{\theta}{2}} - \frac{1}{\sin^2\frac{\theta}{2}\cos^2\frac{\theta}{2}}\right), \quad (12.40)$$

which is named the *Mott formula*.[6] Comparing Eq. (12.40) with the 'classical' *Rutherford-scattering formula*[7] (for scattering of two identical electrons)

$$\frac{d\sigma}{d\Omega} = \frac{\alpha^2}{16m_e^2 v^4}\left(\frac{1}{\sin^4\frac{\theta}{2}} + \frac{1}{\cos^4\frac{\theta}{2}}\right), \quad (12.41)$$

we observe that the last term in Eq. (12.40) is missing in the 'classical' formula.[8] This term is due to the interference between the two electrons, which is a quantum mechanical effect. Classically, the probabilities simply add, whereas quantum mechanically, the two amplitudes add. Thus, the quantum mechanical probability will also contain the interference term.

Finally, in the relativistic limit, i.e. $E \gg m_e$, where $s = 4E^2 \simeq 4p^2$, using the Møller formula (12.39) as well as $t = -2p^2(1 - \cos\theta)$ and $u = -2p^2(1 + \cos\theta)$, we obtain

$$\left(\frac{d\sigma}{d\Omega}\right)_{CM} = \frac{\alpha^2}{E^2}\left(\frac{4}{\sin^4\theta} - \frac{2}{\sin^2\theta} + \frac{1}{4}\right)$$

$$= \frac{\alpha^2}{4E^2}\left(\frac{1}{\sin^4\frac{\theta}{2}} + \frac{1}{\cos^4\frac{\theta}{2}} + 1\right) = \frac{\alpha^2}{4E^2}\frac{(3 + \cos^2\theta)^2}{\sin^4\theta}. \quad (12.42)$$

---

[6] In 1977, N. F. Mott was awarded a third of the Nobel Prize in physics with the motivation 'for their fundamental theoretical investigations of the electronic structure of magnetic and disordered systems'. Mott shared this prize together with P. W. Anderson and J. H. van Vleck.

[7] In 1908, E. Rutherford was awarded the Nobel Prize in chemistry (**not** physics) 'for his investigations into the disintegration of the elements, and the chemistry of radioactive substances'.

[8] Note that the classical Rutherford formula for Coulomb scattering is given by

$$\frac{d\sigma}{d\Omega} = \frac{\alpha^2 Z^2}{4m^2 v^4 \sin^4\frac{\theta}{2}},$$

which is the differential cross-section for protons scattering off nuclei.

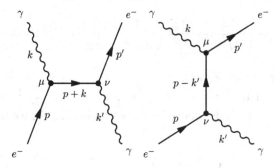

Figure 12.12 Feynman diagrams for Compton scattering.

Again, for small θ, we find that

$$\left(\frac{d\sigma}{d\Omega}\right)_{CM} = \frac{\alpha^2}{E^2}\left[\frac{4}{\theta^4} + \frac{2}{3\theta^2} + \frac{101}{180} + \frac{122}{945}\theta^2 + \mathcal{O}(\theta^4)\right]. \tag{12.43}$$

In addition, in the limits $\theta \to \pi/2$ and $\theta \to \pi/4$, we have $(d\sigma/d\Omega)_{CM} = 9\alpha^2/(4E^2)$ and $(d\sigma/d\Omega)_{CM} = 49\alpha^2/(4E^2)$, respectively.

Exercise 12.4 Derive the averaged amplitude in Eq. (12.38).

Exercise 12.5 Derive the differential cross-section in Eq. (12.39) and show that, in the non-relativistic limit, it is given by Eq. (12.40), whereas in the relativistic limit, it is given by Eq. (12.42).

12.5 $e^- + \gamma \to e^- + \gamma$ and $e^+ + e^- \to 2\gamma$

The reaction $e^- + \gamma \to e^- + \gamma$ is called *Compton scattering*. In physics, Compton scattering or the famous *Compton effect* is the decrease in energy (or the increase in wave-length) of an X-ray or a gamma-ray photon, when it interacts with matter.[9] At tree level, this reaction has two Feynman diagrams, which are presented in Fig. 12.12. Therefore, in the case of Compton scattering, the invariant matrix element has two contributions and is given by

$$i\mathcal{M} = \bar{u}(p')\,(-ie\gamma^\nu)\,\varepsilon_\nu^*(k')\frac{i(\slashed{p}+\slashed{k}+m_e\mathbb{1}_4)}{(p+k)^2 - m_e^2}\,(-ie\gamma^\mu)\,\varepsilon_\mu(k)u(p)$$

$$+ \bar{u}(p')\,(-ie\gamma^\mu)\,\varepsilon_\mu(k)\frac{i(\slashed{p}-\slashed{k'}+m_e\mathbb{1}_4)}{(p-k')^2 - m_e^2}\,(-ie\gamma^\nu)\,\varepsilon_\nu^*(k')u(p), \tag{12.44}$$

---

[9] In 1923, the Compton effect was observed by A. H. Compton, and in 1927, Compton was awarded the Nobel Prize in physics 'for his discovery of the effect named after him'.

where m_e is the electron mass, which can be simplified to

$$i\mathcal{M} = -ie^2 \varepsilon_\nu^*(k')\varepsilon_\mu(k)\bar{u}(p') \left(\frac{\gamma^\nu \not{k}\gamma^\mu + 2\gamma^\nu p^\mu}{2p \cdot k} + \frac{-\gamma^\mu \not{k}'\gamma^\nu + 2\gamma^\mu p^\nu}{-2p \cdot k'} \right) u(p).$$
(12.45)

Note that the fermionic parts of the two Feynman diagrams are the same, which means that the two contributions to the invariant matrix element should have the same relative sign. Now, introducing

$$i\mathcal{M} = i\mathcal{M}^\nu(k)\varepsilon_\nu^*(k),$$
(12.46)

one can show that

$$\sum_\varepsilon \varepsilon_\nu^*(k)\varepsilon_\mu(k)\mathcal{M}^\nu(k)\mathcal{M}^{\mu*}(k) = -g_{\mu\nu}\mathcal{M}^\nu(k)\mathcal{M}^{\mu*}(k),$$
(12.47)

which implies after a tedious calculation that

$$\overline{|\mathcal{M}|^2} = 2e^4 \left[\frac{p \cdot k'}{p \cdot k} + \frac{p \cdot k}{p \cdot k'} + 2m_e^2 \left(\frac{1}{p \cdot k} - \frac{1}{p \cdot k'} \right) + m_e^4 \left(\frac{1}{p \cdot k} - \frac{1}{p \cdot k'} \right)^2 \right].$$
(12.48)

Note that it can be shown that

$$k_\mu \mathcal{M}^\mu(k) = 0,$$
(12.49)

which is called the *Ward identity*. Physically, this identity means that the scalar and longitudinal polarizations of the photon field cancel each other, since they are unphysical (cf. the discussion on the Gupta–Bleuler formalism in Section 8.4).

For Compton scattering, one of the initial particles is a photon, and it is therefore easiest to work with the lab system, since it is difficult to define the centre-of-mass for this reaction. In the lab system (see Fig. 12.13), where the electron is at rest before the reaction, we have the 4-momenta $p = (m_e, \mathbf{0})$, $k = (\omega, \omega \mathbf{e}_z)$, $p' = (E', \mathbf{p}')$, and $k' = (\omega', \omega' \sin\theta, 0, \omega' \cos\theta)$ as well as $v_{ab} = |v_a - v_b| = 1$.[10] Thus, using 4-momentum conservation, we find that $p + k = p' + k'$, where all 4-momenta are known in terms of m_e, ω, and ω', except from the 4-momentum p'. Therefore, we want to remove information on the final electron, since it is described by the 4-momentum p' and only its mass, which is of course m_e, is well known. This is performed by using the fact that p'^2 is a Lorentz invariant and given by $p'^2 = m_e^2$. Now, we obtain

$$m_e^2 = p'^2 = (p+k-k')^2 = p^2 + 2p \cdot (k-k') + (k-k')^2 = m_e^2 + 2p \cdot (k-k') - 2k \cdot k',$$
(12.50)

---

[10] Especially, inserting $k = (\omega, \omega \mathbf{e}_z)$ into Eq. (12.49), it immediately follows that $\mathcal{M}^0 = \mathcal{M}^3$.

Before:

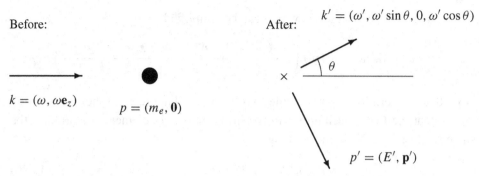

$k = (\omega, \omega e_z)$

$p = (m_e, 0)$

After:

$k' = (\omega', \omega' \sin \theta, 0, \omega' \cos \theta)$

$p' = (E', \mathbf{p}')$

Figure 12.13 The kinematics in the lab system for the reaction $e^- + \gamma \to e^- + \gamma$.

which means that

$$p \cdot (k - k') - k \cdot k' = 0. \tag{12.51}$$

Inserting the expressions for the 4-momenta in the lab system into Eq. (12.51), we have

$$m_e(\omega - \omega') - \omega \omega'(1 - \cos \theta) = 0, \tag{12.52}$$

which leads to

$$\frac{1}{\omega'} - \frac{1}{\omega} = \frac{1}{m_e}(1 - \cos \theta), \tag{12.53}$$

i.e. *Compton's formula* for the shift in the photon wave-length. Thus, using Eqs. (11.182) and (12.48) as well as inserting Eq. (12.53) and $v_{ab} = 1$, the differential cross-section in the lab system for Compton scattering is given by

$$-\frac{d\sigma}{d\cos\theta} = \frac{1}{2\omega \cdot 2m_e \cdot 1} \cdot \frac{1}{8\pi} \frac{\omega'^2}{\omega m_e} \overline{|\mathcal{M}|^2} = \frac{\pi \alpha^2}{m_e^2} \left(\frac{\omega'}{\omega}\right)^2 \left(\frac{\omega'}{\omega} + \frac{\omega}{\omega'} - \sin^2 \theta\right), \tag{12.54}$$

which is known as the *(unpolarized) Klein–Nishina formula*.[11] In the limit $\omega \to 0$ (or $\omega'/\omega \to 1$), we obtain

$$-\frac{d\sigma}{d\cos\theta} \to \frac{\pi \alpha^2}{m_e^2}(1 + \cos^2 \theta) = \frac{\pi \alpha^2}{m_e^2}[2 - \theta^2 + \mathcal{O}(\theta^4)], \tag{12.55}$$

$$\sigma = \int_1^{-1} \left(-\frac{d\sigma}{d\cos\theta}\right)(-d\cos\theta) = \int_{-1}^1 \left(-\frac{d\sigma}{d\cos\theta}\right) d\cos\theta \to \frac{8\pi \alpha^2}{3m_e^2}. \tag{12.56}$$

---

[11] For the interested reader, please see: O. Klein and Y. Nishina, Über die Streuung von Strahlung durch freie Elektronen nach der neuen relativistischen Quantendynamik von Dirac, *Z. Phys.* **52**, 853–868 (1929).

Note that Eq. (12.56) is the famous *Thomson cross-section* for scattering of classical electromagnetic radiation by a free electron (or in general by a charged particle).[12]

Another reaction, which is related by crossing symmetry to Compton scattering, is the reaction $e^+ + e^- \rightarrow 2\gamma$, which is normally referred to as *pair annihilation*. This reaction is obtained by making the replacements $p \rightarrow p_1$, $p' \rightarrow -p_2$, $k \rightarrow -k_1$, and $k' \rightarrow k_2$. In the CM system and in the limit $E \gg m_e$, we obtain the differential cross-section

$$-\frac{d\sigma}{d\cos\theta} \rightarrow \frac{2\pi\alpha^2}{s} \frac{1+\cos^2\theta}{\sin^2\theta} = \frac{2\pi\alpha^2}{s}\left[\frac{2}{\theta^2} - \frac{1}{3} + \frac{2}{15}\theta^2 + \mathcal{O}(\theta^4)\right]. \quad (12.57)$$

The total cross-section is given by

$$\sigma = \int_1^0 \left(-\frac{d\sigma}{d\cos\theta}\right)(-d\cos\theta) = \int_0^1 \left(-\frac{d\sigma}{d\cos\theta}\right)d\cos\theta, \quad (12.58)$$

since the two photons are identical, and therefore, we only have to integrate over $0 \le \theta \le \pi/2$.

In Fig. 12.14, we show results from the HRS, OPAL, and ALEPH Collaborations of the differential cross-section for the reaction $e^+ + e^- \rightarrow 2\gamma$ at $\sqrt{s} = 29$ GeV, $\sqrt{s} = 91.16$ GeV, and $\sqrt{s} = 183$ GeV, respectively.

Exercise 12.6 Derive the invariant matrix element in Eq. (12.44) and the differential cross-section in Eq. (12.54).

Problems

(1) Calculate the total cross-section for the annihilation of electrons and positrons into muons and antimuons, i.e. $e^- + e^+ \rightarrow \mu^- + \mu^+$. Perform the calculation in the centre-of-mass system, which is also the laboratory system in a colliding beam accelerator, where this experiment would usually be carried out. When the energy E of the electron becomes very large, show that the unpolarized total cross-section becomes

$$\sigma = \frac{4\pi\alpha^2}{3s},$$

where $s \equiv 4E^2$.

(2) *Electron–antimuon scattering.* Calculate the differential cross-section for the scattering reaction

$$e^- + \mu^+ \rightarrow e^- + \mu^+$$

in the centre-of-mass system. Assume that the initial particles have positive or negative helicity, whereas the spin states of the final particles are arbitrary.

---

[12] In 1906, J. J. Thomson received the Nobel Prize in physics 'in recognition of the great merits of his theoretical and experimental investigations on the conduction of electricity by gases'.

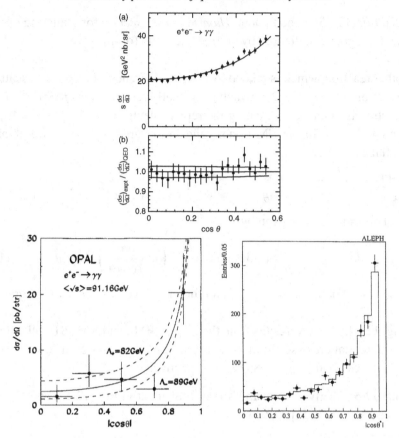

Figure 12.14 *Upper figure:* the differential cross-section as a function of $\cos\theta$ at $\sqrt{s} = 29$ GeV for the reaction $e^+ + e^- \to 2\gamma$. In (a), the data points have been adjusted to remove acceptance losses and to remove the calculated α^3 QED effects. The curve is the second-order QED prediction. In (b), the ratio between the observed differential distribution and the full Monte Carlo simulation of the experiment, including α^3 QED contributions. The upper and lower curves, respectively, represent the lower bounds (at 95% C.L.) of the QED-cut-off parameters. Reprinted figure with permission from the HRS Collaboration, M. Derrick *et al.*, *Phys. Rev. D* **34**, 3286–3303 (1986). © 1986 by the American Physical Society. *Lower-left figure:* the differential cross-section as a function of $|\cos\theta|$ at $\sqrt{s} = 91.16$ GeV for the reaction $e^+ + e^- \to 2\gamma$, which is shown as points with error bars, by the OPAL Collaboration, M. Z. Akrawy *et al.*, *Phys. Lett. B* **241**, 133–140 (1990). In addition, the solid curve shows the QED prediction, whereas the dashed curves show the expectations with cut-off parameters. © 1990 by Elsevier Science B.V. *Lower-right figure:* the differential cross-section as a function of $\cos\theta^*$, where θ^* is the production angle, at $\sqrt{s} = 183$ GeV for the reaction $e^+ + e^- \to 2\gamma$ by the ALEPH Collaboration, R. Barate *et al.*, *Phys. Lett. B* **429**, 201–214 (1998). The figure show both the predicted and observed lowest-order distributions, where the predicted distribution includes a small contribution from the Bhabha background. Note that the errors presented are purely statistical. © 1998 by Elsevier Science B.V.

(3) *Bhabha scattering.*

   (a) Calculate the differential cross-section $d\sigma/d\cos\theta$ for Bhabha scattering, i.e. $e^+ + e^- \to e^+ + e^-$, in the limit $E_{CM} \gg m_e$. Introduce the Mandelstam variables s, t, and u, which should fulfil $s + t + u = 0$, since the electron mass can be neglected. The differential cross-section should be

$$\frac{d\sigma}{d\cos\theta} = -\frac{\pi\alpha^2}{s}\left[u^2\left(\frac{1}{s}+\frac{1}{t}\right)^2 + \left(\frac{t}{s}\right)^2 + \left(\frac{s}{t}\right)^2\right].$$

   (b) Rewrite this formula in terms of $\cos\theta$ and plot it.

   (c) What feature of the Feynman diagrams causes this differential cross-section to diverge as $\theta \to 0$?

(4) Consider the annihilation of electron–positron pairs into two photons, i.e. $e^- + e^+ \to 2\gamma$.

   (a) Draw all Feynman diagrams that contribute to this process to order e^2. Let the 4-momenta of the incoming electron and positron be p_- and p_+, respectively, and of the outgoing photons k_1 and k_2, respectively. Label each diagram with these momenta and the momenta of any internal lines.

   (b) Write down the correct Feynman amplitude for each diagram.

Guide to additional recommended reading

The following books (see the indicated pages) and their authors have similar treatments of the content in the present chapter.

- W. Greiner and J. Reinhardt, *Quantum Electrodynamics*, 3rd edn., Springer (2003), pp. 83–257.
- F. Gross, *Relativistic Quantum Mechanics and Field Theory*, Wiley (1993), pp. 287–318.
- F. Halzen and A. D. Martin, *Quarks & Leptons: An Introductory Course in Modern Particle Physics*, Wiley (1984), pp. 119–132, 141–145.
- C. Itzykson and J.-B. Zuber, *Quantum Field Theory*, McGraw-Hill (1985), pp. 276–282.
- F. Mandl and G. Shaw, *Quantum Field Theory*, rev. edn, Wiley (1994), pp. 146–165.
- M. E. Peskin and D. V. Schroeder, *An Introduction to Quantum Field Theory*, Addison-Wesley (1995), pp. 131–174.
- F. Schwabl, *Advanced Quantum Mechanics*, Springer (1999), pp. 346–358.
- F. J. Ynduráin, *Relativistic Quantum Mechanics and Introduction to Field Theory*, Springer (1996), pp. 229–265.
- For the interested reader: OPAL Collaboration, M. Z. Akrawy *et al.*, A study of the reaction $e^+e^- \to \gamma\gamma$ at LEP, *Phys. Lett. B* **241**, 133–140 (1990); KEDR Collaboration, V. V. Anashin *et al.*, New precise determination of the τ lepton mass at KEDR Detector, *Nucl. Phys. B (Proc. Suppl.)* **169**, 125–131 (2007), hep-ex/0611046; DELCO Collaboration, W. Bacino *et al.*, Measurement of the threshold behavior of $\tau^+\tau^-$ production in e^+e^- annihilation, *Phys. Rev. Lett.* **41**, 13–15 (1978); ALEPH Collaboration, R. Barate *et al.*, Single- and multi-photon production in e^+e^- collisions at a centre-of-mass energy of 183 GeV, *Phys. Lett. B* **429**, 201–214 (1998); CELLO Collaboration, H.-J. Behrend *et al.*, Measurement of $e^+e^- \to e^+e^-$ and $e^+e^- \to \gamma\gamma$ at energies up to

36.7 GeV, *Phys. Lett.* **103B**, 148–152 (1981); PLUTO Collaboration, C. Berger *et al.*, Tests of the standard model with lepton pair production in e^+e^- reactions, *Z. Phys. C* **27**, 341–349 (1985); HRS Collaboration, M. Derrick *et al.*, *Phys. Rev. D* **34**, 3286–3303 (1986); and O. Klein and Y. Nishina, Über die Streuung von Strahlung durch freie Elektronen nach der neuen relativistischen Quantendynamik von Dirac, *Z. Phys.* **52**, 853–868 (1929).

13

Introduction to regularization, renormalization, and radiative corrections

In the previous chapter, we investigated elementary processes in QED and calculated cross-sections to lowest order in perturbation theory. Taking higher orders into account, one will obtain corrections of the order of the Sommerfeld fine-structure constant α to the lowest-order results. However, performing such calculations, one will encounter divergent integrals. Nevertheless, we will try to remedy this situation in three steps. First, we want to regularize the theory, which means that we modify the theory in order to keep it finite and well-defined to all orders in perturbation theory. This step is called *regularization* for which there are several methods. Specific methods of regularization include: Pauli–Villars regularization, dimensional regularization, lattice regularization, Riemann's ζ-function regularization, etc. Here, we will mainly investigate Pauli–Villars regularization and dimensional regularization. Second, we recognize that the non-interacting particles (i.e. the free asymptotic fields) are not identical to the real physical particles that interact. Thus, the interactions modify the properties of the particles such as the masses and the charges. Of course, all the relevant predictions of the theory must be expressed in terms of the properties of the physical particles and not the non-interacting (or bare) particles. This step is called *renormalization*. Third, we have to revert from the regularized theory back to QED, and thus, the infinities of QED will appear in the relations between the bare and physical particles. Of course, these relations as well as the bare particles are completely unobservable. However, the observable predictions of the theory remain finite as QED is restored. In particular, so-called *radiative corrections* are finite and of the order of α.

Using Feynman rules, one can construct Feynman diagrams that describe tree-level processes, i.e. Feynman diagrams, which contain no loops, or higher-order contributions that are called radiative corrections, which contain loops. For example, in QED, at tree level, we have for the reaction $e^- + \mu^- \to e^- + \mu^-$ the Feynman diagram presented in Fig. 13.1. However, the reaction $e^- + \mu^- \to e^- + \mu^-$ also has radiative corrections, which are e.g. the Feynman diagrams given in Fig. 13.2.

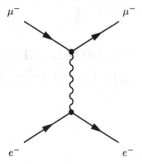

Figure 13.1 Tree-level Feynman diagram for the reaction $e^- + \mu^- \to e^- + \mu^-$.

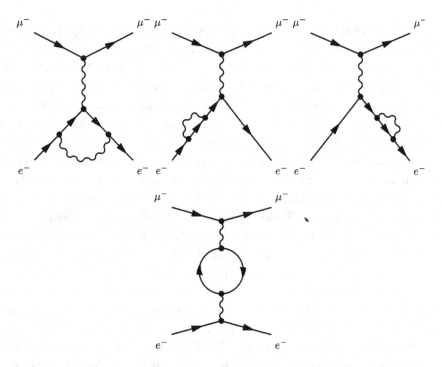

Figure 13.2 Radiative corrections to the reaction $e^- + \mu^- \to e^- + \mu^-$. *Upper-left diagram:* electron vertex correction. *Upper-middle and upper-right diagrams:* external leg corrections. *Lower diagram:* photon self-energy (or vacuum polarization).

In this figure, the upper-left diagram is called the electron *vertex correction* diagram, the upper-middle and upper-right diagrams are called *external leg correction* diagrams, and finally, the lower diagram is called the photon *self-energy* diagram (or the *vacuum polarization* diagram).

Each of the four diagrams in Fig. 13.2 is divergent in the ultraviolet ($k \to \infty$, where k is the loop 4-momentum) region. In addition, the three first diagrams contain infrared divergencies ($k \to 0$). In order to illustrate the two different types of divergencies, consider, for example, the integral

$$\int_{k \le \Lambda} \frac{d^d k}{k^2 - \mu^2 + i\epsilon} \propto \int_0^\Lambda \frac{k^{d-1}\, dk}{k^2 - \mu^2 + i\epsilon}, \tag{13.1}$$

where the quantity Λ is an ultraviolet cut-off in momentum space. The integrand of this integral symbolizes a propagator in d dimensions. In the case when $\mu = 0$, it resembles a photon propagator, whereas in the case when $\mu \ne 0$, it describes a Klein–Gordon propagator. First, for large Λ/μ, the integral is proportional to Λ^{d-2}, unless $d = 2$ when it is proportional to $\log \Lambda$. In quantum field theory and in the continuum limit $\Lambda \to \infty$, the integral diverges for $d \ge 2$. This is known as an *ultraviolet divergency*. Second, on the other hand, if $d \le 2$ and $\mu = 0$ (zero mass), then the integral diverges when $k \to 0$. This is known as an *infrared divergency*. An infrared divergency can be remedied by introducing a fictitious mass $\mu \ne 0$ for the photon. At the end of the calculation, one has to take the limit $\mu \to 0$. This is an example of regularization.

In this chapter, we will investigate the three following radiative corrections:

(1) the electron vertex correction,
(2) the electron self-energy, and
(3) the photon self-energy.

In general, the scheme for the calculational method for all loop diagrams is the following.

- Draw the Feynman diagram(s) and write down the invariant matrix element $\mathcal{M}$.
- Introduce Feynman parameters in order to combine the denominators of the propagators.
- Complete the square in the new denominator by introducing a new loop 4-momentum variable ℓ.
- Write the numerator in terms of ℓ, drop odd powers of ℓ, and rewrite even powers of ℓ.
- Perform the 4-momentum integral using a so-called Wick rotation and four-dimensional spherical coordinates.

Note that the 4-momentum integral will often be divergent. In that case, we have to define (or regularize) the integral using some regularization method.

13.1 The electron vertex correction

The electron vertex correction function $\Gamma^\mu(p', p)$ describes the correction to electron scattering coming from a *virtual* photon. The first-order correction to this function is due to the upper-left Feynman diagram in Fig. 13.2, which is a Feynman diagram with a loop. Note that the zeroth order is given by the Feynman diagram in Fig. 13.1. However, to all orders in perturbation theory, the correction to the function should be given by the sum of electron vertex diagrams $-ie\Gamma^\mu(p, p')$, which is illustrated in Fig. 13.3. In QED, using Feynman rules, the invariant matrix element of the sum of electron vertex diagrams is given by

$$i\mathcal{M} = \bar{u}(p')\left[-ie\Gamma^\mu(p', p)\right]u(p)\frac{-ig_{\mu\nu}}{q^2}\bar{u}(k')\left(-ie\gamma^\nu\right)u(k)$$

$$= ie^2\bar{u}(p')\Gamma^\mu(p', p)u(p)\frac{1}{q^2}\bar{u}(k')\gamma_\mu u(k), \qquad (13.2)$$

where $q = p - p' = k - k'$. Now, what is the structure of the electron vertex correction function $\Gamma^\mu(p', p)$? To lowest (zeroth) order in perturbation theory, we have

$$\Gamma^\mu(p', p) = \gamma^\mu. \qquad (13.3)$$

In general, the electron vertex correction function $\Gamma^\mu(p', p)$ involves the 4-momenta p and p', the gamma matrix γ^μ, and constants such as the coupling constant e, the electron mass m_e, and pure numbers. Then, Lorentz invariance implies that

$$\Gamma^\mu(p', p) = A\gamma^\mu + B\left(p'^\mu + p^\mu\right) + C\left(p'^\mu - p^\mu\right), \qquad (13.4)$$

where A, B, and C are Lorentz scalars. Note that $\Gamma^\mu(p', p)$ transforms as a 4-vector. In addition, using the Ward identity $q_\mu\Gamma^\mu(p', p) \equiv \left(p'_\mu - p_\mu\right)\Gamma^\mu(p', p) = 0$, cf. Eq. (12.49), we find that the Lorentz scalar C is equal to zero, i.e. $C = 0$.

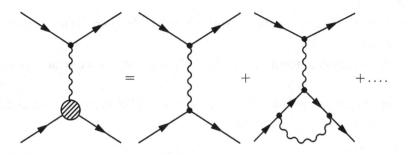

Figure 13.3 The sum of electron vertex diagrams.

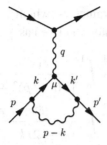

Figure 13.4 Feynman diagram for the electron vertex correction function.

Thus, we can actually write the electron vertex correction function as

$$\Gamma^\mu(p', p) = F_1(q^2)\gamma^\mu + F_2(q^2)\frac{i\sigma^{\mu\nu}q_\nu}{2m_e}, \tag{13.5}$$

where the functions $F_1(q^2)$ and $F_2(q^2)$ are unknown *form factors*. Of course, to lowest (zeroth) order in perturbation theory, we have

$$F_1(q^2) = 1 \quad \text{and} \quad F_2(q^2) = 0. \tag{13.6}$$

Note that the form factors $F_1(q^2)$ and $F_2(q^2)$ can be computed to **any** order in perturbation theory and they contain the complete information about the influence of the electromagnetic field on the electron.

Next, we will evaluate the electron vertex correction function $\Gamma^\mu(p', p)$. To order $\alpha = e^2/(4\pi)$ in perturbation theory, we can write

$$\Gamma^\mu(p', p) = \gamma^\mu + \delta\Gamma^\mu(p', p), \tag{13.7}$$

where $\delta\Gamma^\mu(p', p)$ is the order α correction. Using Feynman rules for the loop diagram shown in Fig. 13.4, we obtain this correction as

$$\bar{u}(p')\delta\Gamma^\mu(p', p)u(p)$$

$$= \int \frac{d^4k}{(2\pi)^4} \frac{-ig_{\nu\rho}}{(k-p)^2 + i\epsilon} \bar{u}(p')(-ie\gamma^\nu)\frac{i(\slashed{k}' + m_e\mathbb{1}_4)}{k'^2 - m_e^2 + i\epsilon}\gamma^\mu \frac{i(\slashed{k} + m_e\mathbb{1}_4)}{k^2 - m_e^2 + i\epsilon}(-ie\gamma^\rho)u(p)$$

$$= 2ie^2 \int \frac{d^4k}{(2\pi)^4} \frac{\bar{u}(p')\left[\slashed{k}\gamma^\mu\slashed{k}' + m_e^2\gamma^\mu - 2m_e(k+k')^\mu\mathbb{1}_4\right]u(p)}{[(k-p)^2 + i\epsilon]\left(k'^2 - m_e^2 + i\epsilon\right)\left(k^2 - m_e^2 + i\epsilon\right)}. \tag{13.8}$$

It is a difficult exercise to compute this integral. Nevertheless, in order to compute the integral, we will introduce so-called *Feynman parameters*. First, we investigate the relation

$$\frac{1}{AB} = \int_0^1 \frac{dx}{[xA + (1-x)B]^2} = \int_0^1\int_0^1 \frac{\delta(x+y-1)\,dx\,dy}{(xA + yB)^2}, \tag{13.9}$$

where x and y are Feynman parameters, which can be used to rewrite the factor $\left[(k-p)^2\left(k^2-m_e^2\right)\right]^{-1}$ in the integral (13.8) as

$$\frac{1}{(k-p)^2\left(k^2-m_e^2\right)} = \int_0^1\int_0^1\frac{\delta(x+y-1)\,dxdy}{\left[x(k-p)^2+y\left(k^2-m_e^2\right)\right]^2}$$

$$= \int_0^1\int_0^1\frac{\delta(x+y-1)\,dxdy}{\left(k^2-2xk\cdot p+xp^2-ym_e^2\right)^2}. \tag{13.10}$$

However, this is not enough in order to rewrite the full denominator in the integral, since it contains three factors. Second, in general, one can prove that

$$\frac{1}{A_1^{m_1}A_2^{m_2}\ldots A_n^{m_n}}$$

$$= \int_0^1\int_0^1\cdots\int_0^1\frac{\delta\left(\sum_{i=1}^n x_i-1\right)\prod_{i=1}^n x_i^{m_i-1}}{\left(\sum_{i=1}^n x_iA_i\right)^{\sum_{i=1}^n m_i}}\frac{\Gamma\left(\sum_{i=1}^n m_i\right)}{\prod_{i=1}^n\Gamma(m_i)}\,dx_1dx_2\ldots dx_n,$$

$$\tag{13.11}$$

where $\Gamma(x)$ is the gamma function. Thus, we have

$$\frac{1}{\left[(k-p)^2+i\epsilon\right]\left(k'^2-m_e^2+i\epsilon\right)\left(k^2-m_e^2+i\epsilon\right)}$$

$$= \int_0^1\int_0^1\int_0^1\frac{2}{D^3}\delta(x+y+z-1)\,dxdydz, \tag{13.12}$$

where the function D is given by

$$D = x\left(k^2-m_e^2\right)+y\left(k'^2-m_e^2\right)+z(k-p)^2+(x+y+z)i\epsilon. \tag{13.13}$$

Now, it holds that

$$x+y+z=1 \quad\text{and}\quad k'=k+q, \tag{13.14}$$

which implies that the function D can be simplified to

$$D = k^2+2k\cdot(yq-zp)+yq^2+zp^2-(x+y)m_e^2+i\epsilon. \tag{13.15}$$

Then, we introduce the variable $\ell=k+yq-zp$, which means that

$$D = \ell^2-\Delta+i\epsilon, \tag{13.16}$$

where

$$\Delta = -xyq^2+(1-z)^2m_e^2 > 0, \tag{13.17}$$

since for scattering we have $q^2 < 0$. Next, express the numerator in terms of the variable ℓ. Thus, the integral (13.8) can be written after a long derivation as

$$
\bar{u}(p')\delta\Gamma^\mu(p', p)u(p) = 2ie^2 \int \frac{d^4\ell}{(2\pi)^4} \int_0^1 \frac{2}{D^3}\delta(x + y + z - 1)\,dxdydz
$$

$$
\times \bar{u}(p')\left\{\gamma^\mu \left[-\tfrac{1}{2}\ell^2 + (1-x)(1-y)q^2 + (1 - 4z + z^2)m_e^2\right] \right.
$$

$$
\left. + \frac{i\sigma^{\mu\nu}q_\nu}{2m_e}2m_e^2 z(1-z)\right\}u(p). \tag{13.18}
$$

Some useful integrals are the following:

$$
\int \frac{d^4\ell}{(2\pi)^4}\frac{1}{(\ell^2 - \Delta)^m} = \frac{i(-1)^m}{(4\pi)^2}\frac{1}{(m-1)(m-2)}\frac{1}{\Delta^{m-2}}, \tag{13.19}
$$

$$
\int \frac{d^4\ell}{(2\pi)^4}\frac{\ell^2}{(\ell^2 - \Delta)^m} = \frac{i(-1)^{m-1}}{(4\pi)^2}\frac{2}{(m-1)(m-2)(m-3)}\frac{1}{\Delta^{m-3}}, \tag{13.20}
$$

where the second integral is only valid when $m > 3$ and not when $m = 3$. However, we need the second integral when $m = 3$. This will lead to a *divergency*. Thus, we have to regularize the integral, which will be performed by introducing a *cut-off* in the photon propagator as follows

$$
\frac{1}{(k - p)^2 + i\epsilon} \rightarrow \frac{1}{(k - p)^2 + i\epsilon} - \frac{1}{(k - p)^2 - \Lambda^2 + i\epsilon}, \tag{13.21}
$$

where Λ is the cut-off and it is assumed to be very large. In fact, Λ acts as a large 'mass' in the second part of the photon propagator that comes with the wrong sign. This regularization procedure is called *Pauli–Villars regularization*. For $(k - p)^2 \gtrsim \Lambda^2$, the two terms on the right-hand side in Eq. (13.21) almost cancel each other, i.e. there is basically no photon propagator, whereas for $(k - p)^2 \lesssim \Lambda^2$, the photon propagator is, in principle, unaffected. In the limit $\Lambda \to \infty$, the propagator on the right-hand side in Eq. (13.21) is equal to the original propagator on the left-hand side in Eq. (13.21). By the introduction of the cut-off, we also need to transform the function Δ in the following way

$$\Delta \to \Delta_\Lambda = -xyq^2 + (1-z)^2 m_e^2 + z\Lambda^2. \tag{13.22}$$

Hence, we find that

$$\int \frac{\mathrm{d}^4\ell}{(2\pi)^4} \left[\frac{\ell^2}{(\ell^2 - \Delta)^3} - \frac{\ell^2}{(\ell^2 - \Delta_\Lambda)^3} \right] = \frac{i}{(4\pi)^2} \log \frac{\Delta_\Lambda}{\Delta}. \tag{13.23}$$

Finally, the electron vertex correction function is given by

$$\delta\Gamma^\mu(p', p) = \frac{\alpha}{2\pi} \int_0^1 \int_0^1 \int_0^1 \delta(x + y + z - 1)\,\mathrm{d}x\mathrm{d}y\mathrm{d}z$$

$$\times \left(\gamma^\mu \left\{ \log \frac{z\Lambda^2}{\Delta} + \frac{1}{\Delta} \left[(1-x)(1-y)q^2 + (1 - 4z + z^2)m_e^2 \right] \right\} \right.$$

$$\left. + \frac{i\sigma^{\mu\nu}q_\nu}{2m_e} \frac{1}{\Delta} 2m_e^2 z(1-z) \right), \tag{13.24}$$

which can be computed numerically. Thus, to first order in α, we obtain the form factors

$$F_1(q^2) = 1 + \frac{\alpha}{2\pi} \int_0^1 \int_0^1 \int_0^1 \delta(x + y + z - 1)\,\mathrm{d}x\mathrm{d}y\mathrm{d}z$$

$$\times \left[\log \frac{m_e^2(1-z)^2}{m_e^2(1-z)^2 - q^2 xy} + \frac{m_e^2(1 - 4z + z^2) + q^2(1-x)(1-y)}{m_e^2(1-z)^2 - q^2 xy + \mu^2 z} \right.$$

$$\left. - \frac{m_e^2(1 - 4z + z^2)}{m_e^2(1-z)^2 + \mu^2 z} \right] + \mathcal{O}(\alpha^2), \tag{13.25}$$

$$F_2(q^2) = \frac{\alpha}{2\pi} \int_0^1 \int_0^1 \int_0^1 \delta(x + y + z - 1)\,\mathrm{d}x\mathrm{d}y\mathrm{d}z \frac{2m_e^2 z(1-z)}{m_e^2(1-z)^2 - q^2 xy} + \mathcal{O}(\alpha^2), \tag{13.26}$$

where μ is a pretend small non-zero 'mass' for the photon. Note that the form factor $F_2(q^2)$ is not affected by either the ultraviolet or the infrared divergency. Thus, we can e.g. compute the value of the form factor $F_2(q^2)$ at $q^2 = 0$. We obtain

$$F_2(0) = \frac{\alpha}{\pi} \int_0^1 \int_0^1 \int_0^1 \frac{z}{1-z} \delta(x + y + z - 1)\,\mathrm{d}x\mathrm{d}y\mathrm{d}z$$

$$= \frac{\alpha}{\pi} \int_0^1 \int_0^{1-z} \frac{z}{1-z}\,\mathrm{d}y\mathrm{d}z = \frac{\alpha}{2\pi}. \tag{13.27}$$

Finally, $F_2(0)$ corresponds to the *anomalous magnetic moment* of the electron (cf. the discussion on the anomalous magnetic moment in Section 3.8), which is defined in terms of the *Landé g-factor* as

$$a_e \equiv \frac{g-2}{2} = F_2(0) \simeq \frac{\alpha}{2\pi} \approx 0.001\,1614, \tag{13.28}$$

which was first computed by Schwinger in 1948. Theoretically, to fourth order in α, i.e. α^4, the calculation requires the evaluation of contributions from about 80 Feynman diagrams (each contribution is typically an eleven-dimensional integral with as many as 2000 algebraic terms), which results in $a_e = 0.001\,159\,652\,140 \pm 0.000\,000\,000\,028$, where the error in the last digits comes nearly entirely from the experimental error in the Sommerfeld fine-structure constant α. Experimentally, the value is $a_e = 0.001\,159\,652\,180\,73 \pm 0.000\,000\,000\,000\,28$.[1] Thus, the theoretical prediction for the anomalous magnetic moment of the electron is in remarkable agreement with the experimental value.

Exercise 13.1 Introduce the interaction Hamiltonian

$$\Delta H_I = e \int A_\mu^{\text{cl}}(x)\, j^\mu(x)\, \mathrm{d}^3 x, \tag{13.29}$$

where $j^\mu(x) = \bar{\psi}(x)\gamma^\mu\psi(x)$ is the current and $A_\mu^{\text{cl}}(x)$ is a fixed classical potential, which implies that the invariant matrix element can be written as

$$i\mathcal{M}(2\pi)\delta(p'^0 - p^0) = -ie\bar{u}(p')\Gamma^\mu(p', p)u(p)\tilde{A}_\mu^{\text{cl}}(p' - p), \tag{13.30}$$

where $\tilde{A}_\mu^{\text{cl}}(q)$ is the four-dimensional Fourier transform of $A_\mu^{\text{cl}}(x)$. How does the quantity $F_2(0)$ enter in the Hamiltonian (13.29) as an effective interaction?

13.2 The electron self-energy

In QED, the complete two-point correlation function is given by

$$\tau(x, y) = \langle\Omega|T\psi(x)\bar{\psi}(y)|\Omega\rangle = y \longrightarrow \bigotimes \longrightarrow x$$

$$= y \longrightarrow x + y \longrightarrow \overset{\frown}{\longrightarrow} x + \cdots, \tag{13.31}$$

which contains all connected Feynman diagrams with two external points, i.e. the full electron propagator. Note that the first term is the free-field electron propagator, whereas the second term is the electron self-energy, which is the first radiative correction to the free-field propagator. In momentum space, the free-field (or bare) electron propagator is

$$\xrightarrow{\hspace{1cm}}_{p} = \frac{i(\not{p} + m_0\mathbb{1}_4)}{p^2 - m_0^2 + i\epsilon}, \tag{13.32}$$

[1] Particle Data Group, K. Nakamura *et al.*, The 2010 Review of Particle Physics, *J. Phys. G* **37**, 075021 (2010), pdg.lbl.gov.

where m_0 is the bare electron mass. Note that this propagator has a pole at $p^2 = m_0^2$. Furthermore, the electron self-energy is

$$\frac{i(\not{p} + m_0 \mathbb{1}_4)}{p^2 - m_0^2} \left[-i\Sigma_2(p)\right] \frac{i(\not{p} + m_0 \mathbb{1}_4)}{p^2 - m_0^2}, \qquad (13.33)$$

where $-i\Sigma_2(p)$ is the first radiative correction.

In general, the full (or dressed) electron propagator, which is the Fourier transform of Eq. (13.31), can be expressed as a series

$$iS_F'(p) = iS_F(p) + iS_F(p)\Sigma(p)S_F(p) + iS_F(p)\Sigma(p)S_F(p)\Sigma(p)S_F(p) + \cdots$$
$$= iS_F(p) + iS_F(p)\Sigma(p)\left[S_F(p) + S_F(p)\Sigma(p)S_F(p) + \cdots\right], \quad (13.34)$$

where

$$S_F(p) = \frac{1}{\not{p} - m_0 \mathbb{1}_4} = \frac{\not{p} + m_0 \mathbb{1}_4}{p^2 - m_0^2} \qquad (13.35)$$

and $\Sigma(p)$ denotes the sum of all one-particle irreducible diagrams[2] with two external electron lines and is sometimes referred to as the proper electron self-energy. In perturbation theory, to leading order in α, we, of course, have $\Sigma(p) = \Sigma_2(p)$. Writing Eq. (13.34) in closed form, we obtain

$$iS_F'(p) = iS_F(p) + iS_F(p)\Sigma(p)S_F'(p), \qquad (13.36)$$

which is called the *Dyson equation* with the solution

$$S_F'(p) = \frac{1}{S_F(p)^{-1} - \Sigma(p)}. \qquad (13.37)$$

This full electron propagator has a simple pole, which is shifted away from m_0 by $\Sigma(p)$. The location of this pole, the *physical* (or *renormalized*) *electron mass* m, is the solution to the equation

$$\left[S_F(p)^{-1} - \Sigma(p)\right]\Big|_{\not{p}=m\mathbb{1}_4} = 0. \qquad (13.38)$$

Thus, the mass of the electron is modified by emission and reabsorption of virtual photons and we can write the full electron propagator as

$$iS_F'(p) = \frac{iZ_2(\not{p} + m\mathbb{1}_4)}{p^2 - m^2 + i\epsilon}, \qquad (13.39)$$

where the quantity Z_2 is given by

$$Z_2^{-1} = 1 - \frac{d\Sigma(p)}{d\not{p}}(\not{p} = m\mathbb{1}_4) \qquad (13.40)$$

and is known as the *electron wave function renormalization constant*.

---

[2] A *one-particle irreducible diagram* is a diagram, which cannot be divided into two diagrams by removing a single line.

In particular, using the Feynman rules for QED, we obtain

$$-i\Sigma_2(p) = (-ie_0)^2 \int \frac{d^4k}{(2\pi)^4} \gamma^\mu \frac{i(\not{k} + m_0 \mathbb{1}_4)}{k^2 - m_0^2 + i\epsilon} \gamma_\mu \frac{-i}{(p-k)^2 - \mu^2 + i\epsilon}, \quad (13.41)$$

where e_0 is the bare electron charge. Note that the integral $\Sigma_2(p)$ has an infrared divergency (when $k \to 0$), which has been regularized by introducing an artificial small non-zero photon 'mass' μ that eventually will be set to zero. In addition, $\Sigma_2(p)$ is ultraviolet divergent, since the integrand falls off only as k^{-3}. Thus, $\Sigma_2(p)$ should naively diverge linearly at the upper limit, i.e. in the limit $k \to \infty$. Introducing a Feynman parameter and performing Pauli–Villars regularization, we find that

$$\Sigma_2(p) = \frac{\alpha}{2\pi} \int_0^1 (2m_0 \mathbb{1}_4 - x\not{p}) \log \frac{x\Lambda^2}{(1-x)m_0^2 + x\mu^2 - x(1-x)p^2} \, dx, \quad (13.42)$$

and therefore, we observe that $\Sigma_2(p)$ diverges logarithmically only. Now, to order α in perturbation theory, the mass shift (or the mass of the mass counter-term $-\delta m \bar{\psi}\psi$) is given by

$$\delta m = m - m_0 = \Sigma_2(\not{p} = m\mathbb{1}_4) \simeq \Sigma_2(\not{p} = m_0\mathbb{1}_4), \quad (13.43)$$

where m is the physical electron mass, which implies that

$$\delta m = \frac{\alpha}{2\pi} m_0 \int_0^1 (2-x) \log \frac{x\Lambda^2}{(1-x)^2 m_0^2 + x\mu^2} \, dx, \quad (13.44)$$

where the cut-off Λ is large. In fact, note that the mass shift is ultraviolet divergent. Thus, in the limit $\Lambda \to \infty$, we obtain

$$\delta m \to \frac{3\alpha}{4\pi} m_0 \log \frac{\Lambda^2}{m_0^2}. \quad (13.45)$$

This is known as *mass renormalization*, which means that the bare mass m_0 is replaced by the physical mass m. Note that the experimentally observed mass of the electron is m and **not** m_0. Thus, predictions from the theory must be expressed in terms of the observable properties of the real interacting particles. Is there a problem that the physical mass m and the bare mass m_0 are different by a divergent quantity δm? No, since the divergent quantity δm occurs only in an unobservable relation $m = m_0 + \delta m$, connecting the physical and bare masses m and m_0, and thus, the divergency is 'absorbed' by δm.

In addition, to order α in perturbation theory, the correction to Z_2 is given by

$$\delta Z_2 = Z_2 - 1 = \frac{\mathrm{d}\Sigma_2}{\mathrm{d}\not{p}}(\not{p} = m\mathbb{1}_4)$$

$$= \frac{\alpha}{2\pi} \int_0^1 \left[-x \log \frac{x\Lambda^2}{(1-x)^2 m^2 + x\mu^2} + 2(2-x)\frac{x(1-x)m^2}{(1-x)^2 m^2 + x\mu^2} \right] \mathrm{d}x,$$

$$(13.46)$$

which is also logarithmically ultraviolet divergent.

Finally, a propagator connects two vertices, each contributing a factor e_0. Thus, Z_2 can be divided into two factors of $\sqrt{Z_2}$, and since two electrons interact at each vertex, the value of the electron charge can be renormalized as

$$e' = Z_2 e_0, \qquad (13.47)$$

which is the *preliminary physical* (or *renormalized*) *electron charge*. Of course, to lowest order in perturbation theory, $Z_2 = 1$, and thus, $e' = e_0$.

13.3 The photon self-energy

In QED, the photon self-energy (or the vacuum polarization) describes a process in which a background electromagnetic field produces a *virtual* electron–positron pair that affects the distribution of charges and currents, which generated the original electromagnetic field. Such a virtual electron–positron pair is short-lived, and therefore, the particles in the pair subsequently annihilate each other. In addition, this pair is itself charged, which means that it will *polarize* the vacuum, and hence, it acts as an *electric dipole*. Thus, the pair is partially counteracting the electromagnetic field, i.e. a dielectric effect that is a *partial screening effect*. Therefore, the effective electromagnetic field will be weaker than without the vacuum fluctuation. Similar to the electron self-energy, the photon self-energy leads to infinities, which will be removed by renormalization.

We will now investigate the effects of the photon self-energy on the photon propagator. In QED, the complete photon two-point correlation function is given by

$$\tau_{\mu\nu}(x,y) = \langle \Omega | T A_\mu(x) A_\nu(y) | \Omega \rangle = y,\nu \ \text{〰〰}\bigotimes\text{〰〰} \ x,\mu$$

$$= y,\nu \ \text{〰〰〰〰} \ x,\mu + y,\nu \ \text{〰〰}\bigcirc\text{〰〰} \ x,\mu + \cdots, \qquad (13.48)$$

which contains all connected Feynman diagrams with two external points, i.e. the full photon propagator. Note that the first term is the free-field photon propagator, whereas the second term is the photon self-energy, which is the first radiative correction to the free-field propagator.

In momentum space, the free-field (or bare) photon propagator is

$$\nu \,\wwwwwww\, \mu = \frac{-ig_{\mu\nu}}{q^2 + i\epsilon}. \qquad (13.49)$$

Furthermore, the photon self-energy is

$$\nu \,\wwww\,\bigcirc\,\wwww\, \mu = \frac{-ig_{\mu\alpha}}{q^2 + i\epsilon}\left[-i\Pi_2^{\alpha\beta}(q)\right]\frac{-ig_{\beta\nu}}{q^2 + i\epsilon}, \qquad (13.50)$$

where $-i\Pi_2^{\alpha\beta}(q)$ is the first radiative correction.

In general, the full (or dressed) photon propagator is given by

$$iD'_{F\mu\nu}(q) = iD_{F\mu\nu}(q) + iD_{F\mu\alpha}(q)\Pi^{\alpha\beta}(q)D'_{F\beta\nu}(q)$$

$$= iD_{F\mu\nu}(q) + iD_{F\mu\alpha}(q)\left(\delta_\nu^\alpha - \frac{q^\alpha q_\nu}{q^2}\right)\left[\Pi(q^2) + \Pi^2(q^2) + \cdots\right]$$

$$= \frac{-i}{q^2\left[1 - \Pi(q^2)\right]}\left(g_{\mu\nu} - \frac{q_\mu q_\nu}{q^2}\right) + \frac{-i}{q^2}\frac{q_\mu q_\nu}{q^2}, \qquad (13.51)$$

where

$$D_{F\mu\nu}(q) = \frac{-g_{\mu\nu}}{q^2} \qquad (13.52)$$

and $\Pi^{\alpha\beta}(q) = (q^2 g^{\alpha\beta} - q^\alpha q^\beta)\Pi(q^2)$ denotes the sum of all one-particle irreducible diagrams with two external photon lines and is sometimes referred to as the proper photon self-energy. According to the Ward identity, terms proportional to q_μ or q_ν must vanish. Therefore, we have

$$iD'_{F\mu\nu}(q) = iD_{F\mu\nu}(q)\frac{1}{1 - \Pi(q^2)}. \qquad (13.53)$$

Note that when $\Pi(q^2)$ is regular at $q^2 = 0$, the full photon propagator has a pole at $q^2 = 0$. Thus, the photon remains massless at **all** orders in perturbation theory. Now, the *photon field renormalization constant Z_3* is defined as

$$Z_3 = \frac{1}{1 - \Pi(0)}. \qquad (13.54)$$

For small q, i.e. close to the pole, the full photon propagator takes the form

$$iD'_{F\mu\nu}(q) = \frac{-iZ_3 g_{\mu\nu}}{q^2 + i\epsilon}. \qquad (13.55)$$

In particular, using the Feynman rules for QED, we obtain

$$i\Pi_2^{\alpha\beta}(q) = (-1)(-ie_0)^2\int\frac{d^4k}{(2\pi)^4}\text{tr}\left(\gamma^\alpha\frac{i}{\slashed{k} - m_0\mathbb{1}_4 + i\epsilon}\gamma^\beta\frac{i}{\slashed{k} + \slashed{q} - m_0\mathbb{1}_4 + i\epsilon}\right), \qquad (13.56)$$

where m_0 and e_0 are the bare electron mass and the bare electron charge, respectively. Note that the integral $\Pi_2^{\alpha\beta}(q)$ is quadratically divergent for large 4-momentum k. In order to handle the integral, we have to regularize it, i.e. we must modify it so that it becomes well-defined and finite. However, in this case, Pauli–Villars regularization will not work. Nevertheless, we will investigate the tensor structure of the integral $\Pi_2^{\alpha\beta}(q)$. It follows from Lorentz invariance that $\Pi_2^{\alpha\beta}(q)$ must be of the form

$$\Pi_2^{\alpha\beta}(q) = A(q^2)g^{\alpha\beta} + B(q^2)q^\alpha q^\beta, \tag{13.57}$$

since this is the most general second-rank tensor which can be formed using only the 4-vector q. In addition, the functions $A(q^2)$ and $B(q^2)$ are two arbitrary functions of q^2. Thus, the only tensors that can appear in $\Pi_2^{\alpha\beta}(q)$ are the metric tensor $g^{\alpha\beta}$ and $q^\alpha q^\beta$. However, the Ward identity implies that $q_\alpha \Pi_2^{\alpha\beta}(q) = 0$. This means that $\Pi_2^{\alpha\beta}(q)$ is proportional to the projector $g^{\alpha\beta} - q^\alpha q^\beta/q^2$. In addition, one would expect that $\Pi_2^{\alpha\beta}(q)$ should not have a pole at $q^2 = 0$. Thus, it is convenient to write the tensor structure of $\Pi_2^{\alpha\beta}(q)$ as follows

$$\Pi_2^{\alpha\beta}(q) = \left(q^2 g^{\alpha\beta} - q^\alpha q^\beta\right) \Pi_2(q^2), \tag{13.58}$$

where the function $\Pi_2(q^2)$ is regular at $q^2 = 0$. Therefore, the vacuum polarization is quantified by the vacuum polarization tensor $\Pi_2^{\mu\nu}(q)$, which describes the dielectric effect as a function of the 4-momentum q carried by the photon. Thus, the vacuum polarization depends on the 4-momentum transfer, or in other words, the dielectric constant is scale dependent.

Introducing a Feynman parameter to combine the denominator factors, writing the numerator in terms of $\ell = k + xq$, and performing a *Wick rotation* by substituting $\ell^0 = i\ell_E^0$, $\ell = \ell_E$, and $\ell^2 = -\ell_E^2$, we find that

$$i\Pi_2^{\alpha\beta}(q) = -4ie_0^2 \int_0^1 dx \int \frac{d^4\ell_E}{(2\pi)^4}$$
$$\times \frac{-\frac{1}{2}g^{\alpha\beta}\ell_E^2 + g^{\alpha\beta}\ell_E^2 - 2x(1-x)q^\alpha q^\beta + g^{\alpha\beta}\left[m_0^2 + x(1-x)q^2\right]}{\left(\ell_E^2 + \Delta\right)^2},$$
$$\tag{13.59}$$

where $\Delta = m_0^2 - x(1-x)q^2$. This integral is ultraviolet divergent. If we use Pauli–Villars regularization with a cut-off Λ, then we would find for the leading term $i\Pi_2^{\alpha\beta}(q) \propto e_0^2\Lambda^2 g^{\alpha\beta}$ with no compensating $q^\alpha q^\beta$ term. Thus, this would violate the Ward identity, since it would give the photon an infinite mass $\mu \propto e_0\Lambda$. Therefore, we need to use another regularization method. We will use *dimensional regularization*, which idea is very simple, i.e. evaluate the Feynman diagram as an analytic function of the dimensionality of spacetime d. For sufficiently small d, any loop 4-momentum integral will be convergent, and therefore, the Ward identity will hold. After renormalization, the final expression for any observable quantity will have a well-defined limit when $d \to 4$.

Some useful d-dimensional integrals are the following:

$$\int \frac{d^d \ell_E}{(2\pi)^d} \frac{1}{\left(\ell_E^2 + \Delta\right)^n} = \frac{1}{(4\pi)^{d/2}} \frac{\Gamma\left(n - \frac{d}{2}\right)}{\Gamma(n)} \left(\frac{1}{\Delta}\right)^{n - \frac{d}{2}}, \tag{13.60}$$

$$\int \frac{d^d \ell_E}{(2\pi)^d} \frac{\ell_E^2}{\left(\ell_E^2 + \Delta\right)^n} = \frac{1}{(4\pi)^{d/2}} \frac{d}{2} \frac{\Gamma\left(n - \frac{d}{2} - 1\right)}{\Gamma(n)} \left(\frac{1}{\Delta}\right)^{n - \frac{d}{2} - 1}. \tag{13.61}$$

In addition, for the gamma matrices in d dimensions, we have the identities (cf. Section 3.3.2)

$$g^{\mu\nu} g_{\mu\nu} = \delta^\mu_\mu = d, \tag{13.62}$$

$$\gamma^\mu \gamma_\mu = d \mathbb{1}_d, \tag{13.63}$$

$$\gamma^\mu \gamma^\nu \gamma_\mu = -(d - 2)\gamma^\nu, \tag{13.64}$$

$$\gamma^\mu \gamma^\nu \gamma^\rho \gamma_\mu = 4g^{\nu\rho} \mathbb{1}_d - (4 - d)\gamma^\nu \gamma^\rho, \tag{13.65}$$

$$\gamma^\mu \gamma^\nu \gamma^\rho \gamma^\sigma \gamma_\mu = -2\gamma^\sigma \gamma^\rho \gamma^\nu + (4 - d)\gamma^\nu \gamma^\rho \gamma^\sigma. \tag{13.66}$$

Using Eq. (13.59) together with the integral formulas (13.60) and (13.61), we obtain

$$\begin{aligned}
i\Pi_2^{\alpha\beta}(q) &= -4ie_0^2 \int_0^1 dx \frac{1}{(4\pi)^{d/2}} \frac{\Gamma\left(2 - \frac{d}{2}\right)}{\Delta^{2 - d/2}} \\
&\quad \times \left\{ g^{\alpha\beta}\left[-m_0^2 + x(1-x)q^2\right] + g^{\alpha\beta}\left[m_0^2 + x(1-x)q^2\right] - 2x(1-x)q^\alpha q^\beta \right\} \\
&= -4ie_0^2 \int_0^1 dx \frac{1}{(4\pi)^{d/2}} \frac{\Gamma\left(2 - \frac{d}{2}\right)}{\Delta^{2 - d/2}} 2x(1-x)\left(q^2 g^{\alpha\beta} - q^\alpha q^\beta\right) \\
&= \left(q^2 g^{\alpha\beta} - q^\alpha q^\beta\right) i\Pi_2(q^2), \tag{13.67}
\end{aligned}$$

which implies that

$$\Pi_2(q^2) = -\frac{8e_0^2}{(4\pi)^{d/2}} \int_0^1 x(1-x) \frac{\Gamma\left(2 - \frac{d}{2}\right)}{\Delta^{2 - d/2}} dx. \tag{13.68}$$

Finally, in the limit $d \to 4$, we obtain

$$\Pi_2(q^2) \to -\frac{2\alpha}{\pi} \int_0^1 x(1-x) \left[\frac{2}{\epsilon} - \log \Delta - \gamma + \log(4\pi)\right] dx, \tag{13.69}$$

where $\epsilon = 4 - d$, since the Γ-function can be written as

$$\Gamma\left(2 - \frac{d}{2}\right) = \Gamma\left(\frac{\epsilon}{2}\right) = \frac{2}{\epsilon} - \gamma + \mathcal{O}(\epsilon), \quad \text{for small } \epsilon, \tag{13.70}$$

where $\gamma \simeq 0.577\,215\,664\,901\,532\,861$ is the *Euler–Mascheroni constant*, which will always cancel out in observable quantities. Thus, the integral is still divergent. In fact, it can be shown that the $1/\epsilon$ pole in dimensional regularization corresponds to a logarithmic divergency in Pauli–Villars regularization.

Finally, again, a propagator connects two vertices, each contributing a factor e_0. Thus, Z_3 can be divided into two factors of $\sqrt{Z_3}$, and since one photon interacts at each vertex, the value of the electron charge can be renormalized as

$$e'' = \sqrt{Z_3}e' = \sqrt{Z_3}Z_2 e_0, \tag{13.71}$$

where e'' is the *preliminary renormalized electron charge*. Again, to lowest order in perturbation theory, $Z_2 = 1$ and $Z_3 = 1$, and thus, $e'' = e_0$.

Now, to order α in perturbation theory, the *electric charge shift* is given by

$$\delta Z_3 = \frac{e^2 - e_0^2}{e_0^2} \simeq \Pi_2(0) \approx -\frac{2\alpha}{3\pi\epsilon}. \tag{13.72}$$

Thus, the bare charge is infinitely larger than the physical charge. However, this difference is **not** observable, but the q^2 dependence of the effective electric charge is observable, which depends on the difference

$$\Pi_2(q^2) - \Pi_2(0) = -\frac{2\alpha}{\pi} \int_0^1 x(1-x) \log \frac{m_0^2}{m_0^2 - x(1-x)q^2} \, dx. \tag{13.73}$$

Note that this difference is independent of ϵ. In particular, for electromagnetism, we can write the Sommerfeld fine-structure constant α as an effective 4-momentum-transfer-dependent quantity. To first order in perturbation theory, we have

$$\alpha_{\text{eff.}}(q^2) = \frac{\alpha}{1 - \left[\Pi_2(q^2) - \Pi(0)\right]}. \tag{13.74}$$

13.4 The renormalized electron charge

In order to complete the electron charge renormalization, we have also to take into account the contribution from the electron vertex correction. Let us therefore consider the electron vertex correction function $\delta\Gamma^\mu$ multiplied from the left and the right by two spinors corresponding to the renormalized (or physical) electron mass m, which leads to

$$\bar{u}(P)\delta\Gamma^\mu(P, P)u(P), \tag{13.75}$$

where the 4-momentum P of the spinors corresponds to real particles. Again, due to Lorentz invariance [cf. Eq. (13.5)], Eq. (13.75) can only be proportional to γ^μ and P^μ. Using the so-called *Gordon identity*

$$\bar{u}_s(p)\gamma^\mu u_{s'}(q) = \frac{1}{2m}\bar{u}_s(p)\left[(p+q)^\mu \mathbb{1}_4 + i\sigma^{\mu\nu}(p-q)_\nu\right]u_{s'}(q), \tag{13.76}$$

where $\bar{u}_s(k)u_{s'}(k) = \delta_{ss'}$, we can replace the 4-momentum P^μ by γ^μ, and thus, we obtain

$$\bar{u}(P)\delta\Gamma^\mu(P, P)u(P) = N\bar{u}(P)\gamma^\mu u(P), \tag{13.77}$$

where N is a constant to be determined. For arbitrary 4-momenta p and p', we have

$$\delta\Gamma^\mu(p', p) = N\gamma^\mu + \Gamma_f^\mu(p', p), \tag{13.78}$$

where the quantity $\Gamma_f^\mu(p', p)$ is finite in the limit $\Lambda \to \infty$, whereas the constant N diverges.

Actually, Eq. (13.5) should correctly read

$$Z_2\Gamma^\mu(p', p) = F_1(q^2)\gamma^\mu + F_2(q^2)\frac{i\sigma^{\mu\nu}q_\nu}{2m}, \tag{13.79}$$

which can be used to reevaluate the form factors to order α in perturbation theory. Since the electron wave function renormalization constant $Z_2 = 1 + \mathcal{O}(\alpha)$ and the form factor $F_2 = \mathcal{O}(\alpha)$, the previous calculation of F_2 is correct. In order to calculate the form factor F_1, we write Eq. (13.79) as

$$Z_2\Gamma^\mu = (1 + \delta Z_2)(\gamma^\mu + \delta\Gamma^\mu) = \gamma^\mu + \delta\Gamma^\mu + \delta Z_2\gamma^\mu. \tag{13.80}$$

Thus, F_1 receives an additional contribution, which is equal to δZ_2. Now, $\delta F_1(q^2)$ denotes the correction to $F_1(q^2)$ and $\delta Z_2 = -\delta F_1(0)$. Then, we have

$$F_1(q^2) = 1 + \delta F_1(q^2) + \delta Z_2 = 1 + \left[\delta F_1(q^2) - \delta F_1(0)\right]. \tag{13.81}$$

Thus, the first term in Eq. (13.78), together with γ^μ, leads to the replacement

$$\gamma^\mu \to (1 + N)\gamma^\mu, \tag{13.82}$$

which produces a final renormalization of the electron charge, i.e.

$$e = (1 + N)e'' \equiv Z_1^{-1}e'', \tag{13.83}$$

where the renormalization constant Z_1 is defined by the relation

$$\Gamma^\mu(q = 0) = Z_1^{-1}\gamma^\mu. \tag{13.84}$$

Note that in order to find $F_1(0) = 1$, we have to show the identity $Z_1 = Z_2$, which means that the electron vertex renormalization constant exactly compensates the electron wave function renormalization constant. This can be shown to all orders in perturbation theory and will be discussed below. Thus, all various renormalization constants for the electron charge yield

$$e_0 \to e = Z_1^{-1}e'' = Z_1^{-1}\sqrt{Z_3}Z_2e_0, \tag{13.85}$$

where the factor $\sqrt{Z_3}$ comes from the vacuum polarization, the factor Z_2 comes from the electron wave function renormalization, and finally, the factor Z_1^{-1} comes from the electron vertex normalization. However, as just stated, in QED, the renormalization constants Z_1 and Z_2 are equal to each other. This can be shown as follows. Differentiating the identity $S_F(p)S_F(p)^{-1} = \mathbb{1}_4$ with respect to the 4-momentum p_μ, we obtain

$$\frac{\partial S_F}{\partial p_\mu} S_F^{-1}(p) + S_F(p) \frac{\partial}{\partial p_\mu} (\not{p} - m\mathbb{1}_4) = 0. \tag{13.86}$$

Multiplying by $S_F(p)$ from the right, we find that

$$\frac{\partial S_F(p)}{\partial p_\mu} = -S_F(p)\gamma^\mu S_F(p), \tag{13.87}$$

which means that differentiation of the electron propagator $S_F(p)$ with respect to p_μ is equivalent to the insertion of a vertex γ^μ in an internal electron line without 4-momentum transfer. Using this identity, we can write the electron vertex correction function

$$\delta\Gamma^\mu(p', p) = ie^2 \int \frac{d^4k}{(2\pi)^4} D_{F\nu\rho}(k)\gamma^\nu S_F(p' - k)\gamma^\mu S_F(p - k)\gamma^\rho \tag{13.88}$$

in the limit of equal 4-momenta as

$$N\gamma^\mu = \lim_{p' \to p} \delta\Gamma^\mu(p', p)\Big|_{\not{p}'=m\mathbb{1}_4, \not{p}=m\mathbb{1}_4}$$

$$= -ie^2 \int \frac{d^4k}{(2\pi)^4} D_{F\nu\rho}(k)\gamma^\nu \frac{\partial S_F(p - k)}{\partial(p - k)_\mu}\gamma^\rho$$

$$= -ie^2 \int \frac{d^4k}{(2\pi)^4} D_{F\nu\rho}(k)\gamma^\nu \frac{\partial S_F(p - k)}{\partial p_\mu}\gamma^\rho, \tag{13.89}$$

whereas from the definition of the renormalization constant Z_2, and using Eqs. (13.41) and (13.89), we have

$$\bar{u}_s(p)\left(Z_2^{-1} - 1\right)\gamma^\mu u_{s'}(p) = \bar{u}_s(p)\left(-\frac{\partial\Sigma}{\partial p_\mu}\right)u_{s'}(p)$$

$$= \left\{\Sigma(p) = ie^2 \int \frac{d^4k}{(2\pi)^4} D_{F\nu\rho}(k)\gamma^\nu S_F(p - k)\gamma^\rho\right\}$$

$$= \bar{u}_s(p)\left[-ie^2 \int \frac{d^4k}{(2\pi)^4} D_{F\nu\rho}(k)\gamma^\nu \frac{\partial S_F(p - k)}{\partial p_\mu}\gamma^\nu\right]u_{s'}(p)$$

$$= \bar{u}_s(p)N\gamma^\mu u_{s'}(p) = \bar{u}_s(p)\left(Z_1^{-1} - 1\right)u_{s'}(p). \tag{13.90}$$

Thus, it follows that $Z_1 = Z_2$. Therefore, the electron charge renormalization simplifies to

$$e_0 \rightarrow e = \sqrt{Z_3}e_0. \qquad (13.91)$$

Note that the renormalized electron charge e is equal to the experimentally measured electron charge $e^2 \equiv 4\pi\alpha \simeq 4\pi/137$. In addition, the bare electron charge e_0 is, according to $e^2 = Z_3 e_0^2$, larger than e. Nevertheless, to lowest order in perturbation theory, $Z_3 = 1$ and we have $e = e_0$.

The prediction that the electron charge renormalization only comes from the photon field renormalization is valid to all orders in perturbation theory. The identity $S_F(p)S_F(p)^{-1} = 1_4$ and its generalization to higher orders as well as its implication $Z_1 = Z_2$ are known as the Ward identity, which is a general consequence of gauge invariance. Thus, expressed in the renormalization constants, the Ward identity reads

$$Z_1 = Z_2. \qquad (13.92)$$

In fact, QED in four spacetime dimensions is renormalizable to all orders in perturbation theory, since at each order in perturbation theory, all divergencies can be 'removed' by means of a finite number of renormalization constants.

Problems

(1) *Electron self-energy.* Show that the Feynman amplitude for the electron self-energy diagram

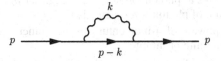

is given by

$$i\mathcal{M}(p) = e^2 \int \frac{d^4k}{(2\pi)^4} D_{F\mu\nu}(k)\bar{u}(p)\gamma^\mu S_F(p-k)\gamma^\nu u(p),$$

where p is the 4-momentum of the electron.
(2) *Photon self-energy (vacuum polarization).* Show that the Feynman amplitude for the photon self-energy diagram

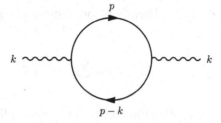

is given by

$$i\mathcal{M}(k) = -e^2 \int \frac{d^4 p}{(2\pi)^4} \text{tr}[\varepsilon_\mu(\mathbf{k})\gamma^\mu S_F(p-k)\varepsilon_\nu(\mathbf{k})^*\gamma^\nu S_F(p)],$$

where $\mathbf{k}$ and $\varepsilon(\mathbf{k})$ are the 3-momentum and polarization vectors of the photon, respectively.

(3) *Photon–photon scattering in QED.*

  (a) Write down the amplitude for the Feynman electron box diagram shown in the following figure:

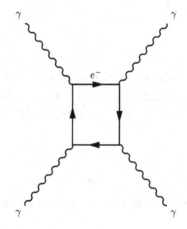

   which contributes to photon–photon scattering. Let p_i and ε_i be the 4-momentum and polarization of photon i, respectively.

  (b) By counting powers of 4-momentum in the numerator and denominator, is the above diagram finite or infinite?

(4) Using dimensional regularization, determine the one-loop self-energy of the Φ particle in Problem 11.4.

(5) Using dimensional regularization, determine the one-loop vertex correction to the interaction in Problem 11.5. In addition, calculate the counter-term δg.

Guide to additional recommended reading

The following books (see the indicated pages) and their authors have similar treatments of the content in the present chapter.

- F. Gross, *Relativistic Quantum Mechanics and Field Theory*, Wiley (1993), pp. 319–372.
- F. Halzen and A. D. Martin, *Quarks & Leptons: An Introductory Course in Modern Particle Physics*, Wiley (1984), pp. 154–158, 159–162, 163–168.
- F. Mandl and G. Shaw, *Quantum Field Theory*, rev. edn, Wiley (1994), pp. 175–219, 221–234.
- W. D. McComb, *Renormalization Methods – A Guide for Beginners*, Oxford (2007), pp. 228–234, 248–251.
- M. E. Peskin and D. V. Schroeder, *An Introduction to Quantum Field Theory*, Addison-Wesley (1995), pp. 175–210, 244–256.
- F. Schwabl, *Advanced Quantum Mechanics*, Springer (1999), pp. 358–373.
- F. J. Ynduráin, *Relativistic Quantum Mechanics and Introduction to Field Theory*, Springer (1996), pp. 267–290.

Appendix A

A brief survey of group theory and its notation

In this appendix, we summarize the basic definitions, notations, concepts, and properties of group theory. In addition, we give some illuminating examples. The reason for presenting this material as an appendix is twofold. On one hand, for readers, who are familiar with group theory, the material should not hinder the readability of the text, whereas on the other hand, for readers, who are unfamiliar with group theory, they should be able to obtain some knowledge of the basics of group theory.

Knowledge of group theory is important in physics, since groups describe the symmetries which the laws of physics seem to obey. Especially, physicists are interested in 'representations of Lie groups'. Examples of such groups include the Lorentz group, the Poincaré group, orthogonal and unitary groups, and the so-called *gauge group* of the Standard Model of particle physics $SU(3) \otimes SU(2) \otimes U(1)$.

A.1 Groups

Definition A *group* is a set $\mathcal{G}$ of elements together with a binary operation $*$ that combines any two elements a and b to form another element denoted $a * b$. In order to qualify as a group, the set and operation $(\mathcal{G}, *)$ must fulfil four requirements, which are called the *group axioms*.

(1) *Closure.* For all $a, b \in \mathcal{G}$, the result $a * b$ is also in $\mathcal{G}$.
(2) *Associativity.* For all $a, b, c \in \mathcal{G}$, the condition $(a * b) * c = a * (b * c)$ holds.
(3) *Identity (or unit) element.* There exists an element $e \in \mathcal{G}$ such that for every element $a \in \mathcal{G}$, the condition $e * a = a * e = a$ holds.
(4) *Inverse element.* For each $a \in \mathcal{G}$, there exists an element $b \in \mathcal{G}$, usually denoted a^{-1}, such that $a * b = b * a = e$, where e is the identity element.

The *order* of a group $|\mathcal{G}|$ is the cardinality of $\mathcal{G}$. If the order $|\mathcal{G}|$ is (in-)finite, then $\mathcal{G}$ itself is called (in-)finite. A subset $\mathcal{H} \subset \mathcal{G}$ is a *subgroup* if $(\mathcal{H}, *)$ fulfils the four group axioms. In addition, a group is called *Abelian* (or *commutative*) if it holds that

$$a * b = b * a \quad \forall a, b \in \mathcal{G}, \tag{A.1}$$

or in other words, if the *commutator*

$$[a, b] := a^{-1} * b^{-1} * a * b = e, \tag{A.2}$$

where e is the identity element. A *non-Abelian group* is a group that is not Abelian.

A.1.1 Examples

Example A.1   The integers $\mathbb{Z}$ with ordinary addition as the group operation are a group.

Example A.2   The rational numbers $\mathbb{Q}$ with ordinary addition as the group operation is a group.

Example A.3   The rational numbers without zero $\mathbb{Q}\backslash\{0\}$ with ordinary multiplication as the group operation are a group.

Example A.4   The integers $\mathbb{Z}$ with ordinary multiplication as the group operator are **not** a group, since not every element in $\mathbb{Z}$ has an inverse.

A.2 Lie groups

In physics, an important class of groups (with infinite order) are so-called *Lie groups* (or *continuous groups*). The Lie groups are characterized by: (i) they can be parametrized by a finite number of parameters n, and (ii) the quantity $a * b^{-1}$ is a continuous mapping with respect to these parameters. The parameters span an n-dimensional parameter space that in a neighbourhood of every point can be made Euclidean. Mathematically, the definition of a Lie group is the following:

Definition   A real Lie group is a group and a finite-dimensional real smooth manifold, in which the group operations of multiplication and inversion are smooth mappings, where the smoothness of the group multiplication

$$m : \mathcal{G} \times \mathcal{G} \to \mathcal{G}, \quad (x, y) \mapsto m(x, y) = xy \tag{A.3}$$

means that m is a smooth mapping of the product manifold $\mathcal{G} \times \mathcal{G}$ into $\mathcal{G}$. Note that the two requirements (on the smoothness of the group multiplication and inversion) can be combined into a single requirement, which means that the mapping

$$M : \mathcal{G} \times \mathcal{G} \to \mathcal{G}, \quad (x, y) \mapsto M(x, y) = x^{-1}y \tag{A.4}$$

is a smooth mapping of the product manifold $\mathcal{G} \times \mathcal{G}$ into $\mathcal{G}$.

A *group representation* is a so-called *homomorphism* from a group to a general linear group (see Example A.7). Basically, one 'represents' a given abstract group as a concrete group consisting of invertible matrices, since such a group is much easier to study.

A.2.1 Examples

Example A.5   The *Euclidean vector space* $\mathbb{R}^n$ with ordinary vector addition as the group operation is an n-dimensional Abelian Lie group.

Example A.6   The *circle group* S^1, consisting of complex numbers with absolute value equal to one, with ordinary multiplication of complex numbers as the group operation. This group is a one-dimensional Abelian Lie group.

Example A.7   The *general linear group* of degree n over a field $\mathbb{F}$, denoted $GL(n, \mathbb{F})$, is the group of $n \times n$ invertible matrices with elements from the field $\mathbb{F}$, where we consider $\mathbb{F} = \mathbb{R}$ (real numbers) or $\mathbb{F} = \mathbb{C}$ (complex numbers), and the group operation being that of matrix multiplication.

Example A.8   The *orthogonal group*, denoted $O(n, \mathbb{F})$, is the group of $n \times n$ orthogonal matrices. This is a subgroup of $GL(n, \mathbb{F})$ given by

$$O(n, \mathbb{F}) = \{A \in GL(n, \mathbb{F}) : A^T A = AA^T = \mathbb{1}_n\}. \tag{A.5}$$

Especially, the *special orthogonal group*, denoted $SO(n, \mathbb{F})$, is given by

$$SO(n, \mathbb{F}) = \{A \in O(n, \mathbb{F}) : \det A = 1\}. \tag{A.6}$$

Note that the *classical orthogonal group* and the *classical special orthogonal group* of real numbers are normally given by $O(n) = O(n, \mathbb{R})$ and $SO(n) = SO(n, \mathbb{R})$, respectively, which have dimension $n(n - 1)/2$.

Example A.9   The *unitary group*, denoted $U(n)$, is the group of $n \times n$ unitary matrices. This is a subgroup of $GL(n, \mathbb{C})$ given by

$$U(n) = \{A \in GL(n, \mathbb{C}) : A^\dagger A = AA^\dagger = \mathbb{1}_n\}. \tag{A.7}$$

Note that the unitary group $U(n)$ is a real Lie group with dimension n^2. In the case that $n = 1$, the group $U(1)$ corresponds to the circle group, consisting of all complex numbers with absolute value equal to one. This group is Abelian but, for $n > 1$, the group $U(n)$ is non-Abelian. Especially, the *special unitary group*, denoted $SU(n)$, is given by

$$SU(n) = \{A \in U(n) : \det A = 1\}, \tag{A.8}$$

which has dimension $n^2 - 1$. In general, the special unitary groups have applications in the Standard Model. In particular, SU(2) is the gauge group of weak interactions and SU(3) is the gauge group of strong interactions.

Example A.10 The Lorentz group and the Poincaré group are groups of linear and so-called *affine isometries* of the Minkowski space. They are Lie groups of dimension six and ten, respectively.

Example A.11 The group SU(3) $\otimes$ SU(2) $\otimes$ U(1) is a Lie group of dimension twelve $(8 + 3 + 1 = 12)$, which is the gauge group of the Standard Model, where the dimensions of the subgroups correspond to the eight gluons, the three vector bosons, and the photon of the Standard Model.

A.3 Lie algebras

Definition A real *Lie algebra* $\mathfrak{g}$ is a linear finite-dimensional vector space V_n over $\mathbb{R}$ with a bilinear operation, i.e. the commutator (or Lie bracket) $[\cdot, \cdot] : V_n \times V_n \to V_n$, such that the following axioms are satisfied.

(1) *Bilinearity.*

$$[\alpha a + \beta b, c] = \alpha[a, c] + \beta[b, c], \qquad (A.9)$$

$$[c, \alpha a + \beta b] = \alpha[c, a] + \beta[c, b], \qquad (A.10)$$

for all $\alpha, \beta \in \mathbb{R}$ and for all $a, b, c \in V_n$.

(2) *Skew-symmetry* (or *anti-commutativity*).

$$[a, b] = -[b, a] \quad \forall a, b \in V_n. \qquad (A.11)$$

In particular, $[a, a] = 0 \quad \forall a \in V_n$.

(3) *The Jacobi identity.* If $a, b, c \in V_n$, then

$$[[a, b], c] + [[b, c], a] + [[c, a], b] = 0. \qquad (A.12)$$

A *subalgebra* of $\mathfrak{g}$ is a subset of V_n that is a Lie algebra under the commutator $[\cdot, \cdot]$. In addition, if $[a, b] = 0 \quad \forall a, b \in V_n$, then $\mathfrak{g}$ is Abelian.

Although Lie algebras are usually investigated in their own right, they were historically developed as a means to study Lie groups.

A.3.1 Examples

Example A.12 The three-dimensional Euclidean vector space $\mathbb{R}^3$ with the commutator given by the cross product of vectors is a three-dimensional Lie algebra.

Example A.13   The real vector space of all complex $n \times n$ skew-Hermitian (or anti-Hermitian) matrices is closed under the commutator and is a real Lie algebra denoted $\mathfrak{u}(n)$, which is the Lie algebra of $U(n)$. The dimension of $\mathfrak{u}(n)$ is n^2.

Example A.14   All traceless complex $n \times n$ anti-Hermitian matrices form a Lie algebra denoted $\mathfrak{su}(n)$, which is the Lie algebra of $SU(n)$. The dimension of $\mathfrak{su}(n)$ is $n^2 - 1$. For example, the Lie algebra $\mathfrak{su}(2)$ is the set of traceless complex 2×2 anti-Hermitian matrices. One can show that $\mathfrak{su}(2)$ is a three-dimensional real algebra given by span$\{i\sigma^1, i\sigma^2, i\sigma^3\}$, where σ^1, σ^2, and σ^3 are the 2×2 Pauli matrices. In addition, $i\sigma^1$, $i\sigma^2$, and $i\sigma^3$ can be viewed as the infinitesimal generators of $SU(2)$.

A.4 Lie algebras of Lie groups

Let $\mathcal{G}$ be a real Lie group that can be parametrized by $\mathbf{t} = (t_1, t_2, \ldots, t_n) \in \mathbb{R}^n$ in a neighbourhood of the identity element $e = g(0, 0, \ldots, 0)$. An arbitrary element of $\mathcal{G}$ is denoted $g = g(\mathbf{t})$. The *infinitesimal generators* $\{x_i\}_{i=1}^n$ of $\mathcal{G}$ are defined by

$$x_i = -\mathrm{i}\, \frac{\partial}{\partial t_i} g(t_1, t_2, \ldots, t_i, \ldots, t_n) \bigg|_{t_1 = t_2 = \cdots = t_n = 0}, \tag{A.13}$$

where $\{x_i\}_{i=1}^n$ spans an n-dimensional vector space, which one can show is a Lie algebra. This Lie algebra corresponding to $\mathcal{G}$ is denoted by $\mathfrak{g}$.

Thus, every Lie group has a unique Lie algebra. In the case of matrix Lie groups, the Lie algebra is given by the exponential mapping. Hence, the representation of the Lie algebra consists of matrices A_i ($i = 1, 2, \ldots, n$) such that $\exp(\mathrm{i}\mathbf{t} \cdot \mathbf{A})$, where $\mathbf{A} = (A_1, A_2, \ldots, A_n)$, belongs to the representation of the Lie group for all parameters $\mathbf{t}$. Note that the quantity $\exp(\mathrm{i}\mathbf{t} \cdot \mathbf{A})$ generates an *n-parameter group*. Conversely, one can show that every Lie algebra is a Lie algebra of a Lie group. However, this Lie group is **not** unique, since in fact two Lie groups can have the same Lie algebra. For example, the Lie groups $SO(3)$ and $SU(2)$ give rise to the same Lie algebra, i.e. they have so-called *isomorphic Lie algebras*. In addition, these Lie algebras are isomorphic to the Lie algebra $\mathbb{R}^3$ with the commutator given by the cross product of vectors (cf. Example A.12). However, although $\mathfrak{so}(3)$ and $\mathfrak{su}(2)$ are isomorphic as Lie algebras, $SO(3)$ and $SU(2)$ are **not** isomorphic as Lie groups. Note that $SU(2)$ is the simply connected double covering group of $SO(3)$, which means that there is a two-to-one homomorphism from $SU(2)$ to $SO(3)$.

If $(\mathcal{G}, *)$ is a Lie group with the Lie algebra $\mathfrak{g}$ and $(\mathcal{H}, *) \subset (\mathcal{G}, *)$ is a subgroup of $(\mathcal{G}, *)$ with the Lie algebra $\mathfrak{h}$, then $\mathfrak{h}$ is a subalgebra of $\mathfrak{g}$.

A.4.1 Examples

Example A.15 The Lie algebra of the Lie group $\mathbb{R}^n$ is just $\mathbb{R}^n$ itself with the commutator given by $[a, b] = 0$ for all $a, b \in \mathbb{R}^n$.

Example A.16 The Lie algebra of the Lie group $GL(n, \mathbb{R})$ is the vector space of real $n \times n$ matrices with the commutator given by $[A, B] = AB - BA$.

Example A.17 If $\mathcal{G}$ is a closed subgroup of $GL(n, \mathbb{R})$, then the Lie algebra of $\mathcal{G}$ can informally be real $n \times n$ matrices m such that $\mathbb{1}_n + \epsilon m \in \mathcal{G}$, where ϵ is an infinitesimal positive number such that $\epsilon^2 = 0$. Of course, note that such ϵ does not exist. For example, consider $\mathcal{G} = O(n)$ with matrices A such that $AA^T = \mathbb{1}_n$, which means that the Lie algebra consists of matrices m such that $(\mathbb{1}_n + \epsilon m)(\mathbb{1}_n + \epsilon m)^T = \mathbb{1}_n$, which is equivalent to $m + m^T = 0$, since $\epsilon^2 = 0$.

Example A.18 The Lie algebra corresponding to the Lie group $SU(n)$ is denoted by $\mathfrak{su}(n)$, cf. Example A.14. In mathematics, the standard representation consists of traceless anti-Hermitian complex $n \times n$ matrices. In physics, the matrices are instead Hermitian, which simply gives rise to a different, but more convenient, representation of the same real Lie algebra. Note that $\mathfrak{su}(n)$ is a real Lie algebra. For example, the 2×2 Pauli matrices, which are Hermitian, give a representation for $\mathfrak{su}(2)$. In addition, the so-called *Gell-Mann matrices*, which are the 3×3 analogues of the Pauli matrices, give a representation for $\mathfrak{su}(3)$.

Example A.19 The set of all translations $x \mapsto x' = x + a$ in Minkowski space is the translation group T^4. Since the translations can be made in arbitrary order, T^4 is Abelian. In addition, T^4 is a Lie group, since it depends continuously on the four parameters a^μ. The only finite-dimensional irreducible unitary representation is given by $U(a) = \exp(ia \cdot P)$, which is the one-dimensional trivial representation, where the infinitesimal generators are the 4-momentum operators which fulfil the corresponding Lie algebra with the commutator $[P^\mu, P^\nu] = 0$.

A.5 The angular momentum algebra

For Lie groups, it is most useful to consider the infinitesimal generators of the group, which structure forms a Lie algebra. In quantum mechanics, the commutation relations among the components of the angular momentum operator ($\mathbf{L} = \mathbf{x} \times \mathbf{p} \leftrightarrow \hat{\mathbf{L}} = -i\mathbf{x} \times \nabla$)

$$L^1 = -i\left(x^2\partial_3 - x^3\partial_2\right), \tag{A.14}$$

$$L^2 = -i\left(x^3\partial_1 - x^1\partial_3\right), \tag{A.15}$$

$$L^3 = -i\left(x^1\partial_2 - x^2\partial_1\right) \tag{A.16}$$

constitute a representation for a complex three-dimensional Lie algebra, which is the complexification of the Lie algebra $\mathfrak{so}(3)$ of the three-dimensional rotation group SO(3) and called the *angular momentum algebra*:

$$[L^i, L^j] = i\epsilon^{ijk} L^k, \tag{A.17}$$

where the components of the angular momentum operators L^i ($i = 1, 2, 3$) are the infinitesimal generators of SO(3).

A specific element of SO(3), which describes rotations around the z-axis, can be written as the matrix

$$R_3(\theta) = \begin{pmatrix} \cos\theta & -\sin\theta & 0 \\ \sin\theta & \cos\theta & 0 \\ 0 & 0 & 1 \end{pmatrix}, \tag{A.18}$$

which must satisfy the requirements $R_3(\theta)^T R_3(\theta) = \mathbb{1}_3$ and $\det R_3(\theta) = 1$, where θ is a real parameter. The infinitesimal generators of SO(3) are introduced by considering elements close to the identity element $\mathbb{1}_3$. In this case, for small values of θ, we have the expansion of $R_3(\theta)$ close to $\mathbb{1}_3$ of the form

$$R_3(\theta) = \mathbb{1}_3 - i\theta L^3 + \mathcal{O}(\theta^2), \tag{A.19}$$

or equivalently,

$$-iL^3 = \left.\frac{dR_3(\theta)}{d\theta}\right|_{\theta=0} = \begin{pmatrix} 0 & -1 & 0 \\ 1 & 0 & 0 \\ 0 & 0 & 0 \end{pmatrix} \quad \Leftrightarrow \quad L^3 = \begin{pmatrix} 0 & -i & 0 \\ i & 0 & 0 \\ 0 & 0 & 0 \end{pmatrix}, \tag{A.20}$$

where L^3 is one of the three generators of SO(3). Note that L^3 is Hermitian and traceless, reflecting the orthogonality and the unit determinant of R_3. The other two generators

$$L^1 = \begin{pmatrix} 0 & 0 & 0 \\ 0 & 0 & -i \\ 0 & i & 0 \end{pmatrix} \quad \text{and} \quad L^2 = \begin{pmatrix} 0 & 0 & i \\ 0 & 0 & 0 \\ -i & 0 & 0 \end{pmatrix} \tag{A.21}$$

can be derived in a similar way.

Now, we can exponentiate the operators L^1, L^2, and L^3 in order to obtain unitary *one-parameter groups*. For example, let us study the one-parameter group generated by the operator L^3. Introducing polar coordinates and taking φ as the polar angle in the $x^1 x^2$-plane, we find that

$$L^3 = -i\partial_\varphi. \tag{A.22}$$

Therefore, we have

$$R_3(\theta) = \exp\left(-i\theta L^3\right) = \sum_{n=0}^{\infty} \frac{(-\theta)^n}{n!} \left(\partial_\varphi\right)^n, \tag{A.23}$$

which means that $R_3(\theta)$ translates the angle φ by the amount $-\theta$ (using Taylor's theorem), i.e. a clock-wise rotation in the $x^1 x^2$-plane by the angle θ. In general, for an arbitrary rotation, we obtain

$$R_n(\theta) = \exp(-i\theta \mathbf{n} \cdot \mathbf{L}), \tag{A.24}$$

where $\mathbf{n}$ is a unit vector.

In fact, the commutation relations of the angular momentum algebra can be realized by 2×2 matrices, which can be chosen as $X^i = \frac{1}{2}\sigma^i$, where σ^i ($i = 1, 2, 3$) are the 2×2 Pauli matrices, which are Hermitian and traceless. Exponentiating the three X^is, we generate the complete group of 2×2 unitary matrices with determinant equal to one, i.e. the group SU(2). In general, a matrix of SU(2) can be written as

$$U = \exp\left(-\frac{i}{2}\boldsymbol{\sigma} \cdot \boldsymbol{\alpha}\right) = \{\boldsymbol{\alpha} = \mathbf{n}\theta\} = \cos\frac{\theta}{2} - i\boldsymbol{\sigma} \cdot \mathbf{n} \sin\frac{\theta}{2}. \tag{A.25}$$

Hence, we have a homomorphism $M : \text{SU}(2) \to \text{SO}(3)$ between the two groups. This mapping is **not** one-to-one, since the so-called *kernel* is non-trivial, which is exactly where the 4π periodicity of SU(2) compared with the 2π periodicity of SO(3) enters. Using $\theta = 0$ and $\theta = 2\pi$, we obtain $U(0) = \mathbb{1}_2$ and $U(2\pi) = -\mathbb{1}_2$, but $R(0) = R(2\pi) = \mathbb{1}_3$. Therefore, two different elements of SU(2), i.e. $U(0)$ and $U(2\pi)$, map onto the identity element of SO(3), which means that the kernel of the homomorphism is non-trivial, consisting of two elements $K = \{\mathbb{1}_2, -\mathbb{1}_2\} = \mathbb{Z}_2$. Thus, the relation between the two groups SU(2) and SO(3) is the following:

$$\text{SO}(3) \simeq \text{SU}(2)/\mathbb{Z}_2, \tag{A.26}$$

which says that the groups have different global properties, despite the fact that they have the same Lie algebra.

Bibliography

I. J. R. Aitchison and A. J. G. Hey, *Gauge Theories in Particle Physics – Volume I: From Relativistic Quantum Mechanics to QED*, 3rd edn, IoP (2003).

D. Bailin and A. Love, *Introduction to Gauge Field Theory*, rev. edn, IOP (1996).

T. Banks, *Modern Quantum Field Theory – A Concise Introduction*, Cambridge (2008).

H. A. Bethe and R. Jackiw, *Intermediate Quantum Mechanics*, 3rd edn, Westview Press (1997).

L. S. Brown, *Quantum Field Theory*, Cambridge (1999).

A. Z. Capri, *Relativistic Quantum Mechanics and Introduction to Quantum Field Theory*, World Scientific (2002).

T.-P. Cheng and L.-F. Li, *Gauge Theory of Elementary Particle Physics*, Oxford (1996).

F. Dyson, transcribed by D. Derbes, *Advanced Quantum Mechanics*, World Scientific (2007).

A. L. Fetter and J. D. Walecka, *Quantum Theory of Many-Particle Systems*, Dover (2003).

W. Greiner, *Relativistic Quantum Mechanics – Wave Equations*, 3rd edn, Springer (2000).

W. Greiner and J. Reinhardt, *Quantum Electrodynamics*, 3rd edn, Springer (2003).

F. Gross, *Relativistic Quantum Mechanics and Field Theory*, Wiley (1993).

F. Halzen and A. D. Martin, *Quarks & Leptons: An Introductory Course in Modern Particle Physics*, Wiley (1984).

K. Huang, *Quarks, Leptons & Gauge Fields*, 2nd edn, World Scientific (1992).

C. Itzykson and J.-B. Zuber, *Quantum Field Theory*, McGraw-Hill (1985).

J. D. Jackson, *Classical Electrodynamics*, 3rd edn, Wiley (1999).

M. Kaku, *Quantum Field Theory – A Modern Introduction*, Oxford (1993).

C. W. Kim and A. Pevsner, *Neutrinos in Physics and Astrophysics*, Harwood Academic Publishers (1993).

R. H. Landau, *Quantum Mechanics II – A Second Course in Quantum Theory*, 2nd edn, Wiley-VCH (2004).

F. Mandl and G. Shaw, *Quantum Field Theory*, rev. edn, Wiley (1994).

W. D. McComb, *Renormalization Methods – A Guide for Beginners*, Oxford (2007).

J. Mickelsson, edited by T. Ohlsson, *Advanced Quantum Mechanics*, KTH (2003).

J. Mickelsson, T. Ohlsson, and H. Snellman, *Relativity Theory*, KTH (2005).

M. E. Peskin and D. V. Schroeder, *An Introduction to Quantum Field Theory*, Addison-Wesley (1995).

H. M. Pilkuhn, *Relativistic Quantum Mechanics*, Springer (2003).

V. Radovanović, *Problem Book in Quantum Field Theory*, 2nd edn, Springer (2008).

L. H. Ryder, *Quantum Field Theory*, 2nd edn, Cambridge (1996).

J. J. Sakurai, *Advanced Quantum Mechanics*, Addison-Wesley (1967).

J. J. Sakurai, *Modern Quantum Mechanics*, rev. edn, Addison-Wesley (1994).

G. Scharf, *Finite Quantum Electrodynamics*, Springer (1989).

F. Schwabl, *Advanced Quantum Mechanics*, Springer (1999).

S. S. Schweber, *An Introduction to Relativistic Quantum Field Theory*, Dover (2005).

H. Snellman, *Elementary Particle Physics*, KTH (2004).

M. Srednicki, *Quantum Field Theory*, Cambridge (2007).

P. Strange, *Relativistic Quantum Mechanics – with Applications in Condensed Matter and Atomic Physics*, Cambridge (1998).

J. D. Walecka, *Advanced Modern Physics – Theoretical Foundations*, World Scientific (2010).

S. Weinberg, *The Quantum Theory of Fields – Volume I: Foundations*, Cambridge (1996).

F. J. Ynduráin, *Relativistic Quantum Mechanics and Introduction to Field Theory*, Springer (1996).

A. Zee, *Quantum Field Theory in a Nutshell*, Princeton (2003).

Index

Printed in the United States
by Baker & Taylor Publisher Services